AF880045

Section 18

Liquid-Solid Operations and Equipment*

Wayne J. Genck, Ph.D. *President, Genck International; consultant on crystallization and precipitation; Member, American Chemical Society, American Institute of Chemical Engineers, Association for Crystallization Technology, International Society of Pharmaceutical Engineers (ISPE) (Section Editor, Crystallization)*

David S. Dickey, Ph.D. *Senior Consultant, MixTech, Inc.; Fellow, American Institute of Chemical Engineers; Member, North American Mixing Forum (NAMF); Member, American Chemical Society; Member, American Society of Mechanical Engineers (Mixing of Viscous Fluids, Pastes, and Doughs)*

Frank A. Baczek, B.S.Ch.E.&Chem. *Manager, Paste and Sedimentation Technology, Dorr-Oliver EIMCO; Member, Society of Metallurgical and Exploration Engineers of the American Institute of Mining, Metallurgical, and Petroleum Engineers (Gravity Sedimentation Operations)*

Daniel C. Bedell, B.S.Ch.E. *Global Market Manager E-CAT & Sedimentation, Dorr-Oliver EIMCO; Member, Society of Metallurgical and Exploration Engineers of the American Institute of Mining, Metallurgical, and Petroleum Engineers (Gravity Sedimentation Operations)*

Kent Brown, B.S.Civ.E. *Sedimentation Product Manager N.A., Dorr-Oliver EIMCO (Gravity Sedimentation Operations)*

Wu Chen, Ph.D. *Fluid/Particle Specialist, Dow Chemical Company; Member, American Filtration and Separations Society, American Institute of Chemical Engineers (Expression)*

Daniel E. Ellis, B.S.Ch.E. *Product Manager, Sedimentation Centrifuges and Belt Presses, Krauss Maffei Process Technology, Inc. (Centrifuges)*

Peter Harriott, Ph.D. *Professor Emeritus, School of Chemical Engineering, Cornell University; Member, American Institute of Chemical Engineers, American Chemical Society (Selection of a Solids-Liquid Separator)*

Tim J. Laros, M.S. *Senior Process Consultant, Dorr-Oliver EIMCO; Member, Society for Mining, Metallurgy, and Exploration (Filtration)*

Wenping Li, Ph.D. *R&D Manager, Agrilectric Research Company; Member, American Filtration and Separations Society, American Institute of Chemical Engineers (Expression)*

James K. McGillicuddy, B.S.M.E. *Product Manager, Filtration Centrifuges and Filters, Krauss Maffei Process Technology, Inc.; Member, American Institute of Chemical Engineers (Centrifuges)*

Terence P. McNulty, Ph.D. *President, T. P. McNulty and Associates, Inc.; Member, National Academy of Engineering; Member, American Institute of Mining, Metallurgical, and Petroleum Engineers; Member, Society for Mining, Metallurgy, and Exploration (Leaching)*

*The contributions of Donald A. Dahlstrom (Section Editor) and Robert C. Emmett, Jr. (Gravity Sedimentation Operations), authors for this section in the Seventh Edition, are acknowledged.

James Y. Oldshue, Ph.D. *Deceased; President, Oldshue Technologies International, Inc.; Adjunct Professor of Chemical Engineering at Beijing Institute of Chemical Technology, Beijing, China; Member, National Academy of Engineering, American Chemical Society, American Institute of Chemical Engineers, Traveler Century Club; Member of Executive Committee on the Transfer of Appropriate Technology for the World Federation of Engineering Organizations (Agitation of Low-Viscosity Particle Suspensions)**

Fred Schoenbrunn, B.S.Ch.E. *Product Manager for Minerals Sedimentation, Dorr-Oliver EIMCO; Member, Society of Metallurgical and Exploration Engineers of the American Institute of Mining, Metallurgical, and Petroleum Engineers; Registered Professional Engineer (Gravity Sedimentation Operations)*

Julian C. Smith, B.Chem.&Ch.E. *Professor Emeritus, School of Chemical Engineering, Cornell University; Member, American Chemical Society, American Institute of Chemical Engineers (Selection of a Solids-Liquid Separator)*

Donald C. Taylor, B.S.Eng.Geol., M.S.Civ.E. *Process Manager Industrial Water & Wastewater Technology, Dorr-Oliver EIMCO; Member, Water Environment Federation; Registered Professional Engineer (Gravity Sedimentation Operations)*

Daniel R. Wells, B.S.Ind.E., MBA *Product Manager Sedimentation Products, Dorr-Oliver EIMCO (Gravity Sedimentation Operations)*

Todd W. Wisdom, M.S.Ch.E. *Global Filtration Product Manager, Dorr-Oliver EIMCO; Member, American Institute of Chemical Engineer's (Filtration)*

PHASE CONTACTING AND LIQUID-SOLID PROCESSING: AGITATION OF LOW-VISCOSITY PARTICLE SUSPENSIONS

MIXING OF VISCOUS FLUIDS, PASTES, AND DOUGHS

*The contribution of the late Dr. J. Y. Oldshue, who authored part of this and many editions, is acknowledged.

CRYSTALLIZATION FROM SOLUTION

LEACHING

GRAVITY SEDIMENTATION OPERATIONS\

FILTRATION

CENTRIFUGES

SELECTION OF A SOLIDS-LIQUID SEPARATOR

Nomenclature

Symbol	Definition	SI units	U.S. customary units
c	Specific heat	J/(kg·k)	Btu/(lb·°F)
C	Constant		
C_o	Orifice coefficient	Dimensionless	Dimensionless
d_o	Orifice diameter	m	in
$d_{p,\,\max}$	Drop diameter	m	ft
d_t	Pipe diameter	m	in
d_t	Tube diameter	m	ft
D	Impeller diameter	m	ft
D_a	Impeller diameter	m	ft
D_j	Diameter of jacketed vessel	m	ft
D_T	Tank diameter	m	ft
g	Acceleration	m/s^2	ft/s^2
g_c	Dimensional constant	$g_c = 1$ when using SI units	32.2 (ft·lb)/(lbf·s^2)
h	Local individual coefficient of heat transfer, equals $dq/(dA)(\Delta T)$	J/(m^2·s·K)	Btu/(h·ft^2·°F)
H	Velocity head	m	ft
k	Thermal conductivity	J/(m·s·K)	(Btu·ft)/(h·ft^2·°F)
L_p	Diameter of agitator blade	m	ft
N	Agitator rotational speed	s^{-1}, (r/s)	s^{-1}, (r/s)
N_{JS}	Agitator speed for just suspension	s^{-1}	s^{-1}
N_{Re}	$D_a^2N\rho/\mu$ impeller Reynolds number	Dimensionless	Dimensionless
N_p	Power number = $(q_cP)/\rho N^3D_a^5$	Dimensionless	Dimensionless
N_Q	Impeller pumping coefficient = Q/ND_a^3	Dimensionless	Dimensionless
N_r	Impeller speed	s^{-1}	s^{-1}
N_t	Impeller speed	s^{-1}	s^{-1}
P	Power	(N·m/s)	ft·lb$_f$/s
Q	Impeller flow rate	m^3/s	ft^3/s
T	Tank diameter	m	ft
v	Average fluid velocity	m/s	ft/s
v'	Fluid velocity fluctuation	m/s	ft/s
V	Bulk average velocity	m/s	ft/s
Z	Liquid level in tank	m	ft
	Greek Symbols		
γ	Rate of shear	s^{-1}	s^{-1}
Δp	Pressure drop across orifice		lbf/ft^2
μ	Viscosity of liquid at tank temperature	Pa·s	lb/(ft·s)
μ	Stirred liquid viscosity	Pa·s	lb/(ft·s)
μ_b	Viscosity of fluid at bulk temperature	Pa·s	lb/(ft·s)
μ_c	Viscosity, continuous phase	Pa·s	lb/(ft·s)
μ_D	Viscosity of dispersed phase	Pa·s	lb/(ft·s)
μ_f	Viscosity of liquid at mean film temperature	Pa·s	lb/(ft·s)
μ_{wt}	Viscosity at wall temperature	Pa·s	lb/(ft·s)
ρ	Stirred liquid density	g/m^3	lb/ft^3
ρ	Density of fluid	kg/m^3	lb/ft^3
ρ_{av}	Density of dispersed phase	kg/m^3	lb/ft^3
ρ_c	Density	kg/m^3	lb/ft^3
σ	Interfacial tension	N/m	lbf/ft
Φ_D	Average volume fraction of discontinuous phase	Dimensionless	Dimensionless

PHASE CONTACTING AND LIQUID-SOLID PROCESSING: AGITATION OF LOW-VISCOSITY PARTICLE SUSPENSIONS

General References: Harnby, N., M. F. Edwards, and A. W. Neinow (eds.), *Mixing in the Process Industries,* Butterworth, Stoneham, Mass., 1986. Lo, T. C., M. H. I. Baird, and C. Hanson, *Handbook of Solvent Extraction,* Wiley, New York, 1983. Nagata, S., *Mixing: Principles and Applications,* Kodansha Ltd., Tokyo, Wiley, New York, 1975. Oldshue, J. Y., *Fluid Mixing Technology,* McGraw-Hill, New York, 1983. Tatterson, G. B., *Fluid Mixing and Gas Dispersion in Agitated Tanks,* McGraw-Hill, New York, 1991. Uhl, V. W., and J. B. Gray (eds.), *Mixing,* vols. I and II, Academic Press, New York, 1966; vol. III, Academic Press, Orlando, Fla., 1992. Ulbrecht, J. J., and G. K. Paterson (eds.), *Mixing of Liquids by Mechanical Agitation,* Godon & Breach Science Publishers, New York, 1985.

Proceedings: *Fluid Mixing,* vol. I, Inst. Chem. Eng. Symp., Ser. No. 64 (Bradford, England), The Institute of Chemical Engineers, Rugby, England, 1984. *Mixing—Theory Related to Practice,* AIChE, Inst. Chem. Eng. Symp. Ser. No. 10 (London), AIChE and The Institute of Chemical Engineers, London, 1965. *Proc. First (1974), Second (1977), Third (1979), Fourth (1982), Fifth (1985), and Sixth (1988) European Conf. on Mixing,* N. G. Coles (ed.), (Cambridge, England) BHRA Fluid Eng., Cranfield, England. *Process Mixing, Chemical and Biochemical Applications,* G. B. Tatterson, and R. V. Calabrese (eds.), AIChE Symp. Ser. No. 286, 1992.

FLUID MIXING TECHNOLOGY

Fluid mixers cut across almost every processing industry including the chemical process industry; minerals, pulp, and paper; waste and water treating and almost every individual process sector. The engineer working with the application and design of mixers for a given process has three basic sources for information. One is published literature, consisting of several thousand published articles and several currently available books, and brochures from equipment vendors. In addition, there may be a variety of in-house experience which may or may not be cataloged, categorized, or usefully available for the process application at hand. Also, short courses are currently available in selected locations and with various course objectives, and a large body of experience and information lies in the hands of equipment vendors.

In the United States, it is customary to design and purchase a mixer from a mixing vendor and purchase the vessel from another supplier. In many other countries, it is more common to purchase the vessel and mixer as a package from one supplier.

In any event, the users of the mixer can issue a mechanical specification and determine the speed, diameter of an impeller, and power with in-house expertise. Or they may issue a process specification which describes the engineering purpose of the mixing operation and the vendor will supply a description of the mixer process performance as well as prepare a mechanical design.

This section describes fluid mixing technology and is referred to in other sections in this handbook which discuss the use of fluid mixing equipment in their various operating disciplines. This section does not describe paste and dough mixing, which may require planetary and extruder-type mixers, nor the area of dry solid-solid mixing.

It is convenient to divide mixing into five pairs (plus three triplets and one quadruplicate combination) of materials, as shown in Table 18-1. These five pairs are blending (miscible liquids), liquid-solid, liquid-gas, liquid-liquid (immiscible liquids), and fluid motion. There are also four other categories that occur, involving three or four phases. One concept that differentiates mixing requirements is the difference between physical criteria listed on the left side of Table 18-1, in which some degree of sampling can be used to determine the character of the mixture in various parts in the tank, and various definitions of mixing requirements can be based on these physical descriptions. The other category on the right side of Table 18-1 involves chemical and mass-transfer criteria in which rates of mass transfer or chemical reaction are of interest and have many more complexities in expressing the mixing requirements.

The first five classes have their own mixing technologies. Each of these 10 areas has its own mixing technology. There are relationships for the optimum geometry of impeller types, D/T ratios, and tank geometry. They each often have general, overall mixing requirements and different scale-up relationships based on process definitions. In addition, there are many subclassifications, some of which are based on the viscosity of fluids. In the case of blending, we have blending in the viscous region, the transition region, and the turbulent region. Since any given mixer designed for a process may be required to do several different parts of these 10 categories, it must be a compromise of the geometry and other requirements for the total process result and may not optimize any one particular process component. If it turns out that one particular process requirement is so predominant that all the other requirements are satisfied as a consequence, then it is possible to optimize that particular process step. Often, the only process requirement is in one of these 10 areas, and the mixer can be designed and optimized for that one step only.

As an example of the complexity of fluid mixing, many batch processes involve adding many different materials and varying the liquid level over wide ranges in the tank, have different temperatures and shear rate requirements, and obviously need experience and expert attention to all of the requirements. Superimposing the requirements for sound mechanical design, including drives, fluid seals, and rotating shafts, means that the concepts presented here are merely a beginning to the overall, final design.

A few general principles are helpful at this point before proceeding to the examination of equipment and process details. For any given impeller geometry, speed, and diameter, the impeller draws a certain amount of power. This power is 100 percent converted to heat. In low-viscosity mixing (defined later), this power is used to generate a *macro-scale* regime in which one typically has the visual observation of flow pattern, swirls, and other surface phenomena. However, these flow patterns are primarily energy transfer agents that transfer the power down to the micro scale. The macro-scale regime involves the pumping capacity of the impeller as well as the total circulating capacity throughout the tank and it is an important part of the overall mixer design. The micro-scale area in which the power is dissipated does not care much which impeller is used to generate the energy dissipation. In contrast, in high-viscosity processes, there is a continual progress of energy dissipation from the macro scale down to the micro scale.

There is a wide variety of impellers using *fluidfoil principles,* which are used when flow from the impeller is predominant in the process requirement and macro- or micro-scale shear rates are a subordinate issue.

TABLE 18-1 Classification System for Mixing Processes

Physical	Components	Chemical, mass transfer
Blending	Blending	Chemical reactions
Suspension	Solid-liquid	Dissolving, precipitation
Dispersion	Gas-liquid	Gas absorption
	Solid-liquid-gas	
Emulsions	Liquid-liquid	Extraction
	Liquid-liquid-solid	
	Gas-liquid-liquid	
	Gas-liquid-liquid-solid	
Pumping	Fluid motion	Heat transfer

Scale-up involves selecting mixing variables to give the desired performance in both pilot and full scale. This is often difficult (sometimes impossible) using geometric similarity, so that the use of nongeometric impellers in the pilot plant compared to the impellers used in the plant often allows closer modeling of the mixing requirements to be achieved.

Computational fluid mixing allows the modeling of flow patterns in mixing vessels and some of the principles on which this is based in current techniques are included.

INTRODUCTORY FLUID MECHANICS

The fluid mixing process involves three different areas of viscosity which affect flow patterns and scale-up, and two different scales within the fluid itself: macro scale and micro scale. Design questions come up when looking at the design and performance of mixing processes in a given volume. Considerations must be given to proper impeller and tank geometry as well as the proper speed and power for the impeller. Similar considerations come up when it is desired to scale up or scale down, and this involves another set of mixing considerations.

If the fluid discharge from an impeller is measured with a device that has a high-frequency response, one can track the velocity of the fluid as a function of time. The velocity at a given point in time can then be expressed as an average velocity v plus fluctuating component v'. Average velocities can be integrated across the discharge of the impeller, and the pumping capacity normal to an arbitrary discharge plane can be calculated. This arbitrary discharge plane is often defined as the plane bounded by the boundaries of the impeller blade diameter and height. Because there is no casing, however, an additional 10 to 20 percent of flow typically can be considered as the primary flow from an impeller.

The velocity gradients between the average velocities operate only on larger particles. Typically, these larger-size particles are greater than 1000 μm. This is not a proven definition, but it does give a feel for the magnitudes involved. This defines macro-scale mixing. In the turbulent region, these macro-scale fluctuations can also arise from the finite number of impeller blades. These set up velocity fluctuations that can also operate on the macro scale.

Smaller particles see primarily only the fluctuating velocity component. When the particle size is much less than 100 μm, the turbulent properties of the fluid become important. This is the definition of the physical size for micro-scale mixing.

All of the power applied by a mixer to a fluid through the impeller appears as heat. The conversion of power to heat is through viscous shear and is approximately 2542 Btu/h/hp. Viscous shear is present in turbulent flow only at the micro-scale level. As a result, the power per unit volume is a major component of the phenomena of micro-scale mixing. At a 1-μm level, in fact, it doesn't matter what specific impeller design is used to supply the power.

Numerous experiments show that power per unit volume in the zone of the impeller (which is about 5 percent of the total tank volume) is about 100 times higher than the power per unit volume in the rest of the vessel. Making some reasonable assumptions about the fluid mechanics parameters, the root-mean-square (rms) velocity fluctuation in the zone of the impeller appears to be approximately 5 to 10 times higher than in the rest of the vessel. This conclusion has been verified by experimental measurements.

The ratio of the rms velocity fluctuation to the average velocity in the impeller zone is about 50 percent with many open impellers. If the rms velocity fluctuation is divided by the average velocity in the rest of the vessel, however, the ratio is on the order of 5 percent. This is also the level of rms velocity fluctuation to the mean velocity in pipeline flow. There are phenomena in micro-scale mixing that can occur in mixing tanks that do not occur in pipeline reactors. Whether this is good or bad depends upon the process requirements.

Figure 18-1 shows velocity versus time for three different impellers. The differences between the impellers are quite significant and can be important for mixing processes.

All three impellers are calculated for the same impeller flow Q and the same diameter. The A310 (Fig. 18-2) draws the least power and has the least velocity fluctuations. This gives the lowest micro-scale turbulence and shear rate. The A200 (Fig. 18-3) shows increased velocity fluctuations and draws more power. The R100 (Fig. 18-4) draws the most power and has the highest micro-scale shear rate. The proper impeller should be used for each individual process requirement.

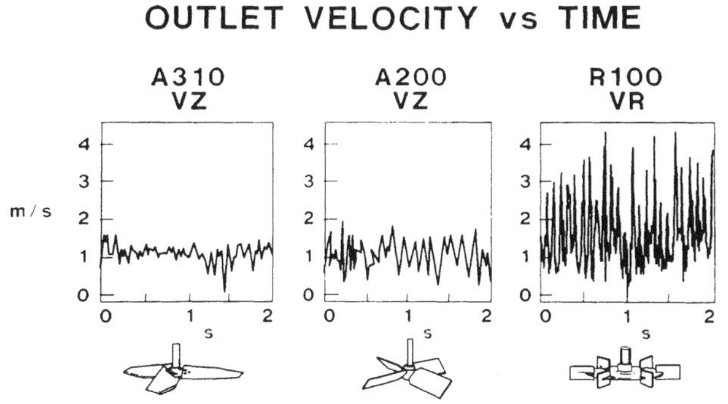

FIG. 18-1 Velocity fluctuations versus time for equal total pumping capacity from three different impellers.

Scale-up/Scale-down Two aspects of scale-up frequently arise. One is building a model based on pilot-plant studies that develop an understanding of the process variables for an existing full-scale mixing installation. The other is taking a new process and studying it in the pilot plant in such a way that pertinent scale-up variables are worked out for a new mixing installation.

There are a few principles of scale-up that can indicate which approach to take in either case. Using geometric similarity, the macro-scale variables can be summarized as follows:

- Blend and circulation times in the large tank will be much longer than in the small tank.

FIG. 18-2 An A310 impeller.

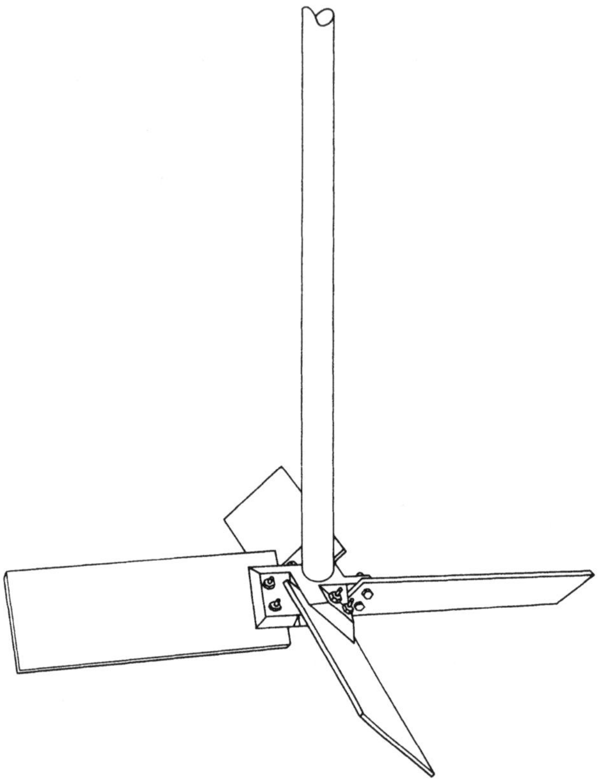

FIG. 18-3 Pitched-blade turbine.

- Maximum impeller zone shear rate will be higher in the larger tank, but the average impeller zone shear rate will be lower; therefore, there will be a much greater variation in shear rates in a full-scale tank than in a pilot unit.
- Reynolds numbers in the large tank will be higher, typically on the order of 5 to 25 times higher than those in a small tank.
- Large tanks tend to develop a recirculation pattern from the impeller through the tank back to the impeller. This results in a behavior similar to that for a number of tanks in a series. The net result is that the mean circulation time is increased over what would be predicted from the impeller pumping capacity. This also increases the standard deviation of the circulation times around the mean.
- Heat transfer is normally much more demanding on a large scale. The introduction of helical coils, vertical tubes, or other heat-transfer devices causes an increased tendency for areas of low recirculation to exist.
- In gas-liquid systems, the tendency for an increase in the gas superficial velocity upon scale-up can further increase the overall circulation time.

What about the micro-scale phenomena? These are dependent primarily on the energy dissipation per unit volume, although one must also be concerned about the energy spectra. In general, the energy dissipation per unit volume around the impeller is approximately 100 times higher than in the rest of the tank. This results in an rms velocity fluctuation ratio to the average velocity on the order of 10:1 between the impeller zone and the rest of the tank.

Because there are thousands of specific processes each year that involve mixing, there will be at least hundreds of different situations requiring a somewhat different pilot-plant approach. Unfortunately, no set of rules states how to carry out studies for any specific program, but here are a few guidelines that can help one carry out a pilot-plant program.

- For any given process, one takes a qualitative look at the possible role of fluid shear stresses. Then one tries to consider pathways related to fluid shear stress that may affect the process. If there are none, then this extremely complex phenomenon can be dismissed and the process design can be based on such things as uniformity, circulation time, blend time, or velocity specifications. This is often the case in the blending of miscible fluids and the suspension of solids.
- If fluid shear stresses are likely to be involved in obtaining a process result, then one must qualitatively look at the scale at which the shear stresses influence the result. If the particles, bubbles, droplets, or fluid clumps are on the order of 1000 μm or larger, the variables are macro scale and average velocities at a point are the predominant variable.

When macro-scale variables are involved, every geometric design variable can affect the role of shear stresses. They can include such items as power, impeller speed, impeller diameter, impeller blade shape, impeller blade width or height, thickness of the material used to make the impeller, number of blades, impeller location, baffle location, and number of impellers.

Micro-scale variables are involved when the particles, droplets, baffles, or fluid clumps are on the order of 100 μm or less. In this case, the critical parameters usually are power per unit volume, distribution of power per unit volume between the impeller and the rest of the tank, rms velocity fluctuation, energy spectra, dissipation length, the smallest micro-scale eddy size for the particular power level, and viscosity of the fluid.

- The overall circulating pattern, including the circulation time and the deviation of the circulation times, can never be neglected. No matter what else a mixer does, it must be able to circulate fluid throughout an entire vessel appropriately. If it cannot, then that mixer is not suited for the task being considered.

Qualitative and, hopefully, quantitative estimates of how the process result will be measured must be made in advance. The evaluations must allow one to establish the importance of the different steps in a process, such as gas-liquid mass transfer, chemical reaction rate, or heat transfer.

- It is seldom possible, either economically or timewise, to study every potential mixing variable or to compare the performance of many impeller types. In many cases, a process needs a specific fluid regime that is relatively independent of the impeller type used to generate it. Because different impellers may require different geometries to achieve an optimum process combination, a random choice of only one diameter of each of two or more impeller types may not tell what is appropriate for the fluid regime ultimately required.
- Often, a pilot plant will operate in the viscous region while the commercial unit will operate in the transition region, or alternatively, the pilot plant may be in the transition region and the commercial unit in the turbulent region. Some experience is required to estimate the difference in performance to be expected upon scale-up.
- In general, it is not necessary to model Z/T ratios between pilot and commercial units.

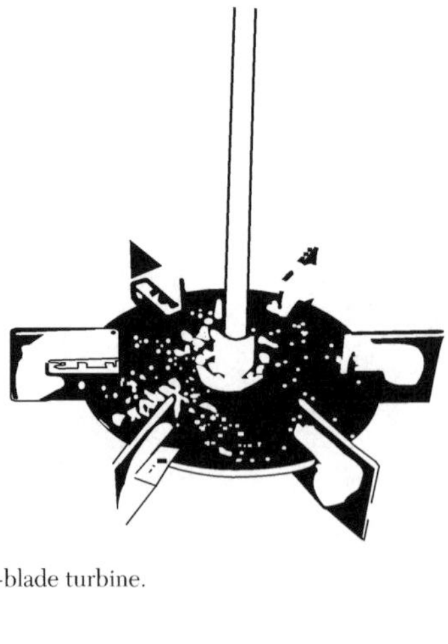

FIG. 18-4 Flat-blade turbine.

- In order to make the pilot unit more like a commercial unit in macro-scale characteristics, the pilot unit impeller must be designed to lengthen the blend time and to increase the maximum impeller zone shear rate. This will result in a greater range of shear rates than is normally found in a pilot unit.

MIXING EQUIPMENT

There are three types of mixing flow patterns that are markedly different. The so-called axial-flow turbines (Fig. 18-3) actually give a flow coming off the impeller of approximately 45°, and therefore have a recirculation pattern coming back into the impeller at the hub region of the blades. This flow pattern exists to an approximate Reynolds number of 200 to 600 and then becomes radial as the Reynolds number decreases. Both the R100 and A200 impellers normally require four baffles for an effective flow pattern. These baffles typically are 1/12 of the tank diameter and width.

Radial-flow impellers include the flat-blade disc turbine, Fig. 18-4, which is labeled an R100. This generates a radial flow pattern at all Reynolds numbers. Figure 18-17 is the diagram of Reynolds number/power number curve, which allows one to calculate the power knowing the speed and diameter of the impeller. The impeller shown in Fig. 18-4 typically gives high shear rates and relatively low pumping capacity.

The current design of fluidfoil impellers includes the A310 (Fig. 18-2), as well as several other impellers of that type commonly referred to as *high-efficiency impellers, hydrofoil,* and other descriptive names to illustrate that they are designed to maximize flow and minimize shear rate. These impellers typically require two baffles, but are normally used with three, since three gives a more stable flow pattern. Since most industrial mixing processes involve pumping capacity and, to a lesser degree, fluid shear rate, the fluidfoil impellers are now used on the majority of the mixer installations. There is now an additional family of these fluidfoil impellers, which depend upon different solidity ratios to operate in various kinds of fluid mixing systems. Figure 18-5 illustrates four of these impellers. The solidity ratio is the ratio of total blade area to a circle circumscribing the impeller and, as viscosity increases, higher values of the solidity ratios are more effective in providing an axial flow pattern rather than a radial flow pattern. Also the A315-type provides an effective area of preventing gas bypassing through the hub of the impeller by having exceptionally wide blades. Another impeller of that type is the Prochem Maxflo T.

Small Tanks For tanks less than 1.8 m in diameter, the clamp or flanged mounted angular, off-center axial-flow impeller without baffles should be used for a wide range of process requirements (refer to Fig. 18-14). The impellers currently used are the fluidfoil type. Since small impellers typically operate at low Reynolds numbers, often in the transition region, the fluidfoil impeller should be designed to give good flow characteristics over a range of Reynolds numbers, probably on the order of 50 to 500. The Z/T ratio should be 0.75 to 1.5. The volume of liquid should not exceed 4 m^3.

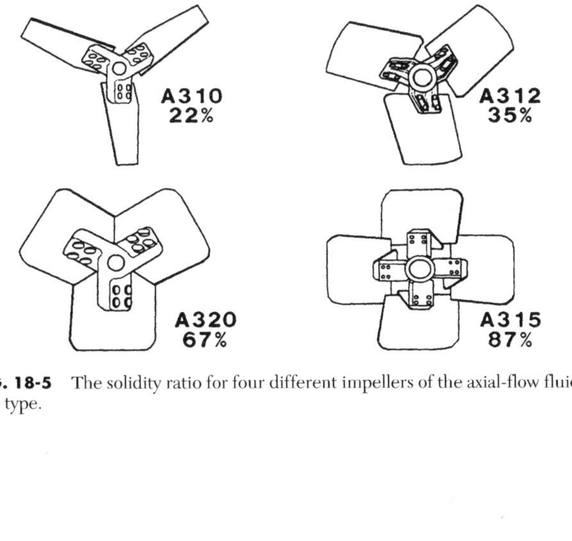

FIG. 18-5 The solidity ratio for four different impellers of the axial-flow fluidfoil type.

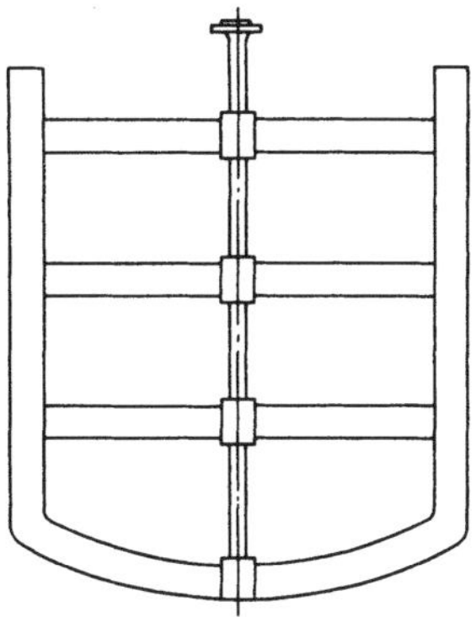

FIG. 18-6 Anchor impeller.

Close-Clearance Impellers There are two close-clearance impellers. They are the *anchor impeller* (Fig. 18-6) and the *helical impeller* (Fig. 18-7), which operate near the tank wall and are particularly effective in pseudoplastic fluids in which it is desirable to have the mixing energy concentrated out near the tank wall where the flow pattern is more effective than with the open impellers that were covered earlier.

Axial-Flow Impellers Axial-flow impellers include all impellers in which the blade makes an angle of less than 90° with the plane of

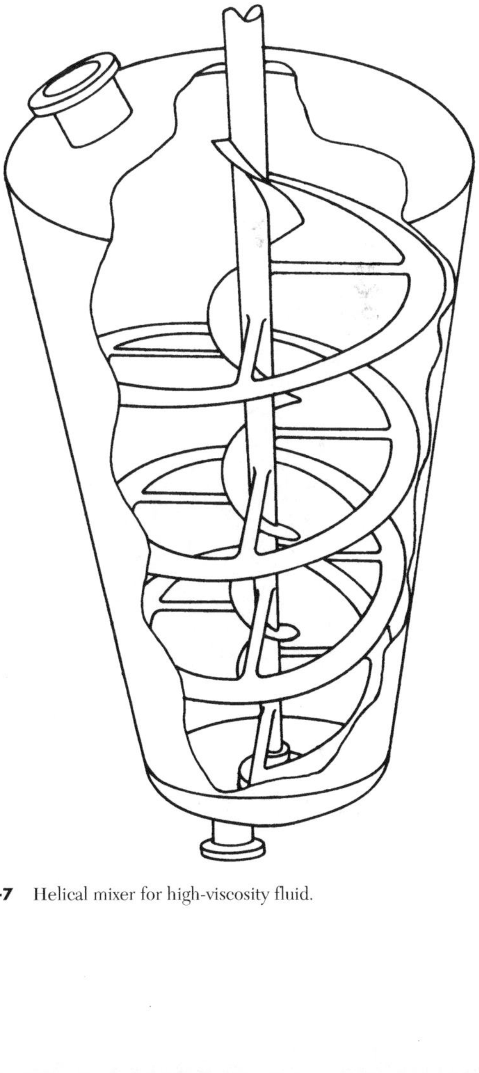

FIG. 18-7 Helical mixer for high-viscosity fluid.

FIG. 18-8 Marine-type mixing impeller.

rotation. Propellers and pitched-blade turbines, as illustrated in Figs. 18-8 and 18-3, are representative axial-flow impellers.

Portable mixers may be clamped on the side of an open vessel in the angular, off-center position shown in Fig. 18-14 or bolted to a flange or plate on the top of a closed vessel with the shaft in the same angular, off-center position. This mounting results in a strong top-to-bottom circulation.

Two basic speed ranges are available: 1150 or 1750 r/min with direct drive and 350 or 420 r/min with a gear drive. The high-speed units produce higher velocities and shear rates (Fig. 18-9) in the impeller discharge stream and a lower circulation rate throughout the vessel than the low-speed units. For suspension of solids, it is common to use the gear-driven units, while for rapid dispersion or fast reactions the high-speed units are more appropriate.

Axial-flow impellers may also be mounted near the bottom of the cylindrical wall of a vessel as shown in Fig. 18-10. Such side-entering agitators are used to blend low-viscosity fluids [<0.1 Pa·s (100 cP)] or to keep slowly settling sediment suspended in tanks as large as some 4000 m^3 (10^6 gal). Mixing of paper pulp is often carried out by side-entering propellers.

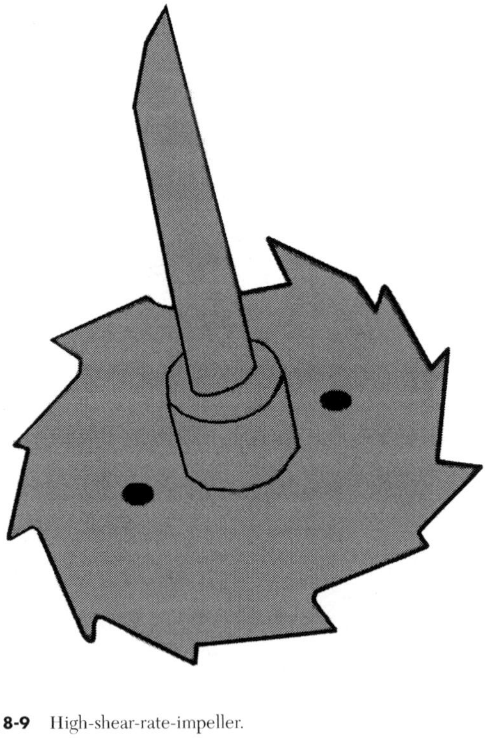

FIG. 18-9 High-shear-rate-impeller.

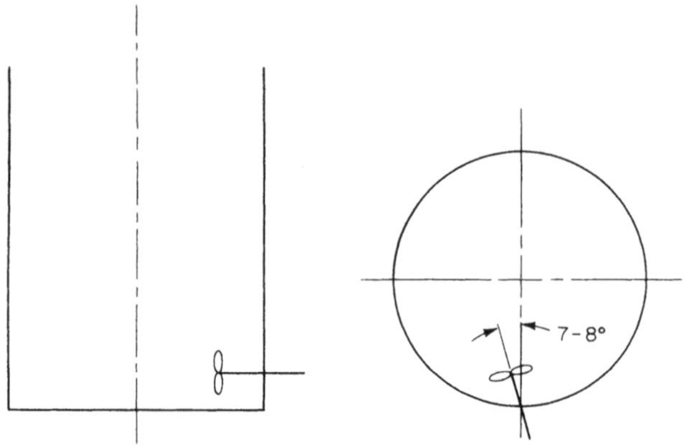

FIG. 18-10 Side-entering propeller mixer.

Pitched-blade turbines (Fig. 18-3) are used on top-entering agitator shafts instead of propellers when a high axial circulation rate is desired and the power consumption is more than 2.2 kW (3 hp). A pitched-blade turbine near the upper surface of liquid in a vessel is effective for rapid submergence of floating particulate solids.

Radial-Flow Impellers Radial-flow impellers have blades which are parallel to the axis of the drive shaft. The smaller multiblade ones are known as *turbines*; larger, slower-speed impellers, with two or four blades, are often called *paddles*. The diameter of a turbine is normally between 0.3 and 0.6 of the tank diameter. Turbine impellers come in a variety of types, such as curved-blade and flat-blade, as illustrated in Fig. 18-4. Curved blades aid in starting an impeller in settled solids.

For processes in which corrosion of commonly used metals is a problem, glass-coated impellers may be economical. A typical modified curved-blade turbine of this type is shown in Fig. 18-11.

Close-Clearance Stirrers For some pseudoplastic fluid systems stagnant fluid may be found next to the vessel walls in parts remote from propeller or turbine impellers. In such cases, an "anchor" impeller may be used (Fig. 18-6). The fluid flow is principally circular or helical (see Fig. 18-7) in the direction of rotation of the anchor. Whether substantial axial or radial fluid motion also occurs depends on the fluid viscosity and the design of the upper blade-supporting spokes. Anchor agitators are used particularly to obtain improved heat transfer in high-consistency fluids.

Unbaffled Tanks If a low-viscosity liquid is stirred in an unbaffled tank by an axially mounted agitator, there is a tendency for a swirling

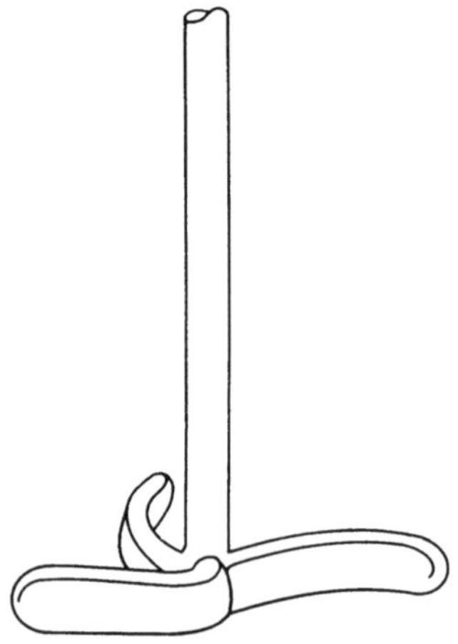

FIG. 18-11 Glass-steel impeller. (*The Pfaudler Company.*)

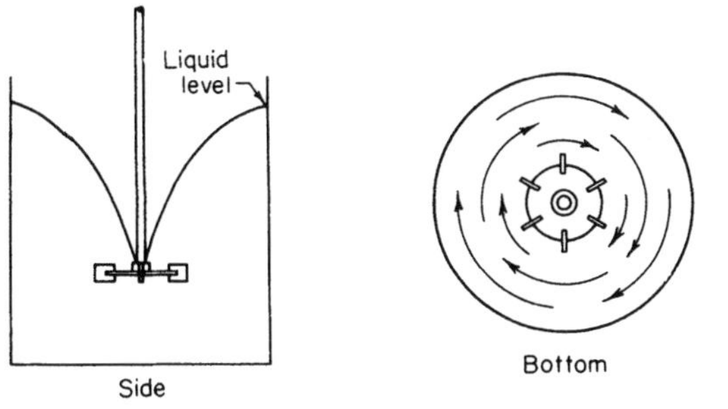

FIG. 18-12 Typical flow pattern for either axial- or radial-flow impellers in an unbaffled tank.

flow pattern to develop regardless of the type of impeller. Figure 18-12 shows a typical flow pattern. A vortex is produced owing to centrifugal force acting on the rotating liquid. In spite of the presence of a vortex, satisfactory process results often can be obtained in an unbaffled vessel. However, there is a limit to the rotational speed that may be used, since once the vortex reaches the impeller, severe air entrainment may occur. In addition, the swirling mass of liquid often generates an oscillating surge in the tank, which coupled with the deep vortex may create a large fluctuating force acting on the mixer shaft.

Vertical velocities in a vortexing low-viscosity liquid are low relative to circumferential velocities in the vessel. Increased vertical circulation rates may be obtained by mounting the impeller off center, as illustrated in Fig. 18-13. This position may be used with either turbines or propellers. The position is critical, since too far or too little off center in one direction or the other will cause greater swirling, erratic vortexing, and dangerously high shaft stresses. Changes in viscosity and tank size also affect the flow pattern in such vessels. Off-center mountings have been particularly effective in the suspension of paper pulp.

With axial-flow impellers, an angular off-center position may be used. The impeller is mounted approximately 15° from the vertical, as shown in Fig. 18-14.

The angular off-center position used with fluidfoil units is usually limited to impellers delivering 2.2 kW (3 hp) or less. The unbalanced fluid forces generated by this mounting can become severe with higher power.

Baffled Tanks For vigorous agitation of thin suspensions, the tank is provided with baffles which are flat vertical strips set radially along the tank wall, as illustrated in Figs. 18-15 and 18-16. Four baffles are almost always adequate. A common baffle width is one-tenth to one-twelfth of the tank diameter (radial dimension). For agitating

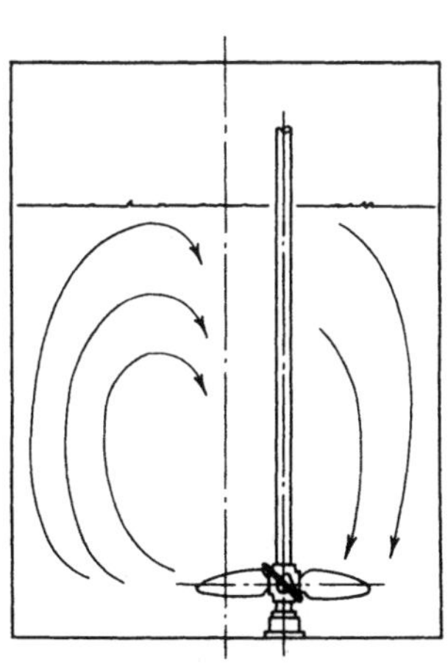

FIG. 18-13 Flow pattern with a paper-stock propeller, unbaffled; vertical off-center position.

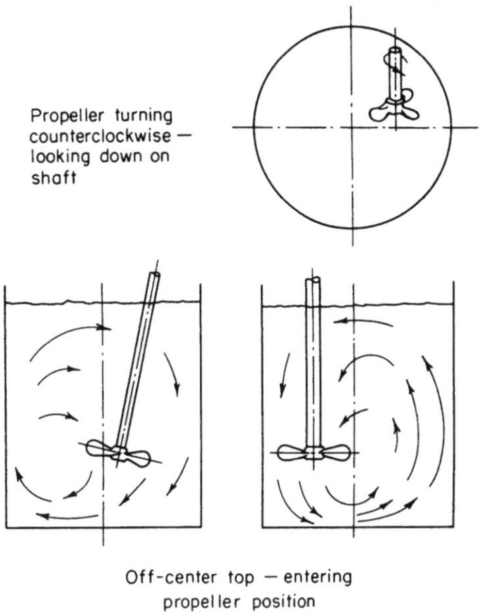

FIG. 18-14 Typical flow pattern with a propeller in angular off-center position without baffles.

slurries, the baffles often are located one-half of their width from the vessel wall to minimize accumulation of solids on or behind them.

For Reynolds numbers greater than 2000 baffles are commonly used with turbine impellers and with on-centerline axial-flow impellers. The flow patterns illustrated in Figs. 18-15 and 18-16 are quite different, but in both cases the use of baffles results in a large top-to-bottom circulation without vortexing or severely unbalanced fluid forces on the impeller shaft.

In the transition region [Reynolds numbers, Eq. (18-1), from 10 to 10,000], the width of the baffle may be reduced, often to one-half of standard width. If the circulation pattern is satisfactory when the tank is unbaffled but a vortex creates a problem, partial-length baffles may be used. These are standard-width and extend downward from the surface into about one-third of the liquid volume.

In the region of laminar flow ($N_{Re} < 10$), the same power is consumed by the impeller whether baffles are present or not, and they are seldom required. The flow pattern may be affected by the baffles, but not always advantageously. When they are needed, the baffles are usually placed one or two widths radially off the tank wall, to allow fluid to circulate behind them and at the same time produce some axial deflection of flow.

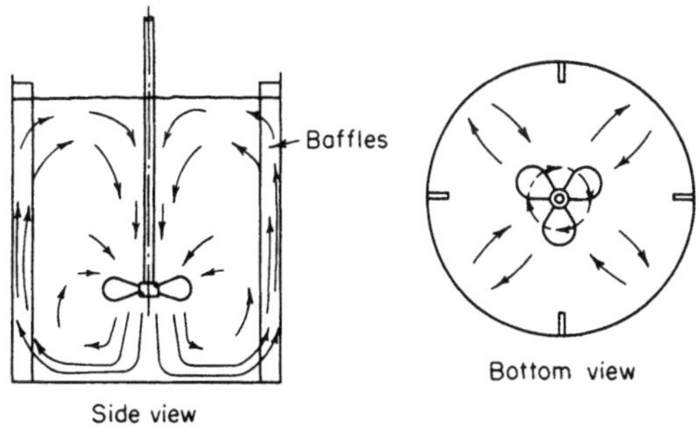

FIG. 18-15 Typical flow pattern in a baffled tank with a propeller or an axial-flow turbine positioned on center.

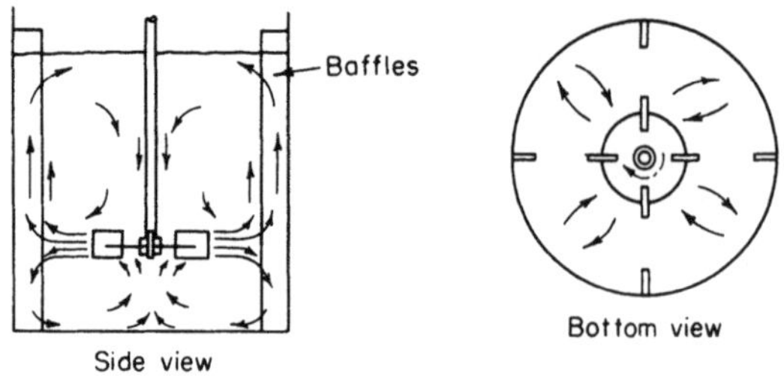

FIG. 18-16 Typical flow pattern in a baffled tank with a turbine positioned on center.

FLUID BEHAVIOR IN MIXING VESSELS

Impeller Reynolds Number The presence or absence of turbulence in an impeller-stirred vessel can be correlated with an impeller Reynolds number defined

$$N_{Re} = \frac{D_a^2 N \rho}{\mu} \tag{18-1}$$

where N = rotational speed, r/s; D_a = impeller diameter, m (ft); ρ = fluid density, kg/m^3 (lb/ft^3); and μ = viscosity, Pa·s [lb/(ft·s)]. Flow in the tank is turbulent when $N_{Re} > 10{,}000$. Thus viscosity alone is not a valid indication of the type of flow to be expected. Between Reynolds numbers of 10,000 and approximately 10 is a transition range in which flow is turbulent at the impeller and laminar in remote parts of the vessel; when $N_{Re} < 10$, flow is laminar only.

Not only is the type of flow related to the impeller Reynolds number, but also such process performance characteristics as mixing time, impeller pumping rate, impeller power consumption, and heat- and mass-transfer coefficients can be correlated with this dimensionless group.

Relationship between Fluid Motion and Process Performance Several phenomena which can be used to promote various processing objectives occur during fluid motion in a vessel.

1. Shear stresses are developed in a fluid when a layer of fluid moves faster or slower than a nearby layer of fluid or a solid surface. In laminar flow, the shear stress is equal to the product of fluid viscosity and velocity gradient or rate of shear. Under laminar-flow conditions, shear forces are larger than inertial forces in the fluid.

With turbulent flow, shear stress also results from the behavior of transient random eddies, including large-scale eddies which decay to small eddies or fluctuations. The scale of the large eddies depends on equipment size. On the other hand, the scale of small eddies, which dissipate energy primarily through viscous shear, is almost independent of agitator and tank size.

The shear stress in the fluid is much higher near the impeller than it is near the tank wall. The difference is greater in large tanks than in small ones.

2. Inertial forces are developed when the velocity of a fluid changes direction or magnitude. In turbulent flow, inertia forces are larger than viscous forces. Fluid in motion tends to continue in motion until it meets a solid surface or other fluid moving in a different direction. Forces are developed during the momentum transfer that takes place. The forces acting on the impeller blades fluctuate in a random manner related to the scale and intensity of turbulence at the impeller.

3. The interfacial area between gases and liquids, immiscible liquids, and solids and liquids may be enlarged or reduced by these viscous and inertia forces when interacting with interfacial forces such as surface tension.

4. Concentration and temperature differences are reduced by bulk flow or circulation in a vessel. Fluid regions of different composition or temperature are reduced in thickness by bulk motion in which velocity gradients exist. This process is called bulk diffusion or Taylor diffusion (Brodkey, in Uhl and Gray, op. cit., vol. 1, p. 48). The turbulent and molecular diffusion reduces the difference between these regions. In laminar flow, Taylor diffusion and molecular diffusion are the mechanisms of concentration- and temperature-difference reduction.

5. Equilibrium concentrations which tend to develop at solid-liquid, gas-liquid, or liquid-liquid interfaces are displaced or changed by molecular and turbulent diffusion between bulk fluid and fluid adjacent to the interface. Bulk motion (Taylor diffusion) aids in this mass-transfer mechanism also.

Turbulent Flow in Stirred Vessels Turbulence parameters such as intensity and scale of turbulence, correlation coefficients, and energy spectra have been measured in stirred vessels. However, these characteristics are not used directly in the design of stirred vessels.

Fluid Velocities in Mixing Equipment Fluid velocities have been measured for various turbines in baffled and unbaffled vessels. Typical data are summarized in Uhl and Gray, op. cit., vol. 1, chap. 4. Velocity data have been used for calculating impeller discharge and circulation rates but are not employed directly in the design of mixing equipment.

Impeller Discharge Rate and Fluid Head for Turbulent Flow When fluid viscosity is low and flow is turbulent, an impeller moves fluids by an increase in momentum from the blades which exert a force on the fluid. The blades of rotating propellers and turbines change the direction and increase the velocity of the fluids.

The pumping rate or discharge rate of an impeller is the flow rate perpendicular to the impeller discharge area. The fluid passing through this area has velocities proportional to the impeller peripheral velocity and velocity heads proportional to the square of these velocities at each point in the impeller discharge stream under turbulent-flow conditions. The following equations relate velocity head, pumping rate, and power for geometrically similar impellers under turbulent-flow conditions:

$$Q = N_Q N D_a^3 \tag{18-2}$$

$$H = \frac{N_p N^2 D_a^2}{N_Q g} \tag{18-3}$$

$$P = N_p \rho N^3 \left(\frac{D_a^5}{g_c} \right) \tag{18-4}$$

$$P = \frac{\rho H Q g}{g_c} \tag{18-5}$$

where Q = impeller discharge rate, m^3/s (ft^3/s); N_Q = discharge coefficient, dimensionless; H = velocity head, m (ft); N_p = power number, dimensionless; P = power, (N·m)/s [(ft·lbf)/s]; g_c = dimensional constant, 32.2 (ft·lb)/(lbf·s^2)(g_c = 1 when using SI units); and g = gravitational acceleration, m/s^2 (ft/s^2).

The discharge rate Q has been measured for several types of impellers, and discharge coefficients have been calculated. The data of a number of investigators are reviewed by Uhl and Gray (op. cit., vol. 1, chap. 4). N_Q is 0.4 to 0.5 for a propeller with pitch equal to diameter at $N_{Re} = 10^5$. For turbines, N_Q ranges from 0.7 to 2.9, depending on the number of blades, blade-height-to-impeller-diameter ratio, and impeller-to-vessel-diameter ratio. The effects of these geometric variables are not well defined.

Power consumption has also been measured and correlated with impeller Reynolds numbers. The velocity head for a mixing impeller can be calculated, then, from flow and power data, by Eq. (18-3) or Eq. (18-5).

The velocity head of the impeller discharge stream is a measure of the maximum force that this fluid can exert when its velocity is changed. Such inertia forces are higher in streams with higher discharge velocities. Shear rates and shear stresses are also higher under these conditions in the smallest eddies. If a higher discharge velocity is desired at the same power consumption, a smaller-diameter impeller must be used at a higher rotational speed. According to Eq. (18-4), at a given power level $N \propto D_a^{-5/3}$ and $ND_a \propto D_a^{-2/3}$. Then, $H \propto D_a^{-4/3}$ and $Q \propto D_a^{4/3}$.

An impeller with a high fluid head is one with high peripheral velocity and discharge velocity. Such impellers are useful for (1) rapid reduction of concentration differences in the impeller discharge stream (rapid mixing), (2) production of large interfacial area and small droplets in gas-liquid and immiscible-liquid systems, (3) solids deagglomeration, and (4) promotion of mass transfer between phases.

The impeller discharge rate can be increased at the same power consumption by increasing impeller diameter and decreasing rotational speed and peripheral velocity so that $N^3D_a^5$ is a constant (Eq. 18-4)]. Flow goes up, velocity head and peripheral velocity go down, but impeller torque T_Q goes up. At the same torque, $N^2D_a^5$ is constant, $P \propto D_a^{-5/2}$, and $Q \propto D_a^{1/2}$. Therefore, increasing impeller diameter at constant torque increases discharge rate at lower power consumption. At the same discharge rate, ND_a^3 is constant, $P \propto D_a^{-4}$, and $T_Q \propto D_a^{-1}$. Therefore, power and torque decrease as impeller diameter is increased at constant Q.

A large-diameter impeller with a high discharge rate is used for (1) short times to complete mixing of miscible liquid throughout a vessel, (2) promotion of heat transfer, (3) reduction of concentration and temperature differences in all parts of vessels used for constant-environment reactors and continuous averaging, and (4) suspension of particles of relatively low settling rate.

Laminar Fluid Motion in Vessels When the impeller Reynolds number is less than 10, the flow induced by the impeller is laminar. Under these conditions, the impeller drags fluid with it in a predominantly circular pattern. If the impeller blades curve back, there is a viscous drag flow toward the tips of these blades. Under moderate-viscosity conditions in laminar flow, centrifugal force acting on the fluid layer dragged in a circular path by the rotating impeller will move fluid in a radial direction. This centrifugal effect causes any gas accumulated behind a rotating blade to move to the axis of impeller rotation. Such radial-velocity components are small relative to tangential velocity.

For turbines at Reynolds numbers less than 100, toroidal stagnant zones exist above and below the turbine periphery. Interchange of liquid between these regions and the rest of the vessel is principally by molecular diffusion.

Suspensions of fine solids may have pseudoplastic or plastic-flow properties. When they are in laminar flow in a stirred vessel, motion in remote parts of the vessel where shear rates are low may become negligible or cease completely. To compensate for this behavior of slurries, large-diameter impellers or paddles are used, with $(D_a/D_T) > 0.6$, where D_T is the tank diameter. In some cases, for example, with some anchors, $D_a > 0.95\ D_T$. Two or more paddles may be used in deep tanks to avoid stagnant regions in slurries.

In laminar flow ($N_{Re} < 10$), $N_p \propto 1/N_{Re}$ and $P \propto \mu N^2 D_a^3$. Since shear stress is proportional to rotational speed, shear stress can be increased at the same power consumption by increasing N proportionally to $D_a^{-3/2}$ as impeller diameter D_a is decreased.

Fluid circulation probably can be increased at the same power consumption and viscosity in laminar flow by increasing impeller diameter and decreasing rotational speed, but the relationship between Q, N, and D_a for laminar flow from turbines has not been determined.

As in the case of turbulent flow, then, small-diameter impellers ($D_a < D_T/3$) are useful for (1) rapid mixing of dry particles into liquids, (2) gas dispersion in slurries, (3) solid-particle deagglomeration, and (4) promoting mass transfer between solid and liquid phases. If stagnant regions are a problem, large impellers must be used and rotational speed and power increased to obtain the required results. Small continuous-processing equipment may be more economical than batch equipment in such cases.

Likewise, large-diameter impellers ($D_a > D_T/2$) are useful for (1) avoiding stagnant regions in slurries, (2) short mixing times to obtain uniformity throughout a vessel, (3) promotion of heat transfer, and (4) laminar continuous averaging of slurries.

Vortex Depth In an unbaffled vessel with an impeller rotating in the center, centrifugal force acting on the fluid raises the fluid level at the wall and lowers the level at the shaft. The depth and shape of such a vortex [Rieger, Ditl, and Novak, *Chem. Eng. Sci.*, **34**, 397 (1978)] depend on impeller and vessel dimensions as well as rotational speed.

Power Consumption of Impellers Power consumption is related to fluid density, fluid viscosity, rotational speed, and impeller diameter by plots of power number ($g_cP/\rho N^3D_a^5$) versus Reynolds number ($D_a^2N\rho/\mu$). Typical correlation lines for frequently used impellers operating in newtonian liquids contained in baffled cylindrical vessels are presented in Fig. 18-17. These curves may be used also

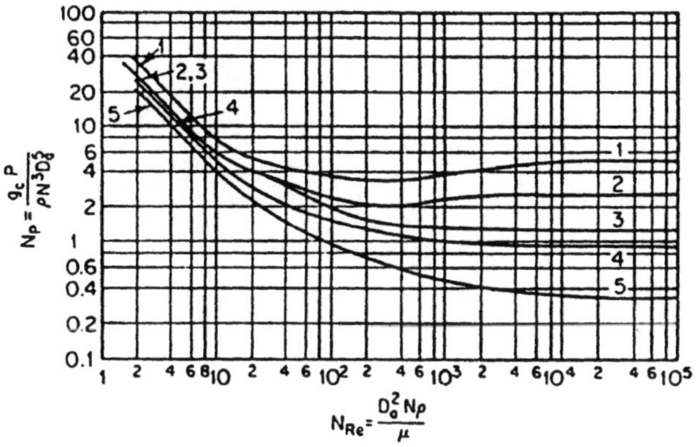

FIG. 18-17 Impeller power correlations: curve 1, six-blade turbine, $D_a/W_i = 5$, like Fig. 18-4 but with six blades, four baffles, each $D_T/12$; curve 2, vertical-blade, open turbine with six straight blades, $D_a/W_i = 8$, four baffles each $D_T/12$; curve 3, 45° pitched-blade turbine like Fig. 18-3 but with six blades, $D_a/W_i = 8$, four baffles, each $D_T/12$; curve 4, propeller, pitch equal to $2D_a$, four baffles, each $0.1D_T$, also same propeller in angular off-center position with no baffles; curve 5, propeller, pitch equal to D_a, four baffles each $0.1D_T$, also same propeller in angular off-center position as in Fig. 18-14 with no baffles. D_a = impeller diameter, D_T = tank diameter, g_c = gravitational conversion factor, N = impeller rotational speed, P = power transmitted by impeller shaft, W_i = impeller blade height, μ = viscosity of stirred liquid, and ρ = density of stirred mixture. Any set of consistent units may be used, but N must be rotations (rather than radians) per unit time. In the SI system, g_c is dimensionless and unity. [Curves 4 and 5 from Rushton, Costich, and Everett, *Chem. Eng. Prog.*, **46**, 395, 467 (1950), by permission; curves 2 and 3 from Bates, Fondy, and Corpstein, *Ind. Eng. Chem. Process Des. Dev.*, **2**, 310 (1963), by permission of the copyright owner, the American Chemical Society.]

for operation of the respective impellers in unbaffled tanks when the Reynolds number is 300 or less. When N_{Re} is greater than 300, however, the power consumption is lower in an unbaffled vessel than indicated in Fig. 18-17. For example, for a six-blade disc turbine with $D_T/D_a = 3$ and $D_a/W_i = 5$, $N_p = 1.2$ when $N_{Re} = 10^4$. This is only about one-fifth of the value of N_p when baffles are present.

Additional power data for other impeller types such as anchors, curved-blade turbines, and paddles in baffled and unbaffled vessels are available in the following references: Holland and Chapman, op. cit., chaps. 2, 4, Reinhold, New York, 1966; and Bates, Fondy, and Fenic, in Uhl and Gray, op. cit., vol. 1, chap. 3.

Power consumption for impellers in pseudoplastic, Bingham plastic, and dilatant non-newtonian fluids may be calculated by using the correlating lines of Fig. 18-17 if viscosity is obtained from viscosity-shear rate curves as described here. For a pseudoplastic fluid, viscosity decreases as shear rate increases. A Bingham plastic is similar to a pseudoplastic fluid but requires that a minimum shear stress be exceeded for any flow to occur. For a dilatant fluid, viscosity increases as shear rate increases.

The appropriate shear rate to use in calculating viscosity is given by one of the following equations when a propeller or a turbine is used (Bates et al., in Uhl and Gray, op. cit., vol. 1, p. 149):

For dilatant liquids,

$$\dot{\gamma} = 13N\left(\frac{D_a}{D_T}\right)^{0.5} \tag{18-6}$$

For pseudoplastic and Bingham plastic fluids,

$$\dot{\gamma} = 10N \tag{18-7}$$

where $\dot{\gamma}$ = average shear rate, s^{-1}.

The shear rate calculated from impeller rotational speed is used to identify a viscosity from a plot of viscosity versus shear rate determined with a capillary or rotational viscometer. Next N_{Re} is calculated, and N_p is read from a plot like Fig. 18-17.

DESIGN OF AGITATION EQUIPMENT

Selection of Equipment The principal factors which influence mixing-equipment choice are (1) the process requirements, (2) the flow properties of the process fluids, (3) equipment costs, and (4) construction materials required.

Ideally, the equipment chosen should be that of the lowest total cost which meets all process requirements. The total cost includes depreciation on investment, operating cost such as power, and maintenance costs. Rarely is any more than a superficial evaluation based on this principle justified, however, because the cost of such an evaluation often exceeds the potential savings that can be realized. Usually optimization is based on experience with similar mixing operations. Often the process requirements can be matched with those of a similar operation, but sometimes tests are necessary to identify a satisfactory design and to find the minimum rotational speed and power.

There are no satisfactory specific guides for selecting mixing equipment because the ranges of application of the various types of equipment overlap and the effects of flow properties on process performance have not been adequately defined. Nevertheless, what is frequently done in selecting equipment is described in the following paragraphs.

Top-Entering Impellers For vessels less than 1.8 m (6 ft) in diameter, a clamp- or flange-mounted, angular, off-center fluidfoil impeller with no baffles should be the initial choice for meeting a wide range of process requirements (Fig. 18-14). The vessel straight-side-height-to-diameter ratio should be 0.75 to 1.5, and the volume of stirred liquid should not exceed 4 m^3 (about 1000 gal).

For suspension of free-settling particles, circulation of pseudoplastic slurries, and heat transfer or mixing of miscible liquids to obtain uniformity, a speed of 350 or 420 r/min should be stipulated. For dispersion of dry particles in liquids or for rapid initial mixing of liquid reactants in a vessel, an 1150- or 1750- r/min propeller should be used at a distance $D_T/4$ above the vessel bottom. A second propeller can be added to the shaft at a depth D_a below the liquid surface if the submergence of floating liquids or particulate solids is otherwise inadequate. Such propeller mixers are readily available up to 2.2 kW (3 hp) for off-center sloped-shaft mounting.

Propeller size, pitch, and rotational speed may be selected by model tests, by experience with similar operations, or, in a few cases, by published correlations of performance data such as mixing time or heat transfer. The propeller diameter and motor power should be the minimum that meets process requirements.

If agitation is required for a vessel less than 1.8 m (6 ft) in diameter and the same operations will be scaled up to a larger vessel ultimately, the equipment type should be the same as that expected in the larger vessel.

Axial-Flow Fluidfoil Impellers For vessel volumes of 4 to 200 m^3 (1000 to 50,000 gal), a turbine mixer mounted coaxially within the vessel with four or more baffles should be the initial choice. Here also the vessel straight-side-height-to-diameter ratio should be 0.75 to 1.5. Four vertical baffles should be fastened perpendicularly to the vessel wall with a gap between baffle and wall equal to $D_T/24$ and a radial baffle width equal to $D_T/12$.

For suspension of rapidly settling particles, the impeller turbine diameter should be $D_T/3$ to $D_T/2$. A clearance of less than one-seventh of the fluid depth in the vessel should be used between the lower edge of the turbine blade tips and the vessel bottom. As the viscosity of a suspension increases, the impeller diameter should be increased. This diameter may be increased to 0.6 D_T and a second impeller added to avoid stagnant regions in pseudoplastic slurries. Moving the baffles halfway between the impeller periphery and the vessel wall will also help avoid stagnant fluid near the baffles.

As has been shown, power consumption is decreased and turbine discharge rate is increased as impeller diameter is increased at constant torque (in the completely turbulent regime). This means that for a stipulated discharge rate, more efficient operation is obtained (lower power and torque) with a relatively large impeller operating at a relatively low speed ($N \propto D_a^{-3}$). Conversely, if power is held constant, decreasing impeller diameter results in increasing peripheral velocity and decreasing torque. Thus at a stipulated power level the rapid, efficient initial mixing of reactants identified with high peripheral velocity can be achieved by a relatively small impeller operating at a relatively high speed ($N \propto D_a^{-5/3}$).

For circulation and mixing to obtain uniformity, the impeller should be located at one-third of the liquid depth above the vessel bottom unless rapidly settling material or a need to stir a nearly empty vessel requires a lower impeller location.

Side-Entering Impellers For vessels greater than 4 m^3 (1000 gal), a side-entering propeller agitator (Fig. 18-9) may be more economical than a top-mounted impeller on a centered vertical shaft. For vessels greater than 38 m^3 (10,000 gal), the economic attractiveness of side-entering impellers increases. For vessels larger than 380 m^3 (100,000 gal), units may be as large as 56 kW (75 hp), and two or even three may be installed in one tank. For the suspension of slow-settling particles or the maintenance of uniformity in a viscous slurry of small particles, the diameter and rotational speed of a side-entering agitator must be selected on the basis of model tests or experience with similar operations.

When abrasive solid particles must be suspended, maintenance costs for the submerged shaft seal of a side-entering propeller may become high enough to make this type of mixer an uneconomical choice.

Jet Mixers Continuous recycle of the contents of a tank through an external pump so arranged that the pump discharge stream appropriately reenters the vessel can result in a flow pattern in the tank which will produce a slow mixing action [Fossett, *Trans. Inst. Chem. Eng.*, **29**, 322 (1951)].

Large Tanks Most large vessels (over 4 m^3) require a heavy-duty drive. About two-thirds of the mixing requirements industrially involve flow, circulation, and other types of pumping capacity requirements, including such applications as blending and solid suspension. There often is no requirement for any marked level of shear rate, so the use of the fluidfoil impellers is most common. If additional shear rate is required over what can be provided by the fluidfoil impeller, the axial-flow turbine (Fig. 18-3) is often used, and if extremely high shear rates are required, the flat-blade turbine (Rushton turbine) (Fig. 18-4) is required. For still higher shear rates, there is an entire variety of high-shear-rate impellers, typified by that shown in Fig. 18-10 that are used.

The fluidfoil impellers in large tanks require only two baffles, but three are usually used to provide better flow pattern asymmetry. These fluidfoil impellers provide a true axial flow pattern, almost as though there was a draft tube around the impeller. Two or three or more impellers are used if tanks with high D/T ratios are involved. The fluidfoil impellers do not vortex vigorously even at relatively low coverage so that if gases or solids are to be incorporated at the surface, the axial-flow turbine is often required and can be used in combination with the fluidfoil impellers also on the same shaft.

BLENDING

If the blending process is between two or more fluids with relatively low viscosity such that the blending is not affected by fluid shear rates, then the difference in blend time and circulation between small and large tanks is the only factor involved. However, if the blending involves wide disparities in the density of viscosity and surface tension between the various phases, then a certain level of shear rate may be required before blending can proceed to the required degree of uniformity.

The role of viscosity is a major factor in going from the turbulent regime, through the transition region, into the viscous regime and the change in the role of energy dissipation discussed previously. The role of non-newtonian viscosities comes into the picture very strongly since that tends to markedly change the type of influence of impellers and determines the appropriate geometry that is involved.

There is the possibility of misinterpretation of the difference between circulation time and blend time. Circulation time is primarily a function of the pumping capacity of the impeller. For axial-flow impellers, a convenient parameter, but not particularly physically accurate, is to divide the pumping capacity of the impeller by the cross-sectional area of the tank to give a superficial liquid velocity.

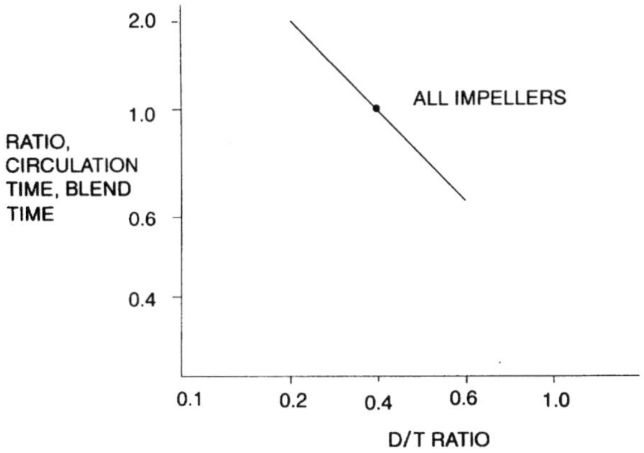

FIG. 18-18 Effect of *D/T* ratio on any impeller on the circulation time and the blend time.

This is sometimes used by using the total volume of flow from the impeller including entrainment of the tank to obtain a superficial liquid velocity.

As the flow from an impeller is increased from a given power level, there will be a higher fluid velocity and therefore a shorter circulation time. This holds true when dealing with any given impeller. This is shown in Fig. 18-18, which shows that circulation time versus *D/T* decreases. A major consideration is when increasing *D/T* becomes too large and actually causes the curve to reverse. This occurs somewhere around 0.45, ± 0.05, so that using impellers of *D/T* ratios of 0.6 to 0.8 is often counterproductive for circulation time. They may be useful for the blending or motion of pseudoplastic fluids.

When comparing different impeller types, an entirely different phenomenon is important. In terms of circulation time, the phenomena shown in Figs. 18-18 and 18-19 still apply with the different impellers shown in Fig. 18-5. When it comes to blending another factor enters the picture. When particles A and B meet each other as a result of shear rates, there has to be sufficient shear stress to cause A and B to blend, react, or otherwise participate in the process.

It turns out that in low-viscosity blending the actual result does depend upon the measuring technique used to measure blend time. Two common techniques, which do not exhaust the possibilities in reported studies, are to use an acid-base indicator and inject an acid or base into the system that will result in a color change. One can also put a dye into the tank and measure the time for color to arrive at uniformity. Another system is to put in a conductivity probe and inject a salt or other electrolyte into the system. With any given impeller type at constant power, the circulation time will increase with the *D/T* ratio of the impeller. Figure 18-18 shows that both circulation time and blend time decrease as *D/T* increases. The same is true for impeller speed. As impeller speed is increased with any impeller, blend time and circulation time are decreased (Fig. 18-19).

However, when comparing different impeller types at the same power level, it turns out that impellers that have a higher pumping capacity will give decreased circulation time, but all the impellers, regardless of their pumping efficiency, give the same blend time at the same power level and same diameter. This means that circulation time must be combined with shear rate to carry out a blending experiment which involves chemical reactions or interparticle mixing (Fig. 18-20).

For other situations in low-viscosity blending, the fluid in tanks may become stratified. There are few studies on that situation, but Oldshue (op. cit.) indicates the relationship between some of the variables. The important difference is that blend time is inversely proportional to power, not impeller flow, so that the exponents are quite different for a stratified tank. This situation occurs more frequently in the petroleum industry, where large petroleum storage tanks become stratified either by filling techniques or by temperature fluctuations.

There is a lot of common usage of the terms *blend time, mixing time,* and *circulation time.* There are differences in concept and interpretation of these different "times." For any given experiment, one must pick a definition of *blend time* to be used. As an example, if one is measuring the fluctuation of concentration after an addition of material to the tank, then one can pick an arbitrary definition of blending such as reducing the fluctuations below a certain level. This often is chosen as a fluctuation equal to 5% of the original fluctuation when the feed material is added. This obviously is a function of the size of the probe used to measure these fluctuations, which often is on the order of 500 to 1000 μm.

At the micro-scale level, there really is no way to measure concentration fluctuations. Resort must be made to other qualitative interpretation of results for either a process or a chemical reaction study.

High-Viscosity Systems All axial-flow impellers become radial flow as Reynolds numbers approach the viscous region. Blending in the transition and low-viscosity system is largely a measure of fluid motion throughout the tank. For close-clearance impellers, the anchor and helical impellers provide blending by having an effective action at the tank wall, which is particularly suitable for pseudoplastic fluids.

Figure 18-21 gives some data on the circulation time of the helical impeller. It has been observed that it takes about three circulation times to get one blend time being the visual uniformity of a dye added to the material. This is a macro-scale blending definition.

Axial-flow turbines are often used in blending pseudoplastic materials, and they are often used at relatively large *D/T* ratios, from 0.5 to 0.7, to adequately provide shear rate in the majority of the batch particularly in pseudoplastic material. These impellers develop a flow

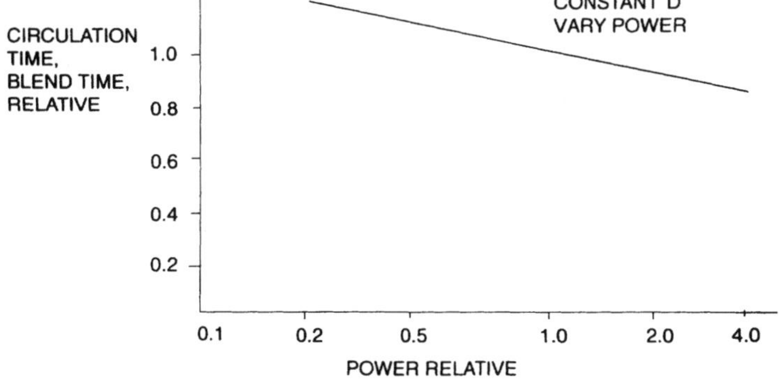

FIG. 18-19 Effect of impeller power for the same diameter on circulation time and blend time for a particular impeller.

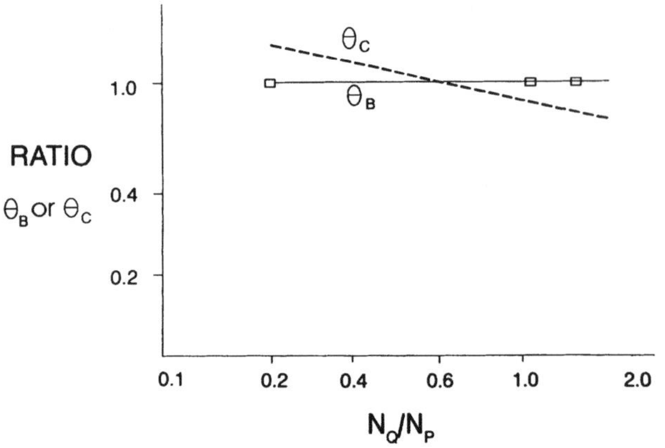

FIG. 18-20 At constant power and constant impeller diameter, three different impellers give the same blend time but different circulation times.

pattern which may or may not encompass an entire tank, and these areas of motion are sometimes referred to as *caverns*. Several papers describe the size of these caverns relative to various types of mixing phenomena. An effective procedure for the blending of pseudoplastic fluids is given in Oldshue (op. cit.).

Chemical Reactions Chemical reactions are influenced by the uniformity of concentration both at the feed point and in the rest of the tank and can be markedly affected by the change in overall blend time and circulation time as well as the micro-scale environment. It is possible to keep the ratio between the power per unit volume at the impeller and in the rest of the tank relatively similar on scale-up, but many details need to be considered when talking about the reaction conditions, particularly where they involve selectivity. This means that reactions can take different paths depending upon chemistry and fluid mechanics, which is a major consideration in what should be examined. The method of introducing the reagent stream can be projected in several different ways depending upon the geometry of the impeller and feed system.

Chemical reactions normally occur in the micro-scale range. In turbulent flow, almost all of the power dissipation occurs eventually in the micro-scale regime because that is the only place where the scale of the fluid fluctuations is small enough that viscous shear stress exists. At approximately 100 μm, the fluid does not know what type of impeller is used to generate the power; continuing down to 10 μm and, even further, to chemical reactions, the actual impeller type is not a major variable as long as the proper macro-scale regime has been provided throughout the entire tank. The intensity of the mixing environment in the micro-scale regime can be related to a series of variables in an increasing order of complexity. Since all of the power is ultimately dissipated in the micro-scale regime, the power per unit volume throughout the tank is one measure of the overall measure of micro-scale mixing and the power dissipation at individual volumes in the tank is another way of expressing the influence. In general, the power per unit volume dissipated around an impeller zone can be 100 times higher than the power dissipated throughout the remainder of the tank.

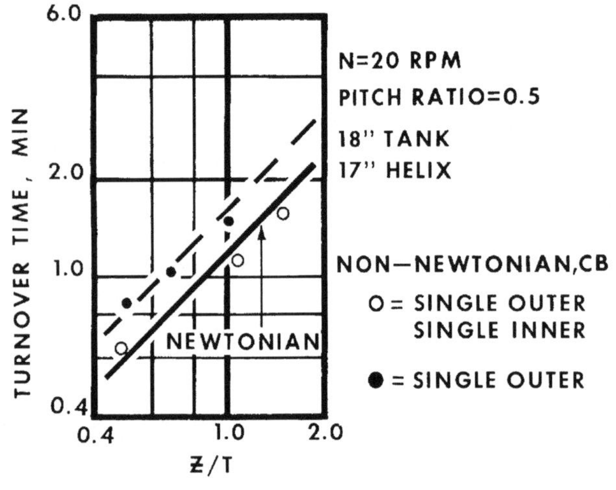

FIG. 18-21 Effect of impeller speed on circulation time for a helical impeller in the Reynolds number arranged less than 10.

The next level of complexity is to look at the rms velocity fluctuation, which is typically 50 percent of the mean velocity around the impeller zone and about 5 percent of the mean velocity in the rest of the vessel. This means that the feed introduction point for either a single reactant or several reactants can be of extreme importance. It seems that the selectivity of competing or consecutive chemical reactions can be a function of the rms velocity fluctuations in the feed point if the chemical reactants remain constant and involve an appropriate relationship to the time between the rms velocity fluctuations. There are three common ways of introducing reagents into a mixing vessel. One is to let them drip on the surface. The second is to use some type of introduction pipe to bring the material into various parts of the vessel. The third is to purposely bring them in and around the impeller zone. Generally, all three methods have to be tried before determining the effect of feed location.

Since chemical reactions are on a scale much below 1 μm, and it appears that the Komolgoroff scale of isotropic turbulence turns out to be somewhere between 10 and 30 μm, other mechanisms must play a role in getting materials in and out of reaction zones and reactants in and out of those zones. One cannot really assign a shear rate magnitude to the area around a micro-scale zone, and it is primarily an environment that particles and reactants witness in this area.

The next level of complexity looks at the kinetic energy of turbulence. There are several models that are used to study the fluid mechanics, such as the Kε model. One can also put the velocity measurements through a spectrum analyzer to look at the energy at various wave numbers.

In the viscous regime, chemical reactants become associated with each other through viscous shear stresses. These shear stresses exist at all scales (macro to micro) and until the power is dissipated continuously through the entire spectrum. This gives a different relationship for power dissipation than in the case of turbulent flow.

SOLID-LIQUID SYSTEMS

The most-used technique to study solid suspension, as documented in hundreds of papers in the literature, is called the *speed for just suspension*, N_{JS}. The original work was done in 1958 by Zwietering and this is still the most extensive range of variables, although other investigators have added to it considerably.

This particular technique is suitable only for laboratory investigation using tanks that are transparent and well illuminated. It does not lend itself to evaluation of the opaque tanks, nor is it used in any study of large-scale tanks in the field. It is a very minimal requirement for uniformity, and definitions suggested earlier are recommended for use in industrial design.

Some Observations on the Use of N_{JS} With D/T ratios of less than 0.4, uniformity throughout the rest of the tank is minimal. In D/T ratios greater than 0.4, the rest of the tank has a very vigorous fluid motion with a marked approach to complete uniformity before N_{JS} is reached.

Much of the variation in N_{JS} can be reduced by using P_{JS}, which is the power in the just-suspended state. This also gives a better feel for the comparison of various impellers based on the energy requirement rather than speed, which has no economic relevance.

The overall superficial fluid velocity, mentioned earlier, should be proportional to the settling velocity of the solids if that were the main mechanism for solid suspension. If this were the case, the requirement for power if the settling velocity were doubled should be eight times. Experimentally, it is found that the increase in power is more nearly four times, so that some effect of the shear rate in macro-scale turbulence is effective in providing uplift and motion in the system.

Picking up the solids at the bottom of the tank depends upon the eddies and velocity fluctuations in the lower part of the tank and is a different criterion from the flow pattern required to keep particles suspended and moving in various velocity patterns throughout the remainder of the vessel. This leads to the variables in the design equation and a relationship that is quite different when these same variables are studied in relation to complete uniformity throughout the mixing vessel.

Another concern is the effect of multiple particle sizes. In general, the presence of fine particles will affect the requirements of suspension of larger particles. The fine particles act largely as a potential viscosity-increasing agent and give a similar result to what would happen if the viscosity of the continuous phase were increased.

Another phenomenon is the increase in power required with percent solids, which makes a dramatic change at approximately 40 percent by volume, and then dramatically changes again as we approach the ultimate weight percent of settled solids. This phenomenon is covered by Oldshue (op. cit.), who describes conditions required for mixing slurries in the 80 to 100 percent range of the ultimate weight percent of settled solids.

Solids suspension in general is not usually affected by blend time or shear-rate changes in the relatively low to medium solids concentration in the range from 0 to 40 percent by weight. However, as solids become more concentrated, the effect of solids concentration on power required gives a change in criterion from the settling velocity of the individual particles in the mixture to the apparent viscosity of the more concentrated slurry. This means that we enter into an area where the blending of non-newtonian fluid regions affects the shear rates and plays a marked role.

The suspension of a single solid particle should depend primarily on the upward velocity at a given point and also should be affected by the uniformity of this velocity profile across the entire tank cross section. There are upward velocities in the tank and there also must be corresponding downward velocities.

In addition to the effect of the upward velocity on a settling particle, there is also the random motion of the micro-scale environment, which does not affect large particles very much but is a major factor in the concentration and uniformity of particles in the transition and micro-scale size range.

Using a draft tube in the tank for solids suspension introduces another, different set of variables. There are other relationships that are very much affected by scale-up in this type of process, as shown in Fig. 18-22. Different scale-up problems exist whether the impeller is pumping up or down within the draft tube.

Solid Dispersion If the process involves the dispersion of solids in a liquid, then we may either be involved with breaking up agglomerates or possibly physically breaking or shattering particles that have a low cohesive force between their components. Normally, we do not think of breaking up ionic bonds with the shear rates available in mixing machinery.

If we know the shear stress required to break up a particle, we can then determine the shear rate required from the machinery by various viscosities with the equation:

$$\text{Shear stress} = \text{viscosity (shear rate)}$$

The shear rate available from various types of mixing and dispersion devices is known approximately and also the range of viscosities in which they can operate. This makes the selection of the mixing equipment subject to calculation of the shear stress required for the viscosity to be used.

In the equation referred to above, it is assumed that there is 100 percent transmission of the shear rate in the shear stress. However, with the slurry viscosity determined essentially by the properties of the slurry, at high concentrations of slurries there is a slippage factor. Internal motion of particles in the fluids over and around each other can reduce the effective transmission of viscosity efficiencies from 100 percent to as low as 30 percent.

Animal cells in biotechnology do not normally have tough skins like those of fungal cells and they are very sensitive to mixing effects. Many approaches have been and are being tried to minimize the effect of increased shear rates on scale-up. These include encapsulating the organism in or on microparticles and/or conditioning cells selectively to shear rates. In addition, traditional fermentation processes have maximum shear-rate requirements in which cells become progressively more and more damaged until they become motile.

Solid-Liquid Mass Transfer There is potentially a major effect of both shear rate and circulation time in these processes. The solids can either be fragile or rugged. We are looking at the slip velocity of the particle and also whether we can break up agglomerates of particles which may enhance the mass transfer. When the particles become small enough, they tend to follow the flow pattern, so the slip velocity necessary to affect the mass transfer becomes less and less available.

What this shows is that, from the definition of off-bottom motion to complete uniformity, the effect of mixer power is much less than from going to on-bottom motion to off-bottom suspension. The initial increase in power causes more and more solids to be in active communication with the liquid and has a much greater mass-transfer rate than that occurring above the power level for off-bottom suspension, in which slip velocity between the particles of fluid is the major contributor (Fig. 18-23).

Since there may well be chemical or biological reactions happening on or in the solid phase, depending upon the size of the process participants, macro- or micro-scale effects may or may not be appropriate to consider.

In the case of living organisms, their access to dissolved oxygen throughout the tank is of great concern. Large tanks in the fermentation industry often have a Z/T ratio of 2:1 to 4:1; thus, top-to-bottom blending can be a major factor. Some biological particles are facultative and can adapt and reestablish their metabolisms at different dissolved-oxygen levels. Other organisms are irreversibly destroyed by sufficient exposure to low dissolved-oxygen levels.

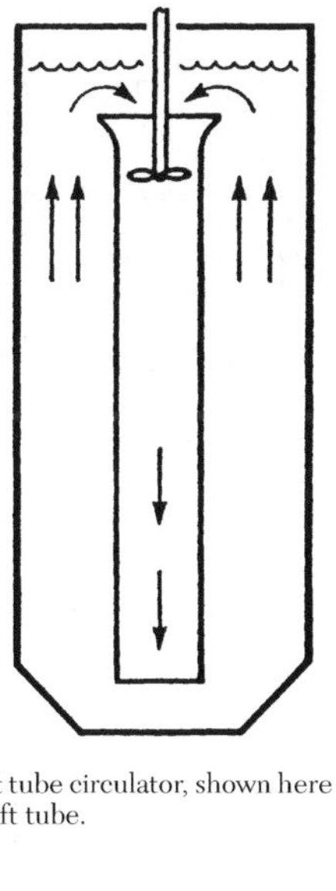

FIG. 18-22 Typical draft tube circulator, shown here for down-pumping mode for the impeller in the draft tube.

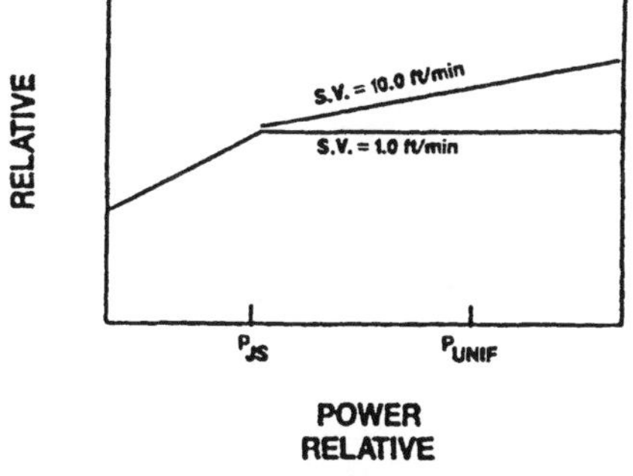

FIG. 18-23 Relative change in solid-liquid mass-transfer ratio with three different suspension levels, i.e., on-bottom motion, off-bottom motion, and complete uniformity.

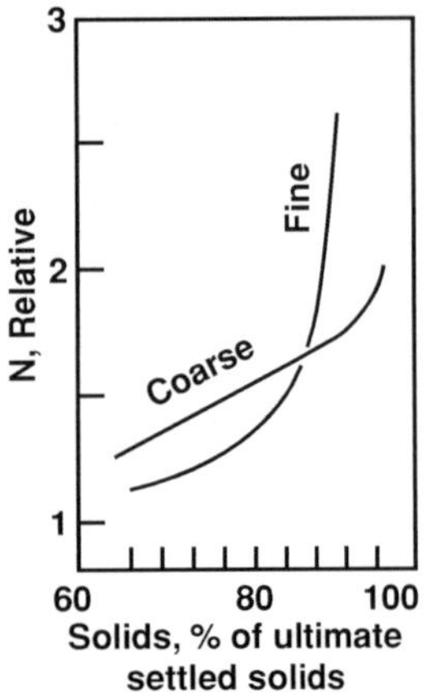

FIG. 18-24 Effect of percent solids on speed required for complete motion throughout the tank on two different grind sizes.

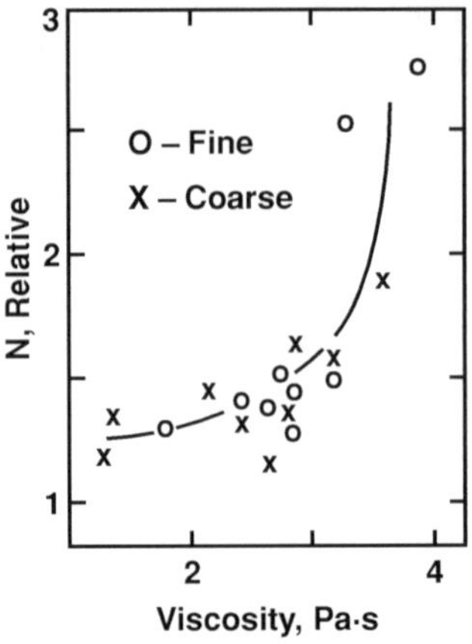

FIG. 18-26 Correlation of impeller speed vs. viscosity of the pulp including both fine and course grind experimental data points.

Leaching and Extraction of Mineral Values from High Concentration of Solids A uranium plant had 10 large slurry tanks for leaching and extraction (approximately 14 m in diameter and 14 m high). They had about 14,000-m^3 capacity.

In a study designed to modify the leaching operation, it was desired to look at two different grind sizes of ore, one labeled five grind and the other labeled coarse grind. Also, the effect of various mixer designs and power levels on the extraction efficiency to arrive at the overall economic optimum was examined. Figure 18-24 shows the results of a pilot study in which the impeller speed for a given impeller and tank geometry was measured for complete overall motion throughout the slurry for both the fine and coarse grinds at various weight percent solids. As can be seen in the figure, the fine material required lower horsepower at low weight percent solids while the coarse grind required less horsepower up near the ultimate settled solids weight percentage.

The interpretation is that at lower percent solids, the viscosity of the fine grind aided suspension whereas at higher percent solids, the higher viscosity of the fine material was detrimental to fluid mixing.

A mixing viscosimeter was used to measure the viscosity of the slurry. Figure 18-25 shows the viscosity of the fine and coarse slurries.

By combining the data from Figs. 18-24 and 18-25 into Fig. 18-26, it is seen that there is a correlation between the impeller speed required and the viscosity of the slurry regardless of whether the material was finely or coarsely ground. This illustrates that viscosity is a key parameter in the process design for solid-liquid slurries.

The overall process economics examined the extraction rate as a function of power, residence time, and grind size. The full-scale design possibilities were represented in the form of Table 18-2, which were accompanied by other charts that gave different heights of suspension in the tank for the three different particle size fractions: fine, medium, and coarse. These various combinations of power levels also gave various blending efficiencies and had different values of the effective residence time used in a system.

By calculating the residence times of the various solids in the tank and relating them to their corresponding extraction curves, the total uranium extraction for the entire train of mixers was estimated. The cost of the various mixer options, the production efficiency net result, and the cost of the installation and tank design could be combined to yield the economic optimum for the plant.

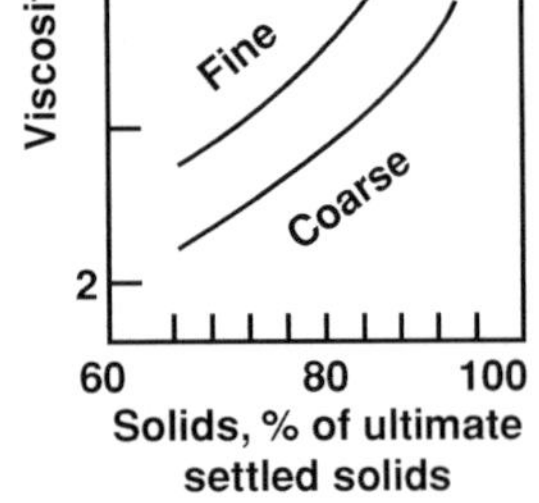

FIG. 18-25 Viscosity of fine grind and coarse grind at various weight percent solids.

GAS-LIQUID SYSTEMS

Gas-Liquid Dispersion This involves physical dispersion of gas bubbles by the impeller, and the effect of gas flow on the impeller.

The observation of the physical appearance of a tank undergoing gas-liquid mass transfer can be helpful but is not a substitute for mass-transfer data on the actual process. The mixing vessel can have four regimes of visual comparisons between gas bubbles and flow patterns. A helpful parameter is the ratio between the power given up by the gas phase and the power introduced by the mixing impeller. In general, if the power in the gas stream (calculated as the expansion energy from the gas expanding from the sparging area to the top of the tank, shown in Fig. 18-27) is greater, there will be considerable blurping and entrainment of liquid drops by a very violent explosion of gas bubbles at the surface. If the power level is more than the expanding gas energy, then the surface action will normally be very coalescent and uniform by comparison, and the gas will be reasonably well distributed throughout the remainder of the tank. With power levels up to 10 to 100 times the

TABLE 18-2 Four Different Selections of Mixers with Different Mixing Characteristics on 14,000 ft^3 of Leach Tanks

Proposal for Revised Installation			
Process factor	Relative torque	Motor kW	D/T (dual)
1.2	1.0	300	0.5
1.0	1.0	250	0.5
0.8	0.9	200	0.5
0.7	1.0	150	0.6

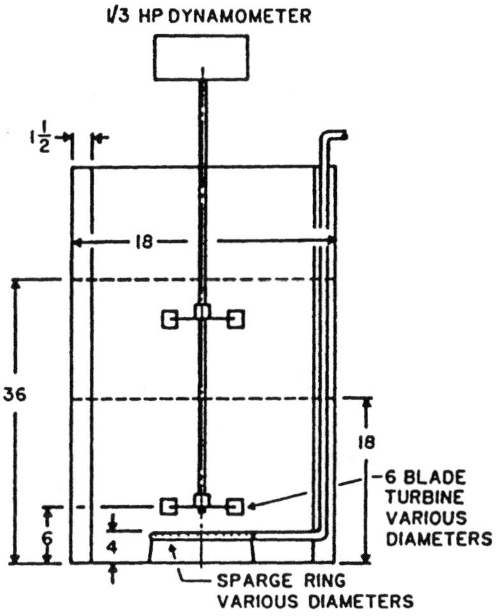

FIG. 18-27 Typical arrangement of Rushton radial-flow R100 flat-blade turbine with typical sparge ring for gas-liquid mass transfer.

gas energy, the impeller will cause a more uniform and vigorous dispersion of the gas bubbles and smaller gas bubbles in the vessel.

In the 1960s and before, most gas-liquid operations were conducted using flat-blade turbines as shown in Fig. 18-4. These impellers required input of approximately three times the energy in the gas stream before they completely control the flow pattern. This was usually the case, and the mass-transfer characteristics were comparable to what would be expected. One disadvantage of the radial-flow impeller is that it is a very poor blending device so blend time is very long compared to that in pilot-scale experiments and compared to the fluidfoil impeller types often used currently. Using curvature of the blades to modify the tendency of gas bubbles to streamline the back of the flat-blade turbine gives a different characteristic to the power drawn by the impeller at a given gas rate compared to no gas rate, but it seems to give quite similar mass transfer at power levels similar to those of the flat-blade design. In order to improve the blending and solid-suspension characteristics, fluidfoil impellers (typified by the A315, Fig. 18-28) have been introduced in recent years and they have many of the advantages and some of the disadvantages of the flat-blade turbine. These impellers typically have a very high solidity ratio, on the order of 0.85 or more, and produce a strong axial downflow at low gas rate. As the gas rate increases, the flow pattern becomes more radial due to the upflow of the gas counteracting the downward flow of the impeller.

Mass-transfer characteristics on large-scale equipment seem to be quite similar, but the fluidfoil impellers tend to release a larger-diameter bubble than is common with the radial-flow turbines. The blend time is one-half or one-third as long, and solid-suspension characteristics are better so that there have been notable improved process results with these impellers. This is particularly true if the process requires better blending and there is solid suspension. If this is not the case, the results from these impellers can be negative compared to radial-flow turbines.

It is very difficult to test these impellers on a small scale, since they provide better blending on a pilot scale where blending is already very effective compared to the large scale. Caution is recommended if it is desirable to study these impellers in pilot-scale equipment.

Gas-Liquid Mass Transfer Gas-liquid mass transfer normally is correlated by means of the mass-transfer coefficient K_ga versus power level at various superficial gas velocities. The superficial gas velocity is

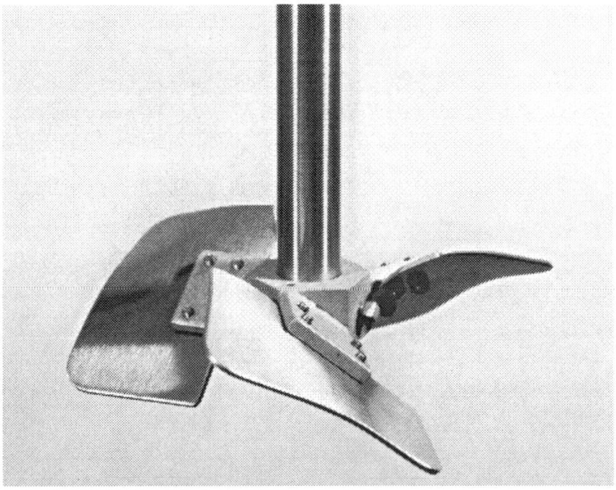

FIG. 18-28 An impeller designed for gas-liquid dispersion and mass transfer of the fluidfoil type, i.e., A315.

the volume of gas at the average temperature and pressure at the midpoint in the tank divided by the area of the vessel. In order to obtain the partial-pressure driving force, an assumption must be made of the partial pressure in equilibrium with the concentration of gas in the liquid. Many times this must be assumed, but if Fig. 18-29 is obtained in the pilot plant and the same assumption principle is used in evaluating the mixer in the full-scale tank, the error from the assumption is limited.

In the plant-size unit, Fig. 18-29 must be translated into a mass-transfer-rate curve for the particular tank volume and operating condition selected. Every time a new physical condition is selected, a different curve similar to that of Fig. 18-30 is obtained.

Typical exponents on the effect of power and gas rate on K_ga tend to be around 0.5 for each variable, ± 0.1.

Viscosity markedly changes the picture and, usually, increasing viscosity lowers the mass-transfer coefficient. For the common application of waste treating and for some of the published data on biological slurries, data for k_La (shown in Fig. 18-31) is obtained in the literature. For a completely new gas or liquid of a liquid slurry system, Fig. 18-29 must be obtained by an actual experiment.

Liquid-Gas-Solid Systems Many gas-liquid systems contain solids that may be the ultimate recipient of the liquid-gas-solid mass transfer entering into the process result. Examples are biological

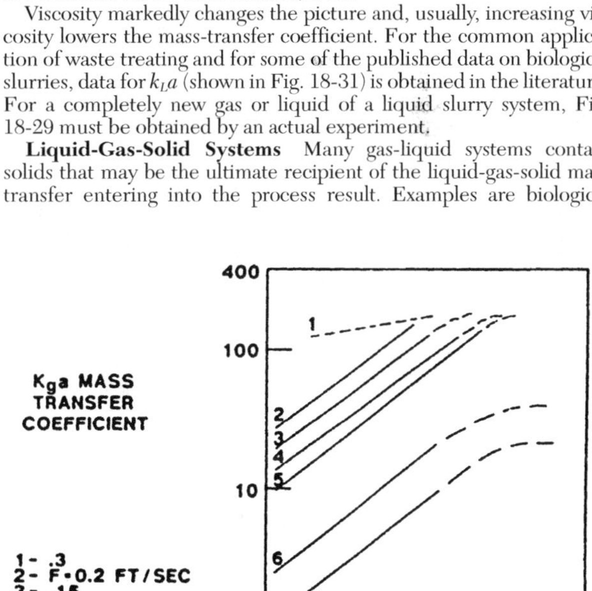

FIG. 18-29 Typical curve for mass transfer coefficient K_ga as a function of mixer power and superficial gas velocity.

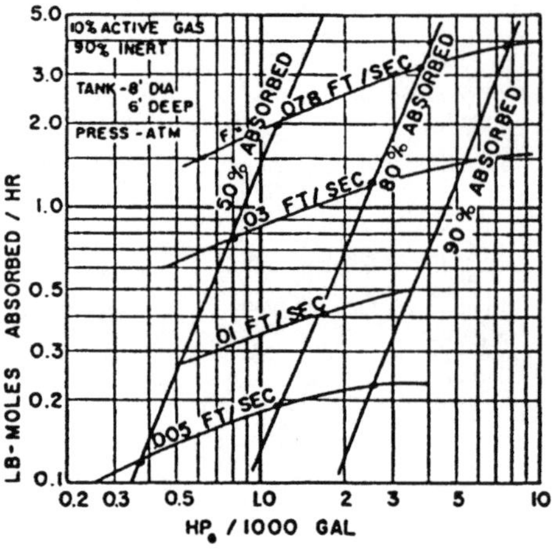

FIG. 18-30 Example of a specific chart to analyze the total mass-transfer rate in a particular tank under a process condition obtained from basic K_ga data shown in Fig. 18-28.

processes in which the biological solids are the user of the mass transfer of the mixing-flow patterns, various types of slurries reactors in which the solids either are being reactive or there may be extraction or dissolving taking place, or there may be polymerization or precipitation of solids occurring.

Normally there must be a way of determining whether the mass-transfer rate with the solids is the key controlling parameter or the gas-liquid mass transfer rate.

In general, introduction of a gas stream to a fluid will increase the blend time because the gas-flow patterns are counterproductive to the typical mixer-flow patterns. In a similar vein, the introduction of a gas stream to a liquid-solid suspension will decrease the suspension uniformity because the gas-flow pattern is normally counterproductive to the mixer-flow pattern. Many times the power needed for the gas-liquid mass transfer is higher than the power needed for solid suspension, and the effect of the gas flow on the solid suspensions are of little concern. On the other hand, if power levels are relatively low and solid-suspension characteristics are critical—examples being the case of activated sludge reactors in the waste-treating field or biological solid reactors in the hydrometallurgical field—then the effect of the gas-flow pattern of the mixing system can be quite critical to the overall design.

Another common situation is batch hydrogenation, in which pure hydrogen is introduced to a relatively high pressure reactor and a decision must be made to recycle the unabsorbed gas stream from the top of the reactor or use a vortexing mode for an upper impeller to incorporate the gas from the surface.

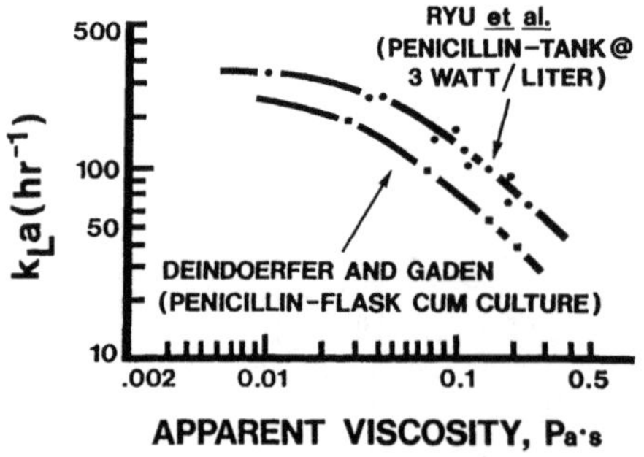

FIG. 18-31 Usually, the gas-liquid mass-transfer coefficient, K_ga, is reduced with increased viscosity. This shows the effect of increased concentration of microbial cells in a fermentation process.

Loop Reactors For some gas-liquid-solid processes, a recirculating loop can be an effective reactor. These involve a relatively high horsepower pumping system and various kinds of nozzles, baffles, and turbulence generators in the loop system. These have power levels anywhere from 1 to 10 times higher than the power level in a typical mixing reactor, and may allow the retention time to be less by a factor of 1 to 10.

LIQUID-LIQUID CONTACTING

Emulsions Almost every shear rate parameter affects liquid-liquid emulsion formation. Some of the effects are dependent upon whether the emulsion is both dispersing and coalescing in the tank, or whether there are sufficient stabilizers present to maintain the smallest droplet size produced for long periods of time. Blend time and the standard deviation of circulation times affect the length of time it takes for a particle to be exposed to the various levels of shear work and thus the time it takes to achieve the ultimate small particle size desired.

The prediction of drop sizes in liquid-liquid systems is difficult. Most of the studies have used very pure fluids as two of the immiscible liquids, and in industrial practice there almost always are other chemicals that are surface-active to some degree and make the prediction of absolute drop sizes very difficult. In addition, techniques to measure drop sizes in experimental studies have all types of experimental and interpretation variations and difficulties so that many of the equations and correlations in the literature give contradictory results under similar conditions. Experimental difficulties include dispersion and coalescence effects, difficulty of measuring actual drop size, the effect of visual or photographic studies on where in the tank you can make these observations, and the difficulty of using probes that measure bubble size or bubble area by light or other sample transmission techniques which are very sensitive to the concentration of the dispersed phase and often are used in very dilute solutions.

It is seldom possible to specify an initial mixer design requirement for an absolute bubble size prediction, particularly if coalescence and dispersion are involved. However, if data are available on the actual system, then many of these correlations could be used to predict relative changes in drop size conditions with changes in fluid properties or impeller variables.

STAGEWISE EQUIPMENT: MIXER-SETTLERS

Introduction Insoluble liquids may be brought into direct contact to cause transfer of dissolved substances, to allow transfer of heat, and to promote chemical reaction. This subsection concerns the design and selection of equipment used for conducting this type of liquid-liquid contact operation.

Objectives There are four principal purposes of operations involving the direct contact of immiscible liquids. The purpose of a particular contact operation may involve any one or any combination of the following objectives:

1. *Separation of components in solution.* This includes the ordinary objectives of liquid extraction, in which the constituents of a solution are separated by causing their unequal distribution between two insoluble liquids, the washing of a liquid with another to remove small amounts of a dissolved impurity, and the like. The theoretical principles governing the phase relationships, material balances, and number of ideal stages or transfer units required to bring about the desired changes are to be found in Sec. 15. Design of equipment is based on the quantities of liquids and the efficiency and operating characteristics of the type of equipment selected.

2. *Chemical reaction.* The reactants may be the liquids themselves, or they may be dissolved in the insoluble liquids. The kinetics of this type of reaction are treated in Sec. 4.

3. *Cooling or heating a liquid by direct contact with another.* Although liquid-liquid-contact operations have not been used widely for heat transfer alone, this technique is one of increasing interest. Applications also include cases in which chemical reaction or liquid extraction occurs simultaneously.

4. *Creating permanent emulsions.* The objective is to disperse one liquid within another in such finely divided form that separation

by settling either does not occur or occurs extremely slowly. The purpose is to prepare the emulsion. Neither extraction nor chemical reaction between the liquids is ordinarily sought.

Liquid-liquid contacting equipment may be generally classified into two categories: **stagewise** and **continuous** (**differential**) contact.

The function of a stage is to contact the liquids, allow equilibrium to be approached, and to make a mechanical separation of the liquids. The contacting and separating correspond to mixing the liquids, and settling the resulting dispersion; so these devices are usually called **mixer-settlers.** The operation may be carried out in batch fashion or with continuous flow. If batch, it is likely that the same vessel will serve for both mixing and settling, whereas if continuous, separate vessels are usually but not always used.

Mixer-Settler Equipment The equipment for extraction or chemical reaction may be classified as follows:

I. Mixers
   A. Flow or line mixers
      1. Mechanical agitation
      2. No mechanical agitation
   B. Agitated vessels
      1. Mechanical agitation
      2. Gas agitation
II. Settlers
   A. Nonmechanical
      1. Gravity
      2. Centrifugal (cyclones)
   B. Mechanical (centrifuges)
   C. Settler auxiliaries
      1. Coalescers
      2. Separator membranes
      3. Electrostatic equipment

In principle, at least, any mixer may be coupled with any settler to provide the complete stage. There are several combinations which are especially popular. Continuously operated devices usually, but not always, place the mixing and settling functions in separate vessels. Batch-operated devices may use the same vessel alternately for the separate functions.

Flow or Line Mixers

Definition Flow or line mixers are devices through which the liquids to be contacted are passed, characterized principally by the very small time of contact for the liquids. They are used only for continuous operations or semibatch (in which one liquid flows continuously and the other is continuously recycled). If holding time is required for extraction or reaction, it must be provided by passing the mixed liquids through a vessel of the necessary volume. This may be a long pipe of large diameter, sometimes fitted with segmental baffles, but frequently the settler which follows the mixer serves. The energy for mixing and dispersing usually comes from pressure drop resulting from flow.

There are many types, and only the most important can be mentioned here. [See also Hunter, in Dunstan (ed.), *Science of Petroleum,* vol. 3, Oxford, New York, 1938, pp. 1779–1797.] They are used fairly extensively in treating petroleum distillates, in vegetable-oil, refining, in extraction of phenol-bearing coke-oven liquors, in some metal extractions, and the like. Kalichevsky and Kobe (*Petroleum Refining with Chemicals,* Elsevier, New York, 1956) discuss detailed application in the refining of petroleum.

Jet Mixers These depend upon impingement of one liquid on the other to obtain a dispersion, and one of the liquids is pumped through a small nozzle or orifice into a flowing stream of the other. Both liquids are pumped. They can be used successfully only for liquids of low interfacial tension. See Fig. 18-32 and also Hunter and Nash [*Ind. Chem.,* **9,** 245, 263, 317 (1933)]. Treybal (*Liquid Extraction,* 2d ed., McGraw-Hill, New York, 1963) describes a more elaborate device. For a study of the extraction of antibiotics with jet mixers, see Anneskova and Boiko, *Med. Prom. SSSR,* **13**(5), 26 (1959). Insonation with ultrasound of a toluene-water mixture during methanol extraction with a simple jet mixer improves the rate of mass transfer, but the energy requirements for significant improvement are large [Woodle and Vilbrandt, *Am. Inst. Chem. Eng. J.,* **6,** 296 (1960)].

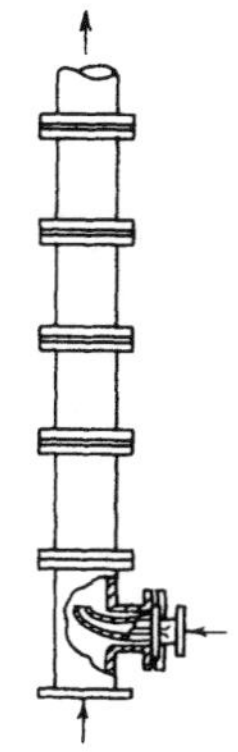

FIG. 18-32 Elbow jet mixer.

Injectors The flow of one liquid is induced by the flow of the other, with only the majority liquid being pumped at relatively high velocity. Figure 18-33 shows a typical device used in semibatch fashion for washing oil with a recirculated wash liquid. It is installed directly in the settling drum. See also Hampton (U.S. Patent 2,091,709, 1933), Sheldon (U.S. Patent 2,009,347, 1935), and Ng (U.S. Patent 2,665,975, 1954). Folsom [*Chem. Eng. Prog.,* **44,** 765 (1948)] gives a good review of basic principles. The most thorough study for extraction is provided by Kafarov and Zhukovskaya [*Zh. Prikl. Khim.,* **31,** 376 (1958)], who used very small injectors. With an injector measuring 73 mm from throat to exit, with 2.48-mm throat diameter, they extracted benzoic acid and acetic acid from water with carbon tetrachloride at the rate of 58 to 106 L/h, to obtain a stage efficiency $E = 0.8$ to 1.0. Data on flow characteristics are also given. Boyadzhiev and Elenkov [*Collect. Czech. Chem. Commun.,* **31,** 4072 (1966)] point out that the presence of surface-active agents exerts a profound influence on drop size in such devices.

Orifices and Mixing Nozzles Both liquids are pumped through constrictions in a pipe, the pressure drop of which is partly utilized to create the dispersion (see Fig. 18-34). Single nozzles or several in series may be used. For the orifice mixers, as many as 20 orifice plates

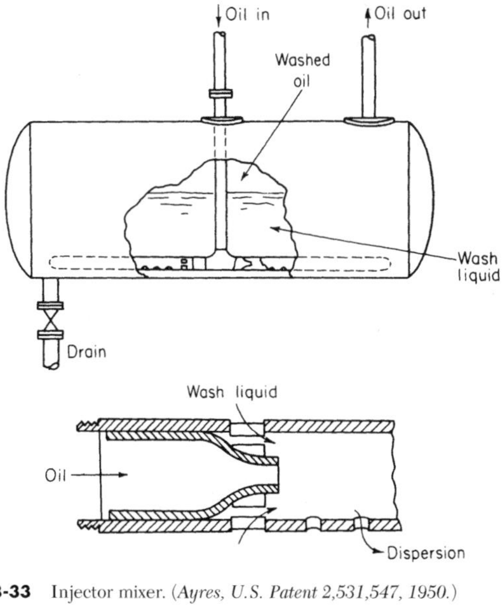

FIG. 18-33 Injector mixer. (*Ayres, U.S. Patent 2,531,547, 1950.*)

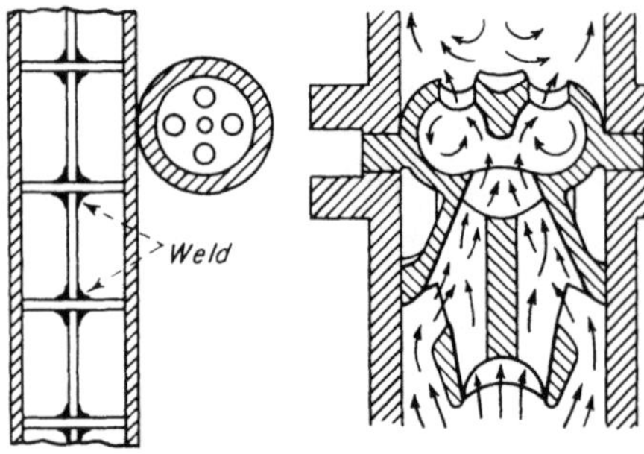

FIG. 18-34 Orifice mixer and nozzle mixer.

each with 13.8-kPa (2-lb/in²) pressure drop may be used in series [Morell and Bergman, *Chem. Metall. Eng.*, **35,** 211 (1928)]. In the Dualayer process for removal of mercaptans from gasoline, 258 m³/h (39,000 bbl/day) of oil and treating solution are contacted with 68.9-kPa (10-lb/in²) pressure drop per stage [Greek et al., *Ind. Eng. Chem.*, **49,** 1938 (1957)]. Holland et al. [*Am. Inst. Chem. Eng. J.*, **4,** 346 (1958); **6,** 615 (1960)] report on the interfacial area produced between two immiscible liquids entering a pipe (diameter 0.8 to 2.0 in) from an orifice, γ_D = 0.02 to 0.20, at flow rates of 0.23 to 4.1 m³/h (1 to 18 gal/min). At a distance 17.8 cm (7 in) downstream from the orifice,

$$a_{av} = \frac{0.179}{\sigma g_c}(C_O^2\,\Delta p)^{0.75}\left(\frac{\sigma\sqrt{g_c\rho_{av}}}{\mu_D}\right)^{0.158}\left[\left(\frac{d_t}{d_O}\right)^4 - 1\right]^{0.117}\gamma_D^{0.878} \quad (18\text{-}8)$$

where a_{av} = interfacial surface, cm²/cm³; C_O = orifice coefficient, dimensionless; d_t = pipe diameter, in; d_O = orifice diameter, in; g_c = gravitational conversion factor, (32.2 lbm·ft)/(lbf·s²); Δp = pressure drop across orifice, lbf/ft²; μ_D = viscosity of dispersed phase, lbm/(ft·s); ρ_{av} = density of dispersed phase, lbm/ft; and σ = interfacial tension, lbf/ft. See also Shirotsuka et al. [*Kagaku Kogaku*, **25,** 109 (1961)].

Valves Valves may be considered to be adjustable orifice mixers. In desalting crude petroleum by mixing with water, Hayes et al. [*Chem. Eng. Prog.*, **45,** 235 (1949)] used a globe-valve mixer operating at 110- to 221-kPa (16- to 32-lb/in²) pressure drop for mixing 66 m³/h (416 bbl/h) oil with 8 m³/h (50 bbl/h) water, with best results at the lowest value. Simkin and Olney [*Am. Inst. Chem. Eng. J.*, **2,** 545 (1956)] mixed kerosine and white oil with water, using 0.35- to 0.62-kPa (0.05- to 0.09-lb/in²) pressure drop across a 1-in gate valve, at 22-m³/h (10-gal/min) flow rate for optimum separating conditions in a cyclone, but higher pressure drops were required to give good extractor efficiencies.

Pumps Centrifugal pumps, in which the two liquids are fed to the suction side of the pump, have been used fairly extensively, and they offer the advantage of providing interstage pumping at the same time. They have been commonly used in the extraction of phenols from coke-oven liquors with light oil [Gollmar, *Ind. Eng. Chem.*, **39,** 596, 1947); Carbone, *Sewage Ind. Wastes*, **22,** 200 (1950)], but the intense shearing action causes emulsions with this low-interfacial-tension system. Modern plants use other types of extractors. Pumps are useful in the extraction of slurries, as in the extraction of uranyl nitrate from acid-uranium-ore slurries [*Chem. Eng.*, **66,** 30 (Nov. 2, 1959)]. Shaw and Long [*Chem. Eng.*, **64**(11), 251 (1957)] obtain a stage efficiency of 100 percent (E = 1.0) in a uranium-ore-slurry extraction with an open impeller pump. In order to avoid emulsification difficulties in these extractions, it is necessary to maintain the organic phase continuous, if necessary by recycling a portion of the settled organic liquid to the mixer

Agitated Line Mixer See Fig. 18-35. This device, which combines the features of orifice mixers and agitators, is used extensively in treating petroleum and vegetable oils. It is available in sizes to fit ½- to 10-in pipe. The device of Fig. 18-36, with two impellers in separate stages, is available in sizes to fit 4- to 20-in pipe.

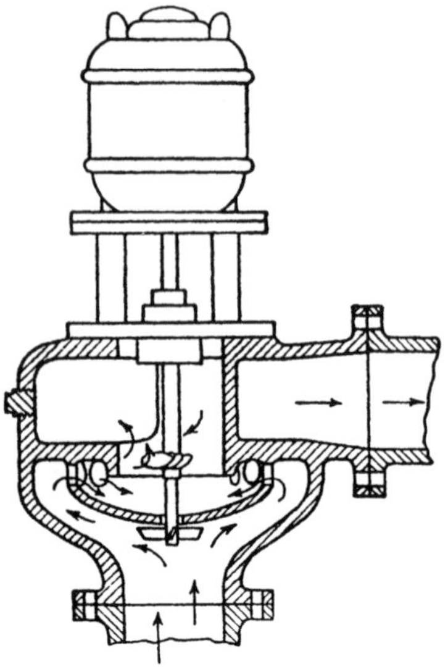

FIG. 18-35 Nettco Corp. Flomix.

Packed Tubes Cocurrent flow of immiscible liquids through a packed tube produces a one-stage contact, characteristic of line mixers. For flow of isobutanol-water* through a 0.5-in diameter tube packed with 6 in of 3-mm glass beads, Leacock and Churchill [*Am. Inst. Chem. Eng. J.*, **7,** 196 (1961)] find

$$k_C a_{av} = c_1 L_C^{0.5} L_D \quad (18\text{-}9)$$

$$k_D a_{av} = c_2 L_C^{0.75} L_D^{0.75} \quad (18\text{-}10)$$

where c_1 = 0.00178 using SI units and 0.00032 using U.S. customary units; and c_2 = 0.0037 using SI units and 0.00057 using U.S. customary units. These indicate a stage efficiency approaching 100 percent. Organic-phase holdup and pressure drop for larger pipes similarly packed are also available [Rigg and Churchill, ibid., **10,** 810 (1964)].

* Isobutanol dispersed: L_D = 3500 to 27,000; water continuous; L_C = 6000 to 32,000 in pounds-mass per hour-square foot (to convert to kilograms per second-square meter, multiply by 1.36 × 10⁻³).

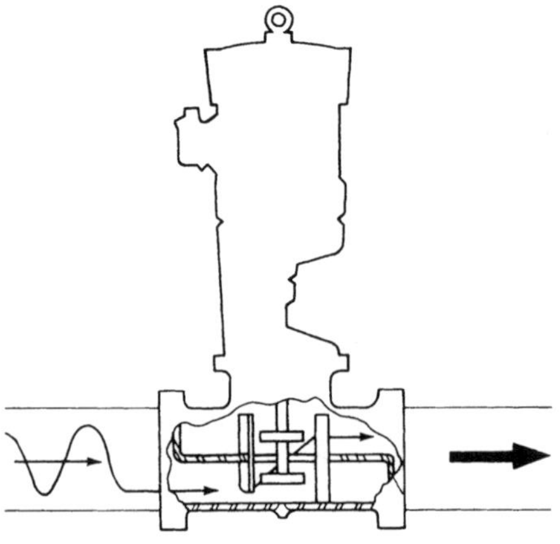

FIG. 18-36 Lightnin line blender. (*Mixing Equipment Co., Inc., with permission.*)

Pipe Lines The principal interest here will be for flow in which one liquid is dispersed in another as they flow cocurrently through a pipe (stratified flow produces too little interfacial area for use in liquid extraction or chemical reaction between liquids). Drop size of dispersed phase, if initially very fine at high concentrations, increases as the distance downstream increases, owing to coalescence [see Holland, loc. cit.; Ward and Knudsen, *Am. Inst. Chem. Eng. J.*, **13,** 356 (1967)]; or if initially large, decreases by breakup in regions of high shear [Sleicher, ibid., **8,** 471 (1962); *Chem. Eng. Sci.*, **20,** 57 (1965)]. The maximum drop size is given by (Sleicher, loc. cit.)

$$\frac{d_{p,\max}\rho_C V^2}{\sigma g_c}\sqrt{\frac{\mu_C V}{\sigma g_c}} = C\left[1 + 0.7\left(\frac{\mu_D V}{\sigma g_c}\right)^{0.7}\right] \quad (18\text{-}11)$$

where $C = 43$ ($d_t = 0.013$ m or 0.0417 ft) or 38 ($d_t = 0.038$ m or 0.125 ft), with $d_{p,\text{av}} = d_{p,\max}/4$ for high flow rates and $d_{p,\max}/13$ for low velocities.

Extensive measurements of the rate of mass transfer between *n*-butanol and water flowing in a 0.008-m (0.314-in) ID horizontal pipe are reported by Watkinson and Cavers [*Can. J. Chem. Eng.*, **45,** 258 (1967)] in a series of graphs not readily reproduced here. Length of a transfer unit for either phase is strongly dependent upon flow rate and passes through a pronounced maximum at an organic-water phase ratio of 0.5. In energy (pressure-drop) requirements and volume, the pipe line compared favorably with other types of extractors. Boyadzhiev and Elenkov [*Chem. Eng. Sci.*, **21,** 955 (1966)] concluded that, for the extraction of iodine between carbon tetrachloride and water in turbulent flow, drop coalescence and breakup did not influence the extraction rate. Yoshida et al. [*Coal Tar* (Japan), **8,** 107 (1956)] provide details of the treatment of crude benzene with sulfuric acid in a 1-in diameter pipe, $N_{\text{Re}} = 37{,}000$ to 50,000. Fernandes and Sharma [*Chem. Eng. Sci.*, **23,** 9 (1968)] used cocurrent flow downward of two liquids in a pipe, agitated with an upward current of air.

The pipe has also been used for the transfer of heat between two immiscible liquids in cocurrent flow. For hydrocarbon oil-water, the heat-transfer coefficient is given by

$$\frac{Ua_{\text{av}}d_t^2}{vk_{to}} = \frac{\gamma_D N_{\text{We},t}^{6/5}}{\dfrac{k_{to}}{0.415k_{tC}} + \dfrac{k_{to}}{0.173k_{tD}}} \quad (18\text{-}12)$$

for $\gamma_D = 0$ to 0.2. Additional data for $\gamma_D = 0.4$ to 0.8 are also given. Data for stratified flow are given by Wilke et al. [*Chem. Eng. Prog.*, **59,** 69 (1963)] and Grover and Knudsen [*Chem. Eng. Prog.*, **51,** *Symp. Ser.* **17,** 71 (1955)].

Mixing in Agitated Vessels Agitated vessels may frequently be used for either batch or continuous service and for the latter may be sized to provide any holding time desired. They are useful for liquids of any viscosity up to 750 Pa·s (750,000 cP), although in contacting two liquids for reaction or extraction purposes viscosities in excess of 0.1 Pa·s (100 cP) are only rarely encountered.

Mechanical Agitation This type of agitation utilizes a rotating impeller immersed in the liquid to accomplish the mixing and dispersion. There are literally hundreds of devices using this principle, the major variations being found when chemical reactions are being carried out. The basic requirements regarding shape and arrangement of the vessel, type and arrangement of the impeller, and the like are essentially the same as those for dispersing finely divided solids in liquids, which are fully discussed in Sec. 18.

The following summary of operating characteristics of mechanically agitated vessels is confined to the data available on liquid-liquid contacting.

Phase Dispersed There is an ill-defined upper limit to the volume fraction of dispersed liquid which may be maintained in an agitated dispersion. For dispersions of organic liquids in water [Quinn and Sigloh, *Can. J. Chem. Eng.*, **41,** 15 (1963)],

$$\gamma_{Do,\max} = \gamma' + \left(\frac{C}{N^3}\right) \quad (18\text{-}13)$$

where γ' is a constant, asymptotic value, and C is a constant, both depending in an unestablished manner upon the systems' physical properties and geometry. Thus, inversion of a dispersion may occur if the agitator speed is increased. With systems of low interfacial tension ($\sigma' = 2$ to 3 mN/m or 2 to 3 dyn/cm), γ_D as high as 0.8 can be maintained. Selker and Sleicher [*Can. J. Chem. Eng.*, **43,** 298 (1965)] and Yeh et al. [*Am. Inst. Chem. Eng. J.*, **10,** 260 (1964)] feel that the viscosity ratio of the liquids alone is important. Within the limits in which either phase can be dispersed, for *batch operation* of baffled vessels, that phase in which the impeller is immersed when at rest will normally be continuous [Rodger, Trice, and Rushton, *Chem. Eng. Prog.*, **52,** 515 (1956); Laity and Treybal, *Am. Inst. Chem. Eng. J.*, **3,** 176 (1957)]. With water dispersed, *dual emulsions* (continuous phase found as small droplets within larger drops of dispersed phase) are possible. In *continuous operation,* the vessel is first filled with the liquid to be continuous, and agitation is then begun, after which the liquid to be dispersed is introduced.

Uniformity of Mixing This refers to the gross uniformity throughout the vessel and not to the size of the droplets produced. For *unbaffled vessels, batch, with an air-liquid interface,* Miller and Mann [*Trans. Am. Inst. Chem. Eng.*, **40,** 709 (1944)] mixed water with several organic liquids, measuring uniformity of mixing by sampling the tank at various places, comparing the percentage of dispersed phase found with that in the tank as a whole. A power application of 200 to 400 W/m^3 [(250 to 500 ft·lb)/(min·ft^3)] gave maximum and nearly uniform performance for all. See also Nagata et al. [*Chem. Eng. (Japan),* **15,** 59 (1951)].

For *baffled vessels operated continuously, no air-liquid interface,* flow upward, light liquid dispersed [Treybal, *Am. Inst. Chem. Eng. J.*, **4,** 202 (1958)], the average fraction of dispersed phase in the vessel $\gamma_{D,\text{av}}$ is less than the fraction of the dispersed liquid in the feed mixture, unless the impeller speed is above a certain critical value which depends upon vessel geometry and liquid properties. Thornton and Bouyatiotis [*Ind. Chem.*, **39,** 298 (1963); *Inst. Chem. Eng. Symp. Liquid Extraction,* Newcastle-upon-Tyne, April 1967] have presented correlations of data for a 17.8-cm (7-in) vessel, but these do not agree with observations on 15.2- and 30.5-cm (6- and 12-in) vessels in Treybal's laboratory. See also Kovalev and Kagan [*Zh. Prikl. Khim.*, **39,** 1513 (1966)] and Trambouze [*Chem. Eng. Sci.*, **14,** 161 (1961)]. Stemerding et al. [*Can. J. Chem. Eng.*, **43,** 153 (1965)] present data on a large mixing tank [15 m^3 (530 ft^3)] fitted with a marine-type propeller and a draft tube.

Drop Size and Interfacial Area The drops produced have a size range [Sullivan and Lindsey, *Ind. Eng. Chem. Fundam.*, **1,** 87 (1962); Sprow, *Chem. Eng. Sci.*, **22,** 435 (1967); and Chen and Middleman, *Am. Inst. Chem. Eng. J.*, **13,** 989 (1967)]. The average drop size may be expressed as

$$d_{p,\text{av}} = \frac{\sum n_i d_{pi}^3}{\sum n_i d_{pi}^2} \quad (18\text{-}14)$$

and if the drops are spherical,

$$a_{\text{av}} = \frac{6\gamma_{D,\text{av}}}{d_{p,\text{av}}} \quad (18\text{-}15)$$

The drop size varies locally with location in the vessel, being smallest at the impeller and largest in regions farthest removed from the impeller owing to coalescence in regions of relatively low turbulence intensity [Schindler and Treybal, *Am. Inst. Chem. Eng. J.*, **14,** 790 (1968); Vanderveen, U.S. AEC UCRL-8733, 1960]. Interfacial area and hence average drop size have been measured by light transmittance, light scattering, direct photography, and other means. Typical of the resulting correlations is that of Thornton and Bouyatiotis (*Inst. Chem. Eng. Symp. Liquid Extraction,* Newcastle-upon-Tyne, April 1967) for a 17.8-cm- (7-in-) diameter baffled vessel, six-bladed flat-blade turbine, $d_i = 6.85$ cm (0.225 ft), operated full, for organic liquids

($\sigma' = 8.5$ to 34, $\rho_D = 43.1$ to 56.4, $\mu_D = 1.18$ to 1.81) dispersed in water, in the absence of mass transfer, and under conditions giving nearly the vessel-average $d_{p,\text{av}}$:

$$\frac{d_{p,\text{av}}}{d_p^0} = 1 + 1.18\phi_D \left(\frac{\sigma^2 g_c^2}{d_p^0 \mu_c^2 g}\right)\left(\frac{\mu_c^4 g}{\Delta\rho\, \sigma^3 g_c^3}\right)^{0.62}\left(\frac{\Delta\rho}{\rho_c}\right)^{0.05} \tag{18-16}$$

where d_p^0 is given by

$$\frac{(d_p^0)^3 \rho_C^2 g}{\mu_c^2} = 29.0\left(\frac{P^3 g_c^3}{\upsilon^3 \rho_c^2 \mu_c g^4}\right)^{-0.32}\left(\frac{\rho_C \sigma^3 g_c^3}{\mu_c^4 g}\right)^{0.14} \tag{18-17}$$

Caution is needed in using such correlations, since those available do not generally agree with each other. For example, Eq. (21-28) gives $d_{p,\text{av}} = 4.78(10^{-4})$ ft for a liquid pair of properties $a' = 30$, $\rho_C = 62.0$, $\rho_D = 52.0$, $\mu_C = 2.42$, $\mu_D = 1.94$, $\gamma_{D,\text{av}} = 0.20$ in a vessel $T = Z = 0.75$, a turbine impeller $d_i = 0.25$ turning at 400 r/min. Other correlations provide $3.28(10^{-4})$ [Thornton and Bouyatiotis, *Ind. Chem.*, **39,** 298 (1963)], $8.58(10^{-4})$ [Calderbank, *Trans. Inst. Chem. Eng. (London),* **36,** 443 (1958)], $6.1(10^{-4})$ [Kafarov and Babinov, *Zh. Prikl. Khim,* **32,** 789 (1959)], and $2.68(10^{-3})$ (Rushton and Love, paper at AIChE, Mexico City, September 1967). See also Vermeulen et al. [*Chem. Eng. Prog.*, **51,** 85F (1955)], Rodgers et al. [ibid., **52,** 515 (1956); U.S. AEC ANL-5575 (1956)], Rodrigues et al. [*Am. Inst. Chem. Eng. J.*, **7,** 663 (1961)], Sharma et al. [*Chem. Eng. Sci.*, **21,** 707 (1966); **22,** 1267 (1967)], and Kagan and Kovalev [*Khim. Prom.*, **42,** 192 (1966)]. For the effect of absence of baffles, see Fick et al. (U.S. AEC UCRL-2545, 1954) and Schindler and Treybal [*Am. Inst. Chem. Eng. J.*, **14,** 790 (1968)]. The latter have observations during mass transfer.

Coalescence Rates The droplets coalesce and redisperse at rates that depend upon the vessel geometry, N, $\gamma_{D,\text{av}}$, and liquid properties. The few measurements available, made with a variety of techniques, do not as yet permit quantitative estimates of the coalescence frequency υ. Madden and Damarell [*Am. Inst. Chem. Eng. J.*, **8,** 233 (1962)] found for baffled vessels that υ varied as $N^{2.2}\gamma_{D,\text{av}}^{0.5}$, and this has generally been confirmed by Groothius and Zuiderweg [*Chem. Eng. Sci.*, **19,** 63 (1964)], Miller et al. [*Am. Inst. Chem. Eng. J.*, **9,** 196 (1963)], and Howarth [ibid., **13,** 1007 (1967)], although absolute values of υ in the various studies are not well related. Hillestad and Rushton (paper at AIChE, Columbus, Ohio, May 1966), on the other hand, find υ to vary as $N^{0.73}\gamma_{D,\text{av}}$ for impeller Weber numbers $N_{We,i}$ below a certain critical value and as $N^{-3.5}\gamma_{D,\text{av}}^{1.58}$ for higher Weber numbers. The influence of liquid properties is strong. There is clear evidence [Groothius and Zuiderweg, loc. cit.; *Chem. Eng. Sci.*, **12,** 288 (1960)] that coalescence rates are enhanced by mass transfer from a drop to the surrounding continuum and retarded by transfer in the reverse direction. See also Howarth [*Chem. Eng. Sci.*, **19,** 33 (1964)]. For a theoretical treatment of drop breakage and coalescence and their effects, see Valentas and Amundsen [*Ind. Eng. Chem. Fundam.*, **5,** 271, 533 (1966); **7,** 66 (1968)], Gal-Or and Walatka [*Am. Inst. Chem. Eng. J.*, **13,** 650 (1967)], and Curl [ibid., **9,** 175 (1963)].

In calculating the power required for mixers, a reasonable estimate of the average density and viscosity for a two-phase system is satisfactory.

Solids are often present in liquid streams either as a part of the processing system or as impurities that come along and have to be handled in the process. One advantage of mixers in differential contact equipment is the fact that they can handle slurries in one or both phases. In many industrial leaching systems, particularly in the minerals processing industry, coming out of the leach circuit is a slurry with a desired material involved in the liquid but a large amount of solids contained in the stream. Typically, the solids must be separated out by filtration or centrifugation, but there has always been a desire to try a direct liquid-liquid extraction with an immiscible liquid contact with this often highly concentrated slurry leach solution. The major problem with this approach is loss of organic material going out with the highly concentrated liquid slurry.

Data are not currently available on the dispersion with the newer fluidfoil impellers, but they are often used in industrial mixer-settler systems to maintain dispersion when additional resonance time holdup is required, after an initial dispersion is made by a radial- or axial-flow turbine.

Recent data by Calabrese[5] indicates that the sauter mean drop diameter can be correlated by equation and is useful to compare with other predictions indicated previously.

As an aside, when a large liquid droplet is broken up by shear stress, it tends initially to elongate into a dumbbell shape, which determines the particle size of the two large droplets formed. Then, the neck in the center between the ends of the dumbbell may explode or shatter. This would give a debris of particle sizes which can be quite different than the two major particles produced.

Liquid-Liquid Extraction The actual configuration of mixers in multistage mixer-settlers and/or multistage columns is summarized in Section 15. A general handbook on this subject is *Handbook of Solvent Extraction* by Lowe, Beard, and Hanson. This handbook gives a comprehensive review of this entire operation as well.

In the liquid-liquid extraction area, in the mining industry, coming out of the leach tanks is normally a slurry, in which the desired mineral is dissolved in the liquid phase. To save the expense of separation, usually by filtration or centrifugation, attempts have been made to use a resident pump extraction system in which the organic material is contacted directly with the slurry. The main economic disadvantage to this proposed system is the fact that considerable amounts of organic liquid are entrained in the aqueous slurry system, which, after the extraction is complete, are discarded. In many systems this has caused an economic loss of solvent into this waste stream.

LIQUID-LIQUID-SOLID SYSTEMS

Many times solids are present in one or more phases of a solid-liquid system. They add a certain level of complexity in the process, especially if they tend to be a part of both phases, as they normally will do. Approximate methods need to be worked out to estimate the density of the emulsion and determine the overall velocity of the flow pattern so that proper evaluation of the suspension requirements can be made. In general, the solids will behave as though they were a fluid of a particular average density and viscosity and won't care much that there is a two-phase dispersion going on in the system. However, if solids are being dissolved or precipitated by participating in one phase and not the other, then they will be affected by which phase is dispersed or continuous, and the process will behave somewhat differently than if the solids migrate independently between the two phases within the process.

FLUID MOTION

Pumping Some mixing applications can be specified by the pumping capacity desired from the impeller with a certain specified geometry in the vessel. As mentioned earlier, this sometimes is used to describe a blending requirement, but circulation and blending are two different things. The major area where this occurs is in draft tube circulators or pump-mix mixer settlers. In draft tube circulators (shown in Fig. 18-22), the circulation occurs through the draft tube and around the annulus and for a given geometry, the velocity head required can be calculated with reference to various formulas for geometric shapes. What is needed is a curve for head versus flow for the impeller, and then the system curve can be matched to the impeller curve. Adding to the complexity of this system is the fact that solids may settle out and change the character of the head curve so that the impeller can get involved in an unstable condition which has various degrees of erratic behavior depending upon the sophistication of the impeller and inlet and outlet vanes involved. These draft tube circulators often involve solids, and applications are often for precipitation or crystallization in these units. Draft tube circulators can either have the impeller pump up in the draft tube and flow down the annulus or just the reverse. If the flow is down the annulus, then the flow has to make a 180° turn where it comes back at the bottom of the tank into the draft tube again. This is a very sensitive area, and special baffles must be used to carefully determine how the fluid will make this turn since many areas of constriction are involved in making this change in direction.

When pumping down the draft tube, flow normally makes a more troublefree velocity change to a flow going up the annulus. Since the area of the draft tube is markedly less than the area of the annulus, pumping up the draft tube requires less flow to suspend solids of a given settling velocity than does pumping down the draft tube.

Another example is to eliminate the interstage pump between mixing and settling stages in the countercurrent mixer-settler system. The radial-flow impeller typically used is placed very close to an orifice at the bottom of the mixing tank and can develop heads from 12 to 18 in. All the head-loss terms in the mixer and settler circuit have to be carefully calculated because they come very close to that 12- to 18-in range when the passages are very carefully designed and streamlined. If the mixing tank gets much above 10 ft in depth, then the heads have to be higher than the 12- to 18-in range and special designs have to be worked on which have the potential liability of increasing the shear rate acting on the dispersed phase to cause more entrainment and longer settling times. In these cases, it is sometimes desirable to put the mixer system outside the actual mixer tank and have it operate in a single phase or to use multiple impellers, each one of which can develop a portion of the total head required.

Heat Transfer In general, the fluid mechanics of the film on the mixer side of the heat transfer surface is a function of what happens at that surface rather than the fluid mechanics going on around the impeller zone. The impeller largely provides flow across and adjacent to the heat-transfer surface and that is the major consideration of the heat-transfer result obtained. Many of the correlations are in terms of traditional dimensionless groups in heat transfer, while the impeller performance is often expressed as the impeller Reynolds number.

The fluidfoil impellers (shown in Fig. 18-2) usually give more flow for a given power level than the traditional axial- or radial-flow turbines. This is also thought to be an advantage since the heat-transfer surface itself generates the turbulence to provide the film coefficient and more flow should be helpful. This is true to a limited degree in jacketed tanks (Fig. 18-37), but in helical coils (Fig. 18-38), the extreme axial flow of these impellers tends to have the first or second turn in the coil at the bottom of the tank blank off the flow from the turns above it in a way that (at the same power level) the increased flow from the fluidfoil impeller is not helpful. It best gives the same coefficient as with the other impellers and on occasion can cause a 5 to 10 percent reduction in the heat-transfer coefficient over the entire coil.

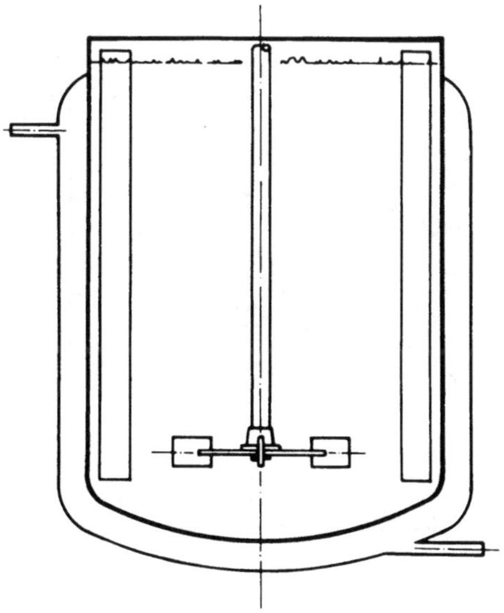

FIG. 18-37 Typical jacket arrangement for heat transfer.

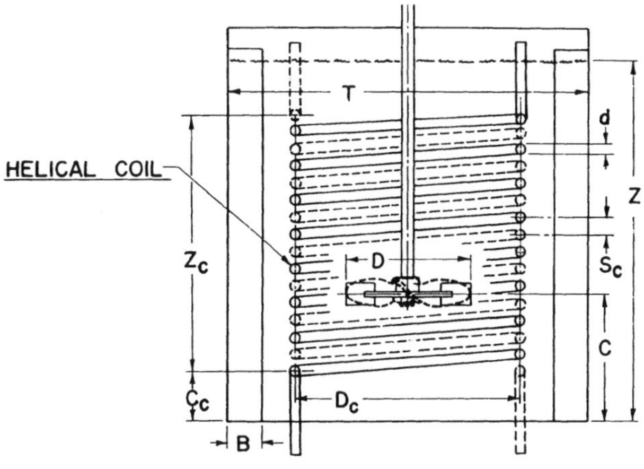

FIG. 18-38 Typical arrangement of helical coil at mixing vessel for heat transfer.

JACKETS AND COILS OF AGITATED VESSELS

Most of the correlations for heat transfer from the agitated liquid contents of vessels to jacketed walls have been of the form:

$$\frac{hD_j}{k} = a\left(\frac{L_p^2 N_r \rho}{\mu}\right)^b \left(\frac{c\mu}{k}\right)^{1/3} \left(\frac{\mu_b}{\mu_w}\right)^m \qquad (18\text{-}18)$$

The film coefficient h is for the inner wall; D_j is the inside diameter of the mixing vessel. The term $L_p^2 N_r \rho/\mu$ is the Reynolds number for mixing in which L_p is the diameter and N_r the speed of the agitator. Recommended values of the constants a, b, and m are given in Table 18-3.

A wide variety of configurations exists for coils in agitated vessels. Correlations of data for heat transfer to helical coils have been of two forms, of which the following are representative:

$$\frac{hD_j}{k} = 0.87\left(\frac{L_p^2 N_t \rho}{\mu}\right)^{0.62} \left(\frac{c\mu}{k}\right)^{1/3} \left(\frac{\mu_b}{\mu_w}\right)^{0.14} \qquad (18\text{-}19)$$

TABLE 18-3 Values of Constants for Use in Eq. (18-18)

Agitator	a	b	m	Range of Reynolds number
Paddle[a]	0.36	⅔	0.21	300–3 × 10⁵
Pitched-blade turbine[b]	0.53	⅔	0.24	80–200
Disc, flat-blade turbine[c]	0.54	⅔	0.14	40–3 × 10⁵
Propeller[d]	0.54	⅔	0.14	2 × 10³ (one point)
Anchor[b]	1.0	½	0.18	10–300
Anchor[b]	0.36	⅔	0.18	300–40,000
Helical ribbon[e]	0.633	½	0.18	8–10⁵

[a]Chilton, Drew, and Jebens, *Ind. Eng. Chem.*, **36,** 510 (1944), with constant *m* modified by Uhl.
[b]Uhl, *Chem. Eng. Progr., Symp. Ser.* 17, **51,** 93 (1955).
[c]Brooks and Su, *Chem. Eng. Progr.*, **55**(10), 54 (1959).
[d]Brown et al., *Trans. Inst. Chem. Engrs.* (*London*), **25,** 181 (1947).
[e]Gluz and Pavlushenko, *J. Appl. Chem. U.S.S.R.*, **39,** 2323 (1966).

where the agitator is a paddle, the Reynolds number range is 300 to 4×10^5 [Chilton, Drew, and Jebens, *Ind. Eng. Chem.*, **36,** 510 (1944)], and

$$\frac{hD_o}{k} = 0.17 \left(\frac{L_p^2 N_r \rho}{\mu}\right)^{0.67} \left(\frac{c\mu}{k}\right)^{0.37} \left(\frac{L_p}{D_j}\right)^{0.1} \left(\frac{D_o}{D_j}\right)^{0.5} \quad (18\text{-}20)$$

where the agitator is a disc flat-blade turbine, and the Reynolds number range is 400 to $(2)(10^5)$ [Oldshue and Gretton, *Chem. Eng. Prog.*, **50,** 615 (1954)]. The term D_o is the outside diameter of the coil tube.

The most comprehensive correlation for heat transfer to vertical baffle-type coils is for a disc flat-blade turbine over the Reynolds number range 10^3 to $(2)(10^6)$:

$$\frac{hD_o}{k} = 0.09 \left(\frac{L_p^2 N_r \rho}{\mu}\right)^{0.65} \left(\frac{c\mu}{k}\right)^{1/3} \left(\frac{L_p}{D_j}\right)^{1/3} \left(\frac{2}{n_b}\right)^{0.2} \left(\frac{\mu}{\mu_f}\right)^{0.4} \quad (18\text{-}21)$$

where n_b is the number of baffle-type coils and μ_f is the fluid viscosity at the mean film temperature [Dunlop and Rushton, *Chem. Eng. Prog. Symp. Ser.* 5, **49,** 137 (1953)].

Chapman and Holland (*Liquid Mixing and Processing in Stirred Tanks,* Reinhold, New York, 1966) review heat transfer to low-viscosity fluids in agitated vessels. Uhl ["Mechanically Aided Heat Transfer," in Uhl and Gray (eds.), *Mixing: Theory and Practice,* vol. I, Academic, New York, 1966, chap. V] surveys heat transfer to low- and high-viscosity agitated fluid systems. This review includes scraped-wall units and heat transfer on the jacket and coil side for agitated vessels.

LIQUID-LIQUID-GAS-SOLID SYSTEMS

This is a relatively unusual combination, and one of the more common times it exists is in the fermentation of hydrocarbons with aerobic microorganisms in an aqueous phase. The solid phase is a microorganism which is normally in the aqueous phase and is using the organic phase for food. Gas is supplied to the system to make the fermentation aerobic. Usually the viscosities are quite low, percent solids is also modest, and there are no special design conditions required when this particular gas-liquid-liquid-solid combination occurs. Normally, average properties for the density of viscosity of the liquid phase are used. In considering that the role the solids play in the system is adequate, there are cases of other processes which consist of four phases, each of which involves looking at the particular properties of the phases to see whether there are any problems of dispersion, suspension, or emulsification.

COMPUTATIONAL FLUID DYNAMICS

There are several software programs that are available to model flow patterns of mixing tanks. They allow the prediction of flow patterns based on certain boundary conditions. The most reliable models use accurate fluid mechanics data generated for the impellers in question and a reasonable number of modeling cells to give the overall tank flow pattern. These flow patterns can give velocities, streamlines, and localized kinetic energy values for the systems. Their main use at the present time is to look at the effect of making changes in mixing variables based on doing certain things to the mixing process. These programs can model velocity, shear rates, and kinetic energy, but probably cannot adapt to the actual chemistry of diffusion or mass-transfer kinetics of actual industrial process at the present time.

Relatively uncomplicated transparent tank studies with tracer fluids or particles can give a similar feel for the overall flow pattern. It is important that a careful balance be made between the time and expense of calculating these flow patterns with computational fluid dynamics compared to their applicability to an actual industrial process. The future of computational fluid dynamics appears very encouraging and a reasonable amount of time and effort put forth in this regard can yield immediate results as well as potential for future process evaluation.

Figures 18-39, 18-40, and 18-41 show some approaches. Figure 18-39 shows velocity vectors for an A310 impeller. Figure 18-40 shows contours of kinetic energy of turbulence. Figure 18-41 uses a particle trajectory approach with neutral buoyancy particles.

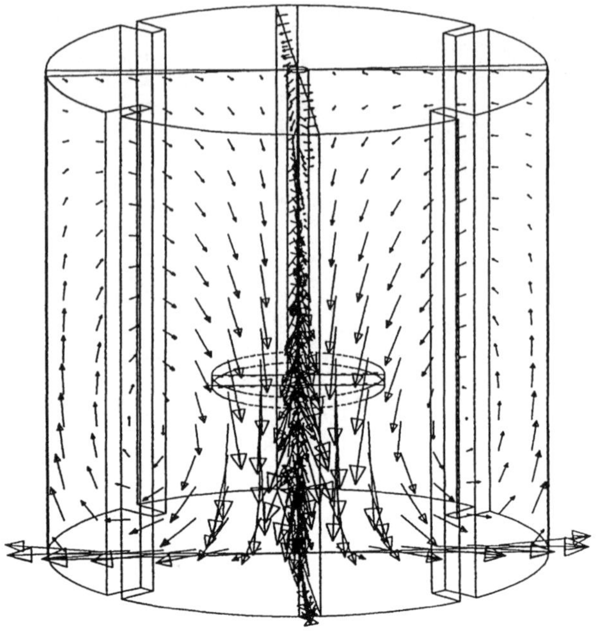

FIG. 18-39 Laser scan.

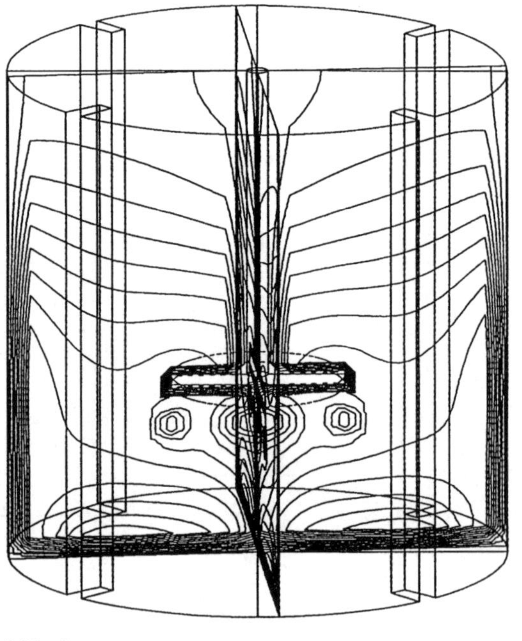

FIG. 18-40 Laser scan.

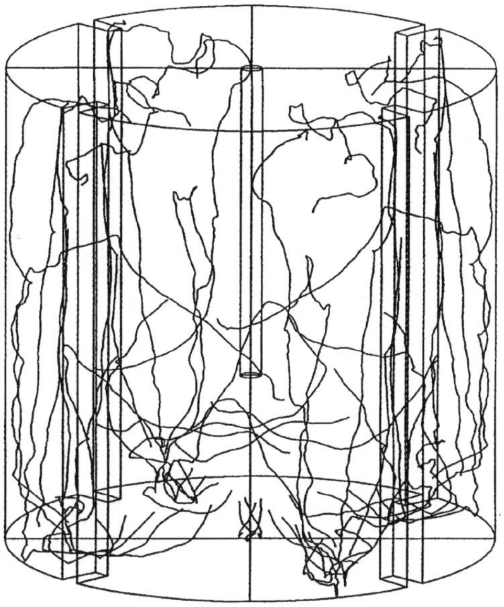

FIG. 18-41 Laser scan.

Numerical fluid mechanics can define many of the fluid mechanics parameters for an overall reactor system. Many of the models break up the mixing tank into small microcells. Suitable material and mass-transfer balances between these cells throughout the reactor are then made. This can involve long and massive computational requirements. Programs are available that can give reasonably acceptable models of experimental data taken in mixing vessels. Modeling the three-dimensional aspect of a flow pattern in a mixing tank can require a large amount of computing power.

Most modeling codes are a time-averaging technique. Depending upon the process, a time-dependent technique may be more suitable. Time-dependent modeling requires much more computing power than does time averaging.

General References

1. J. Y. Oldshue, "Mixing '89," *Chemical Engineering Progress,* **85**(5): 33–42 (1989).
2. J. C. Middleton, *Proc. 3d European Conf. on Mixing,* 4/89, BHRA, pp. 15–36.
3. J. Y. Oldshue, T. A. Post, R. J. Weetman, "Comparison of Mass Transfer Characteristics of Radial and Axial Flow Impellers," *BHRA Proc. 6th European Conf. on Mixing,* 5/88.
4. A. W. Neinow, B. Buckland, R. J. Weetman, *Mixing XII Research Conference,* Potosi, Mo., 8/89.
5. R. Calabrese et al., *AIChE J.* **32:** 657, 677 (1986).
6. T. N. Zwietering, *Chemical Engineering Science,* **8**(3): 244–253 (1958).
7. J. Y. Oldshue, *Chemical Engineering Progress,* "Mixing of Slurries Near the Ultimate Settled Solids Concentration," **77**(5): 95–98 (1981).

MIXING OF VISCOUS FLUIDS, PASTES, AND DOUGHS

General References: Paul, E. L., V. A. Atiemo-Obeng, and S. M. Kresta (eds.), *Handbook of Industrial Mixing, Science and Practice,* Wiley, Hoboken, N.J., 2004. Harnby, N., M. F. Edwards, and A. W. Nienow (eds.), *Mixing in the Process Industries,* 2d ed., Butterworth-Heinemann, Boston, 1992. Oldshue, J. Y., *Fluid Mixing Technology,* McGraw-Hill, New York, 1983. Ottino, J. M., *The Kinematics of Mixing: Stretching, Chaos, and Transport,* Cambridge University Press, New York, 1999. Tatterson, G. B., *Fluid Mixing and Gas Dispersion in Agitated Tanks,* McGraw-Hill, New York, 1991. Zlokarnik, M., *Stirring, Theory and Practice,* Wiley-VCH, New York, 2001.

INTRODUCTION

Even the definition of mixing for viscous fluids, pastes, and doughs is complicated. While mixing can be defined simply as increasing or maintaining uniformity, the devices that cause mixing to take place may also accomplish deagglomeration, dispersion, extrusion, heat transfer, or other process objectives. Fluids with viscosities greater than 10 Pa·s (10,000 cP) can be considered *viscous.* However, non-newtonian fluid properties are often as important in establishing mixing requirements. Viscous fluids can be polymer melts, polymer solutions, and a variety of other high-molecular-weight or low-temperature materials. Many polymeric fluids are shear thinning. Pastes are typically formed when particulate materials are wetted by a fluid to the extent that particle-particle interactions create flow characteristics similar to those of viscous fluids. The particle-particle interactions may cause shear thickening effects. Doughs have the added characteristic of elasticity. Viscous materials often exhibit a combination of other non-newtonian characteristics, such as a yield stress.

One common connection between viscous fluids, pastes, and doughs is the types of equipment used to mix or process them. While often designed for a specific process objective or a certain fluid characteristic, most types of viscous mixing equipment have some common characteristics. The nature of all viscous materials is their resistance to flow. This resistance is usually overcome by a mixer that will eventually contact or directly influence all the material in a container, particularly material near the walls or in corners. Small clearances between rotating and stationary parts of a mixer create regions of high local shear. Intermeshing blades or stators prevent material from rotating as a solid mass. Such equipment provides greater control of fluid motion than equipment used for low-viscosity fluids, but typically at greater cost and complexity.

The one failure common to all mixing equipment is any region of stagnant material. With a shear thinning material, the relative motion between a rotating mixer blade and adjacent fluid will reduce the local viscosity. However, away from the mixer blade, shear will decrease and the viscosity will increase, leading to the possibility of stagnation. With a shear thickening material, high shear near a mixer blade will result in high viscosity, which may reduce either local relative motion or the surrounding bulk motion. Yield stress requires some minimum shear stress to accomplish any motion at all. Viscoelastic characteristics cause motion normal to the applied stresses. Thus all major non-newtonian characteristics reduce effective mixing and increase the possibility of local stagnation.

Blade shape and mixing action can have significant impacts on the mixing process. A scraping action is often necessary to promote heat transfer or prevent adhesion to equipment surfaces. A smearing action can improve dispersion. A combination of actions is necessary to accomplish the random or complicated pattern necessary for complete mixing. No one mixing effect or equipment design is ideal for all applications.

Because of high viscosity, the mixing Reynolds number ($N_{Re} = D^2N\rho/\mu$, where D is impeller diameter, N is rotational speed, ρ is density, and μ is viscosity) may be less than 100. At such viscous conditions, mixing occurs because of laminar shearing and stretching. Turbulence is not a factor, and complicated motion is a direct result of the mixer action. The relative motion between moving parts of the mixer and the walls of the container or other mixer parts creates both shear and bulk motion. The shear effectively creates thinner layers of nonuniform material, which diminishes striations or breaks agglomerates to increase homogeneity. Bulk motion redistributes the effects of the stretching processes throughout the container.

Often as important as or more important than the primary viscosity is the relative viscosity of fluids being mixed. When a high-viscosity material is added to a low-viscosity material, the shear created by the

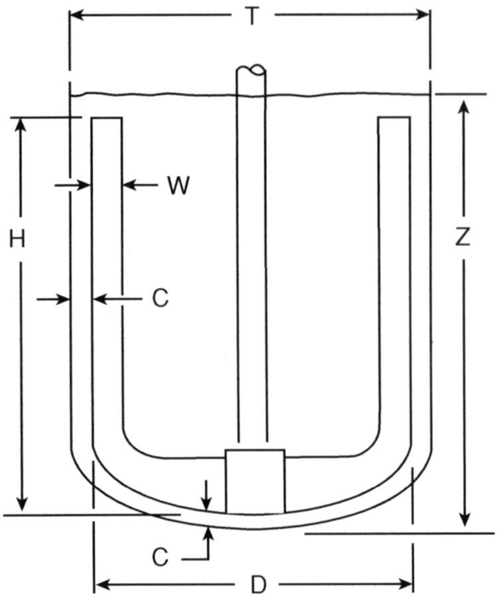

FIG. 18-42 Anchor impeller with nomenclature.

low-viscosity material may not be sufficient to stretch and interact with the high-viscosity material. When a low-viscosity material is added to a high-viscosity material, the low-viscosity material may act as a lubricant, thus allowing slippage between the high-viscosity material and the mixer surfaces. Viscosity differences can be orders of magnitude different. Density differences are smaller and typically less of a problem in viscous mixing.

Besides mixing fluids, pastes, and doughs, the same equipment may be used to create those materials. Viscous fluids such as polymers can be created by reaction from low-viscosity monomers in the same equipment described for viscous mixing. Pastes may be created by either the addition of powders to liquids or the removal of liquids from slurries, again using the same type of equipment as for bulk mixing. Doughs are usually created by the addition of a powder to liquid and the subsequent hydration of the powder. The addition process itself becomes a mixer application, which may fall somewhere between low-viscosity and high-viscosity mixing, but often including both types of mixing.

BATCH MIXERS

Anchor Mixers Anchor mixers are the simplest and one of the more common types of high-viscosity mixers (Fig. 18-42). The diameter of the anchor D is typically 90 to 95 percent of the tank diameter T. The result is a small clearance C between the rotating impeller and the tank wall. Within this gap the fluid is sheared by the relative motion between the rotating blade and the stationary tank wall. The shear near the wall typically reduces the buildup of stagnant material and promotes heat transfer. To reduce buildups further, flexible or spring-loaded scrapers, typically made of polymeric material, can be mounted on the rotating blades to move material physically away from the wall.

The benefits of an anchor mixer are limited by the fact that the vertical blades provide very little fluid motion between the top and bottom of the tank. Ingredient additions at the surface of the fluid may make many rotations before gradually being spread and circulated to the bottom of the tank. To promote top-to-bottom fluid motion, angled blades on the anchor or helical ribbon blades, described in the next subsection, make better mixers for uniform blending. Significant viscosity differences between fluids may extend mixing times to unacceptable limits with the basic anchor.

Anchor mixers may be used in combination with other types of mixers, such as turbine mixers, high-shear mixers, or rotor-stator mixers, which were described in the previous subsection. Such mixers can be placed on a vertical shaft midway between the anchor shaft and blade. A secondary mixer can promote top-to-bottom motion and also limit bulk rotation of the fluid. A stationary baffle is sometimes placed between the anchor shaft and rotating blade to limit fluid rotation and enhance shear.

A dimensionless group called the power number is commonly used to predict the power required to rotate a mixing impeller. The *power number* is defined as $P/(\rho N^3 D^5)$, where P is power, ρ is fluid density, N is rotational speed, and D is impeller diameter. To be dimensionless, the units of the variables must be coherent, such as SI metric; otherwise appropriate conversions factors must be used. The conversion factor for common engineering units gives the following expression for power number:

$$N_P = \frac{1.524 \times 10^{13} P}{\text{sp gr } N^3 D^5} \tag{18-22}$$

where P is power in horsepower, sp gr is fluid specific gravity based on water, N is rotational speed in rpm, and D is impeller diameter in inches. The power number is an empirically measured value that describes geometrically similar impellers. Power number is a function of Reynolds number, which accounts for the effects of fluid properties. Impeller Reynolds number, as defined earlier, is another dimensionless group. A conversion factor is needed for common engineering units:

$$N_{\text{Re}} = \frac{10.4 D^2 N \text{ sp gr}}{\mu} \tag{18-23}$$

where D is the impeller diameter in inches, N is rotational speed in rpm, sp gr is specific gravity based on water, and μ is viscosity in centipoise.

Power can be calculated by rearranging the definition of power number; see the following example. A value for the appropriate power number must be obtained from empirically derived data for geometrically similar impellers. Power number correlations for anchor impellers are shown in Fig. 18-43. The typical anchor impellers have two vertical arms with a blade width W equal to one-tenth of the impeller diameter D, and the arm height H equal to the impeller diameter D. Correlations are shown for typical impellers 95 and 90 percent of the tank diameter. The clearance C is one-half of the difference between the impeller diameter and the tank diameter, or 2.5 and 5.0 percent of the tank diameter for the respective correlations. An additional correlation is shown for an anchor with three vertical arms and a diameter equal to 95 percent of the tank diameter. The correlation for a three-arm impeller which anchors 90 percent of the tank diameter is the same as that for the typical anchor 95 percent of the tank diameter.

The power number and corresponding power of an anchor impeller are proportional to the height of the vertical arm. Thus, an anchor with a height H equal to 75 percent of the impeller diameter would have a power number equal to 75 percent of the typical values shown in Fig. 18-43. Similarly, a partially filled tank with a liquid level Z that covers only 75 percent of the vertical arm will also have a power number that is 75 percent of the typical correlation value. The addition of scrapers will increase the power requirement for an anchor impeller, but the effect depends on the clearance at the wall, design of the scrapers, processed material, and many other factors. Correlations are not practical or available.

Unfortunately, the power number only provides a relationship between impeller size, rotational speed, and fluid properties. The power number does not tell whether a mixer will work for an application. Successful operating characteristics for an anchor mixer usually depend on experience with a similar process or experimentation in a pilot plant. Scale-up of pilot-plant experience is most often done for a geometrically similar impeller and equal tip (peripheral) speed.

Helical Ribbon Mixers Helical ribbon mixers (Fig. 18-44), or simply helix mixers, have major advantages over the anchor mixer, because they force strong top-to-bottom motion even with viscous

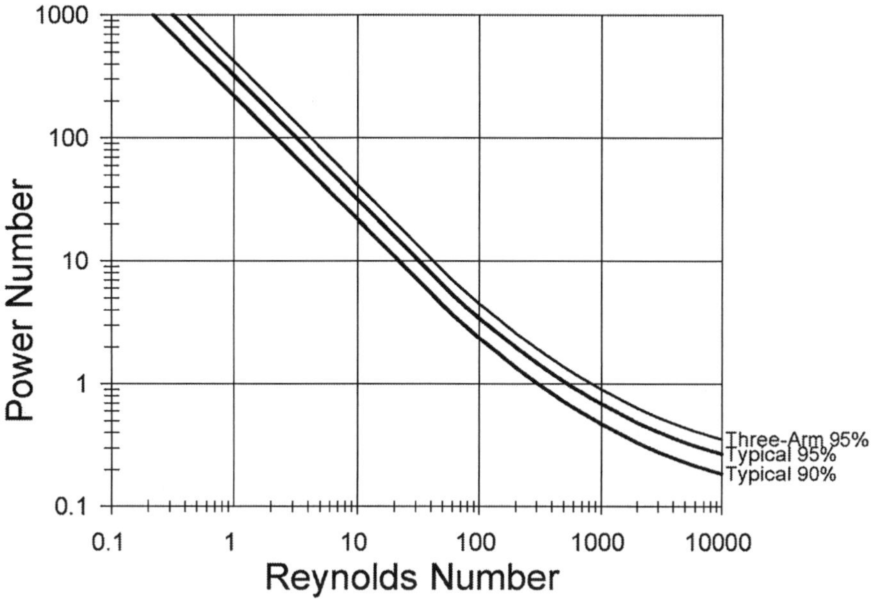

FIG. 18-43 Power numbers for anchor impellers: typical two-arm impeller anchors 95 percent of tank diameter T and 90 percent of T; three-arm impeller anchors 95 percent of T; and three-arm impeller anchors 90 percent of T, similar to two-arm impeller that anchors 95 percent of T.

materials. These impellers are some of the most versatile mixing impellers, but also some of the most expensive. Besides having a formed helical shape, the blades must be rolled the hard way with the thick dimension normal to the direction of the circular rolled shape. Helical ribbon mixers will work with most viscous fluids up to the limits of a flowable material, as high as 4,000,000 cP or more depending on non-newtonian characteristics. While not cost-effective for low-viscosity materials, they will adequately mix, and even suspend solids, in low-viscosity liquids. These characteristics make helical ribbon mixers effective for batch processes, such as polymerization or other processes beginning with low-viscosity materials and changing to high-viscosity products. Helical ribbon mixers will even work with heavy pastes and flowable powders. Usually the helix pumps down at the tank wall with fluids and up at the wall with pastes or powders.

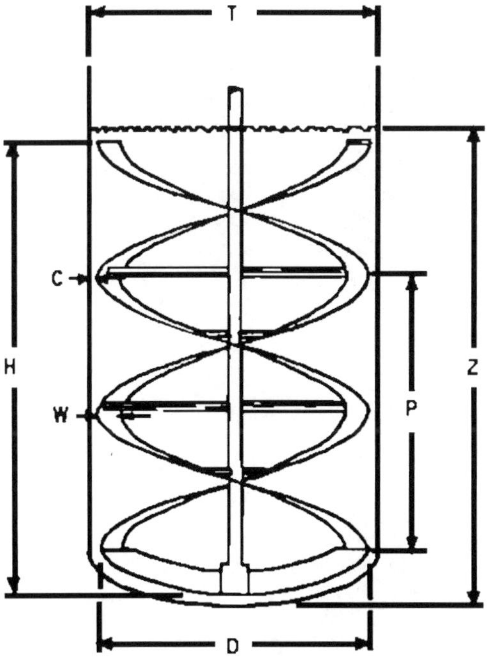

FIG. 18-44 Helical ribbon impeller with nomenclature.

The helical ribbon power numbers are a function of Reynolds number similar to the correlations for anchor impellers. Figure 18-45 shows correlations for some typical helical ribbon power numbers. The upper curve is for a double-flight helix with the blade width W equal to one-tenth the impeller diameter D, the pitch P equal to the impeller diameter, and the impeller diameter at 95 percent of the tank diameter T. The height H for this typical helix is equal to the impeller diameter and pitch, not 15 times the pitch, as shown in Fig. 18-45. A second curve shows the power number correlation for a helical ribbon impeller that is 90 percent of the tank diameter. The curve marked "Single 90%" is for a single flight helix, 90 percent of the tank diameter. Each ribbon beginning at the bottom of the impeller and spiraling around the axis of the impeller is called a flight. Single-flight helixes are theoretically more efficient, but a partially filled tank can cause imbalanced forces on the impeller. The correlation for a 95 percent diameter single-flight helix is the same as the correlation for the double-flight 90 percent diameter helix.

Example 1: Calculate the Power for a Helix Impeller Calculate the power required to rotate a double-flight helix impeller that is 57 in in diameter, 57 in high, with a 57-in pitch operating at 30 rpm in a 60-in-diameter tank. The tank is filled 85 percent full with a 100,000-cP fluid, having a 1.05 specific gravity.

$$N_{\text{Re}} = \frac{10.4\, D^2 N \text{ sp gr}}{\mu} = \frac{10.4(57)^2(30)(1.05)}{100{,}000} = 10.6$$

Referring to Fig. 18-45, the power number N_P for the full-height helix impeller is 27.5 at $N_{\text{Re}} = 10.6$. At 85 percent full, the power number is $0.85 \times 27.5 = 23.4$. Power can be calculated by rearranging Eq. (18-22).

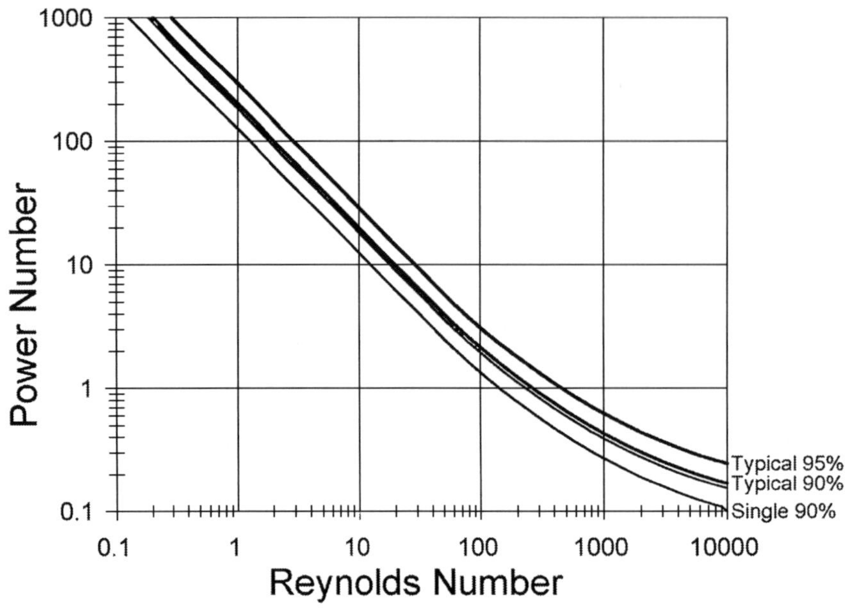

FIG. 18-45 Power numbers for helical-ribbon impeller: typical double-flight helixes 95 percent of tank diameter T and 90 percent of T; single-flight helix 90 percent of T; single-flight 95 percent of T similar to double-flight 90 percent of T.

$$P = \frac{N_P \text{sp gr } N^3 D^5}{1.524 \times 10^{13}} = \frac{23.4(1.05)(30)^3(57)^5}{1.524 \times 10^{13}} = 26.2 \text{ hp}$$

Helical ribbon mixers can also be formed to fit in conical bottom tanks. While not as effective at mixing as in a cylindrical tank, the conical bottom mixer can force material to the bottom discharge. By more effectively discharging, a higher yield of the product can be obtained.

Planetary Mixers A variation on the single anchor mixer is essentially a double anchor mixer with the impellers moving in a planetary pattern. Each anchor impeller rotates on its own axis, while the pair of intermeshing anchors also rotates on the central axis of the tank. The intermeshing pattern of the two impellers gives a kneading action with blades alternately wiping each other. The rotation around the central axis also creates a scraping action at the tank wall and across the bottom. With successive rotations of the impellers, all the tank contents can be contacted directly. A typical planetary mixer is shown in Fig. 18-46.

The intimate mixing provided by the planetary motion means that the materials need not actively flow from one location in the tank to another. The rotating blades cut through the material, creating local shear and stretching. Even thick pastes and viscoelastic and high-viscosity fluids can be mixed with planetary mixers. The disadvantage of poor top-to-bottom motion still exists with conventional planetary mixers. However, some new designs offer blades with a twisted shape to increase vertical motion.

To provide added flexibility and reduce batch-to-batch turnaround or cross-contamination, a change-can feature is often available with planetary and other multishaft mixers. The container (can) in a change-can mixer is a separate part that can be rapidly exchanged between batches. Batch ingredients can even be put in the can before it is placed under the mixing head. Once the mixing or processing is accomplished, the container can be removed from the mixer and taken to another location for packaging and cleaning. After one container is removed from the mixer and the blades of the impeller are cleaned, another batch can begin processing. Because the cans are relatively inexpensive compared with the cost of the mixer head, a change-can mixer can be better utilized and processing costs can be reduced.

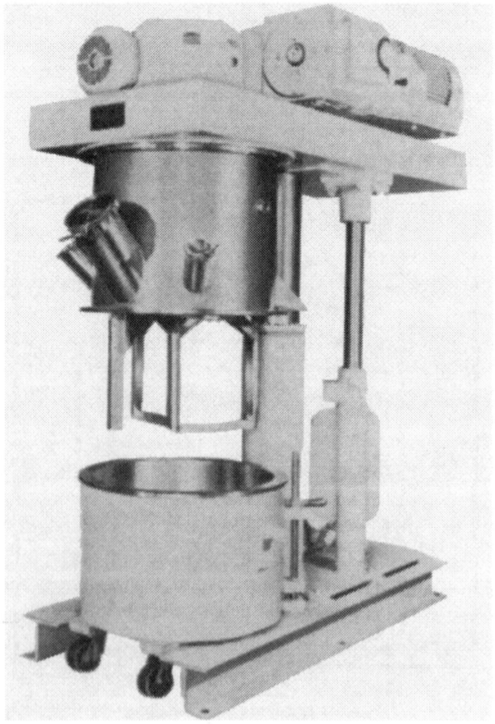

FIG. 18-46 Planetary mixer. *(Charles Ross & Son Company.)*

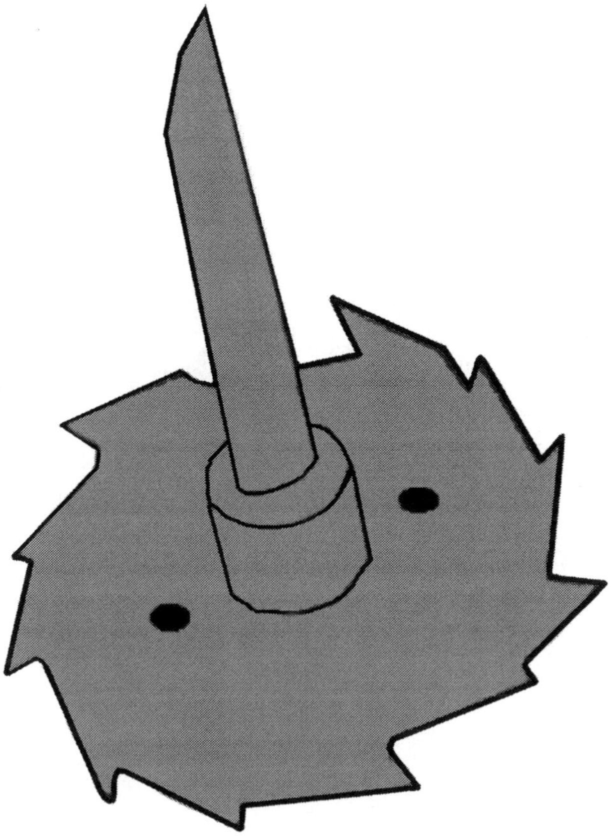

FIG. 18-47 High-shear impeller.

Double- and Triple-Shaft Mixers The planetary mixer is an example of a double shaft mixer. However, many different combinations of mixing actions can be achieved with multi-shaft mixers. One variation on planetary motion involves replacing one anchor-style impeller with a high-shear impeller similar to the one shown in Fig. 18-47. The high-shear mixer can be used to incorporate powdered material effectively or create a stable emulsion leading to a final batch of viscous paste or fluid.

Many types of multishaft mixers do not require planetary motion. Instead the mixers rely on an anchor-style impeller to move and shear material near the tank wall, while another mixer provides a different type of mixing. The second or third mixer shafts may have a pitched-blade turbine, hydrofoil impeller, high-shear blade, rotor-stator mixer, or other type of mixer. The combination of multiple impeller types adds to the flexibility of the total mixer. Many batch processes involve different types of mixing over a range of viscosities. Some mixer types provide the top-to-bottom motion that is missing from the anchor impeller alone.

Double-Arm Kneading Mixers A double-arm kneader consists of two counter-rotating blades in a rectangular trough with the bottom formed like two overlapping or adjacent half-cylinders (Fig. 18-48). The blades are driven by gearing at one or both ends. The older-style kneaders emptied through a door or valve at the bottom. Those mixers are still used where complete discharge or thorough cleaning between batches is not essential. More commonly, double-arm kneaders are tilted for discharge. The tilting mechanism may be manual, mechanical, or hydraulic, depending on the size of the mixer and weight of the material.

A variety of blade shapes have evolved for different applications. The mixing action is a combination of bulk movement, shearing, stretching, folding, dividing, and recombining. The material being mixed is also squeezed and stretched against the blades, bottom, and sidewalls of the mixer. Clearances may be as close as 1 mm (0.04 in). Rotation is usually such that the material is drawn down in the center between the blades and up at the sidewalls of the trough. Most of the blades are pitched to cause end-to-end motion.

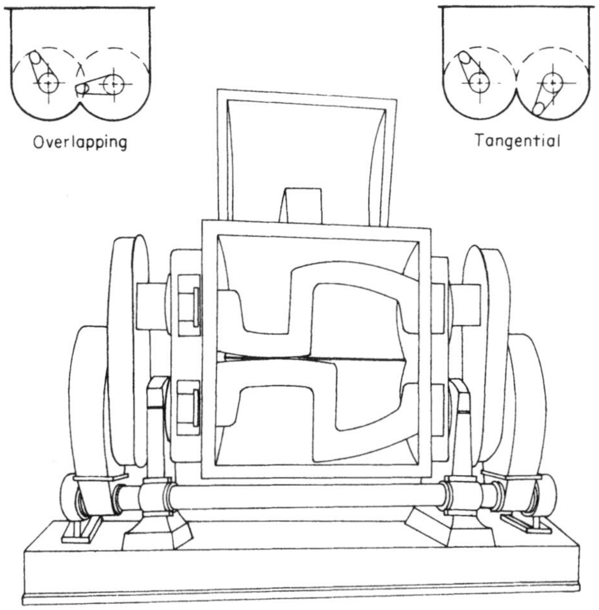

FIG. 18-48 Double-arm kneader. *(APV Baker Invensys.)*

The blades can be tangential or overlapping. Tangential blades can run at different speeds with the advantages of faster mixing caused by changes in the relative position of the blades, greater heat-transfer surface area per unit volume, and less tendency for the material to ride above the blades. Overlapping blades can reduce the buildup of material sticking to the blades.

Because the materials most commonly mixed in kneaders are very viscous, often elastic or rubbery materials, a large amount of energy must be applied to the mixer blades. All that energy is converted to heat within the material. Often the material begins as a semisolid mass, with liquid or powder additives, and the blending process both combines the materials and heats them to create uniform bulk properties.

The blade design most commonly used is the sigma blade (Fig. 18-49*a*). The sigma-blade mixer can start and operate with either liquids or solids, or a combination of both. Modifications to the blade faces have been introduced to increase particular effects, such as shredding or wiping. The sigma blades can handle elastic materials and readily discharge materials that do not stick to the blades. The sigma blades are easy to clean, even with sticky materials.

The dispersion blade in Fig. 18-49*b* was developed to provide higher compressive shear than the standard sigma blade. The blade shape forces material against the trough surface. The compressive action is especially good for dispersing fine particles in a viscous material. Rubbery materials have a tendency to ride the blades, and a dispersion blade is frequently used to keep the material in the mixing zone.

Multiwiping overlapping (MWOL) blades (Fig. 18-49*c*), are commonly used for mixtures that start tough and rubberlike. The blade shape initially cuts the material into small pieces before plasticating it.

The single-curve blade (Fig. 18-49*d*), was developed for incorporating fiber reinforcement into plastics. In this application the individual fibers, e.g., glass, must be wetted with the polymer without undue fiber breakage.

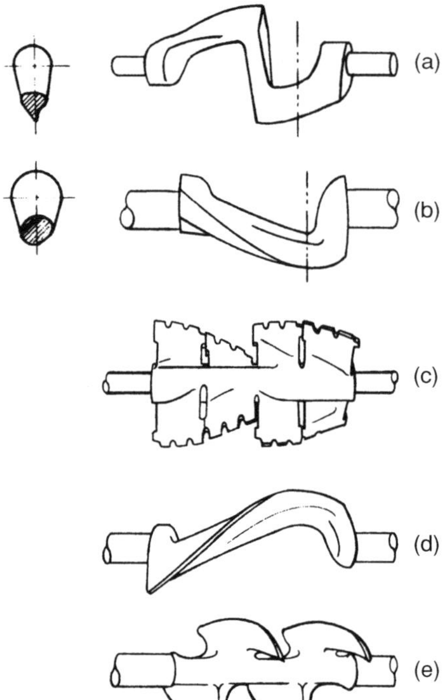

FIG. 18-49 Agitator blades for double-arm kneader: *(a)* Sigma; *(b)* dispersion; *(c)* multiwiping; *(d)* single-curve; *(e)* double-naben. *(APV Baker Invensys.)*

Many other designs have been developed for specific applications. The double-naben blade (Fig. 18-49*e*), is good for mixtures which "ride," meaning they form a lump that bridges across the sigma blade.

Figure 18-50 provides a guide for some typical applications of double-arm mixers. Individual formulations may require more power.

Screw-Discharge Batch Mixers A variant of the sigma-blade mixer has an extrusion-discharge screw located at the center of the trough, just below the rotating blades. During the mixing cycle the screw moves the material within the reach of the mixing blades, thus accelerating the mixing process. At discharge time, the screw extrudes the finished material through a die opening in the end of the machine. The discharge screw is driven independently of the mixer blades.

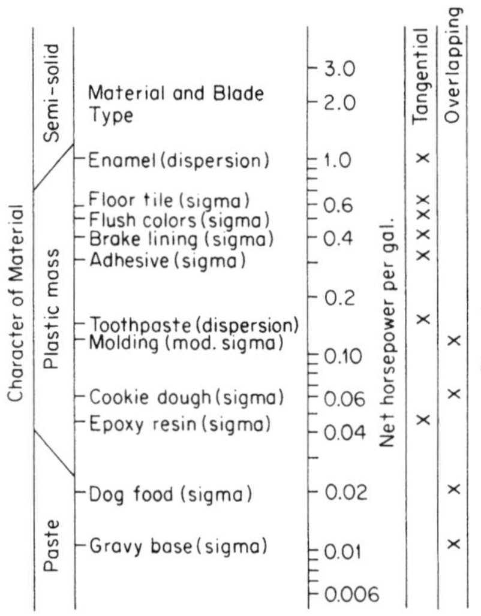

FIG. 18-50 Typical application and power for double-arm kneaders. To convert horsepower per gallon to kilowatts per cubic meter, multiply by 197.3. [*Parker, Chem. Eng.* **72***(18): 125 (1965); excerpted by special permission of the copyright owner, McGraw-Hill, Inc.*]

INTENSIVE MIXERS

Banbury Mixers The dominant high-intensity mixer, with power input up to 6000 kW/m^3 (30 hp/gal), is the Banbury mixer made by Farrel Co. (Fig. 18-51). It is used primarily in the plastics and rubber industries. The batch charge of material is forced into the mixing chamber by an air-operated ram at the top of the mixer. The clearance between the rotors and the walls is extremely small. The mixing action takes place in that small gap. The rotors of the Banbury mixer operate at different speeds, so one rotor can drag material against the rear of the other and thus clean ingredients from behind and between the rotors.

The extremely high power consumption of these machines, which operate at speeds of 40 rpm or less, requires large-diameter shafts. The combination of heavy shafts, stubby blades, close clearances, and

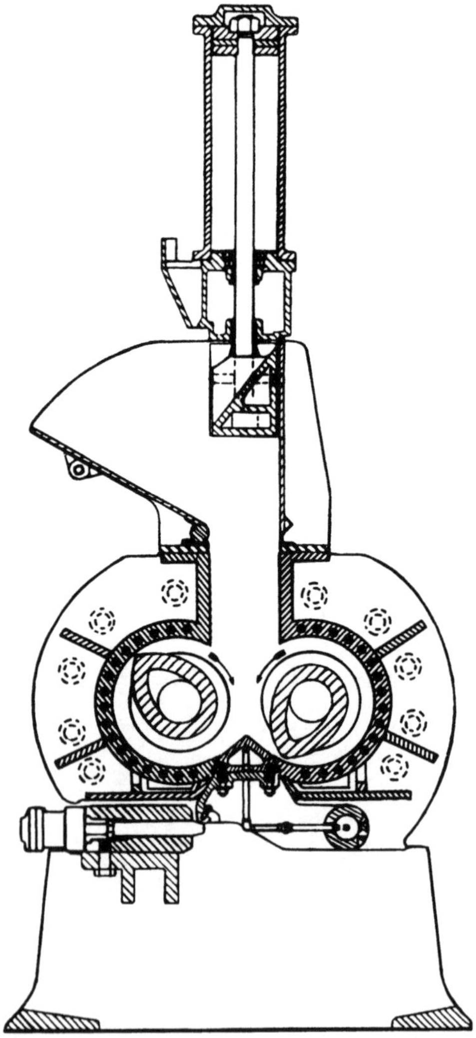

FIG. 18-51 Banbury mixer. *(Farrel Co.)*

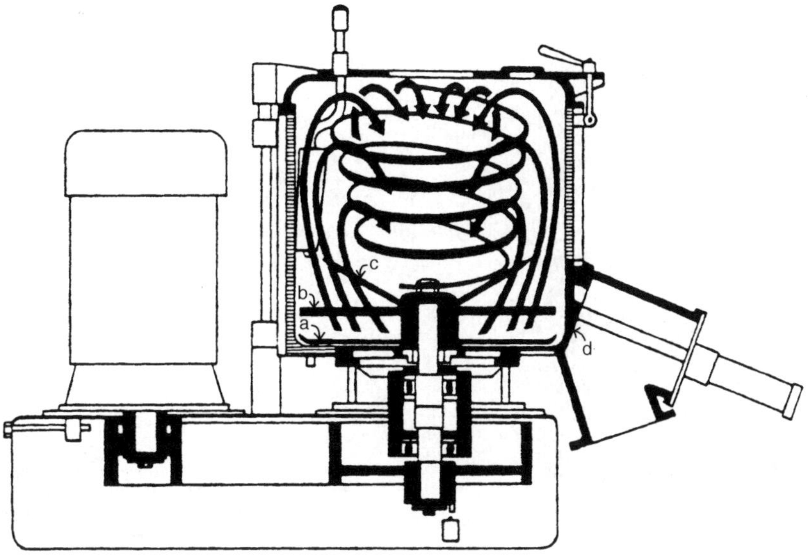

FIG. 18-52 High-intensity mixer: (*a*) bottom scraper; (*b*) fluidizing tool; (*c*) horn tool; (*d*) flush-mounted discharge valve. (*Henschel Mixers America, Inc.*)

a confined charge limits the Banbury mixer to small batches. The production rate is increased as much as possible by using powerful drives and rotating the blades at the highest speed that the material can tolerate without degradation. The heat added by the high-power input often limits operating conditions because of temperature limits on the material being mixed. Equipment is available from laboratory size to a mixer that can handle a 450-kg (1000-lb) charge and applying 2240 kW (3000 hp).

High-Intensity Mixers Mixers, such as the one shown in Fig. 18-52, combine a high-shear zone with a fluidized vortex for mixing of pastes and powders. Blades at the bottom of the vessel scoop the material upward with peripheral speeds of about 40 m/s (130 ft/s). The high-shear stresses between the blade and the bowl, along with blade impact, easily reduce agglomerates and create an intimate dispersion of powders and liquids. Because the energy input is high, 200 kW/m^3 (8 hp/ft^3), even powdery material can heat rapidly.

These mixers are particularly suited for the rapid mixing of powders and granules with liquids, for dissolving resins or solids in liquids, or for removal of volatiles from pastes under a vacuum. Scale-up is usually based on constant peripheral speed of the impeller.

Roll Mills Roll mills can provide extremely high localized shear while retaining extended surface area for temperature control. A typical roll mill has two parallel rolls mounted in a heavy frame with provisions for accurately regulating the pressure and distance between the rolls. Since one pass between the rolls does only a little blending, the mills are usually used as a series of mixers. Only a small amount of material is in the high-shear zone at a time, thus allowing time and exposure for cooling.

To increase the shearing action, the rolls are usually operated at different speeds. The material passing between the rolls can be returned to the feed by the rotation of the rolls. If the rolls are at different temperatures, the material will usually stick to the hotter roll and return to the feed point as a thick layer.

At the end of a period of batch mixing, heavy materials may be discharged by simply dropping from between the rolls. Thin, lighter mixes may be removed by a scraper bar pressing against the descending surface of one of the rolls. Roll mixers are used primarily for preparing color pastes for inks, paints, and coatings. A few applications in heavy-duty blending of rubber stocks use corrugated rolls for masticating the material.

Miscellaneous Batch Mixers Many mixers used for solids blending (Sec. 19 of seventh edition) are suitable for liquid-solids blending. Some solids processing applications involve the addition of liquids, and the same blenders may transition from dry powders to cohesive pastes.

Ribbon blenders typically have multiple helical ribbons with opposing pitches operating in a horizontal trough with a half-cylinder bottom. These mixers can be used for wetting or coating a powder. The final product may have a paste consistency, but must remain at least partially flowable for removal from the blender.

Plowshare mixers have plow-shaped blades mounted at the ends of arms on a horizontal rotating shaft in a cylindrical chamber. The shaft rotates at a sufficient speed to toss the material into the free space in the vessel. The angled surfaces of the plow-shaped blades provide additional intermixing and blending in the bed of solids. High-speed (3600-rpm) chopper blades mounted in the lower side of the mixing chamber can disperse fine particles or break agglomerates. Mixers are available in sizes from 0.03- to 30-m^3 (1.0- to 1000-ft^3) working capacity. Plowshare mixers can be used for either batch or continuous processing.

Conical mixers are also known as Nauta mixers (Fig. 18-53). Material placed in the conical bin is lifted by the rotation of the helical screw, which in turn is rotated around the wall of the cone. The lifting actions of the screw combined with motion around the cone provide bulk mixing for flowable dry powders, paste materials, and even viscous fluids. The specific energy input is relatively small, and the large volume of the mixers can even provide storage capacity. The mixers may have multiple screws, tapered screws, and high-speed dispersers for different applications. At constant speed, both the mixing time and power scale up with the square root of volume. Sizes from 0.1 to 20 m^3 (3.3 to 700 ft^3) are available.

Pan mullers are the modern industrial equivalent of the traditional mortar and pestle. Typical mullers have two broad wheels (M1 and M2) on an axle (Fig. 18-54). The mixer rotates about the approximate midpoint of the axle, so that the wheels both rotate and skid over the bottom of the mixing chamber (A). Plow blades (P1 and P2), which rotate with the mixer, push material from the center (T) and walls (C) of the mixing chamber into the path of the rollers. The mixing action combines both crushing and shearing to break lumps or agglomerates and evenly distribute moisture.

Mullers can be used if the paste is not too fluid or sticky. The main application of muller mixers is now in the foundry industry to mix small amounts of moisture and binder with sand for both core and molding sand. Muller mixers also handle such diverse materials as

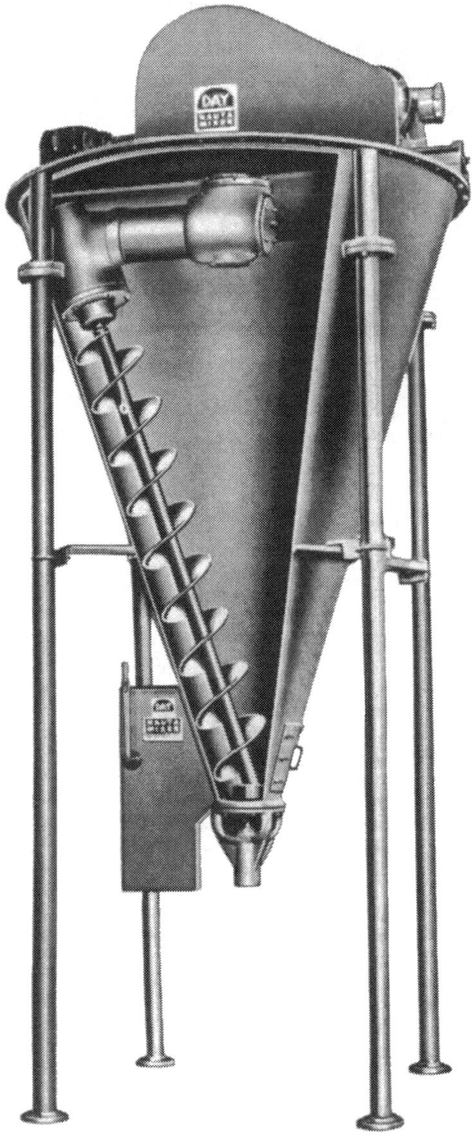

FIG. 18-53 Day Nauta conical mixer. (*Littleford Day, Inc.*)

clay, storage-battery paste, welding-rod coatings, and chocolate coatings. Standard muller mixers range in capacity from 0.01 to 1.7 m^3 (0.4 to 60 ft^3), with power requirements from 0.2 to 56 kW (⅓ to 75 hp).

A continuous muller design employs two intersecting and communicating chambers, each with its own mullers and plows. At the point of intersection of the two chambers, the outside plows give an approximately equal exchange of material from one chamber to the other. Material builds in the first chamber until the feed rate and the discharge rate of the material are equal. The quantity of material in the muller is regulated by adjusting the outlet gate.

CONTINUOUS MIXERS

Some batch mixers previously described can be modified for continuous processing. Product uniformity may be limited because of broad residence time distributions. If ingredients can be accurately metered,

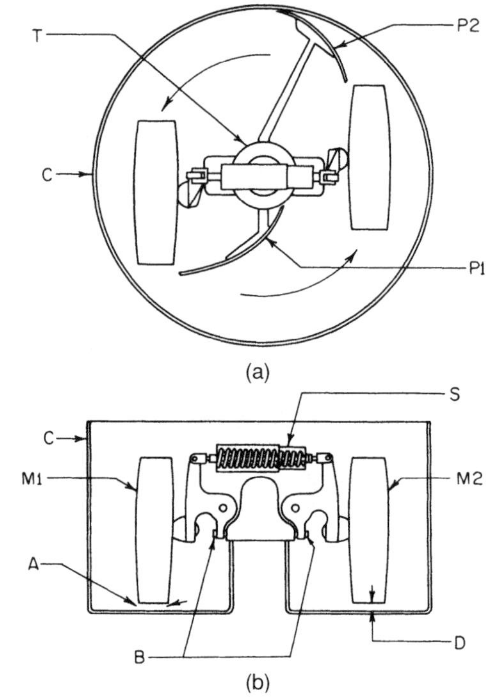

FIG. 18-54 Pan muller: (*a*) plan view; (*b*) sectional elevation. [*Bullock*, Chem. Eng. Prog. ***51:*** *243 (1955); by permission.*]

which can be a problem with powdered or viscous materials, several continuous mixers are available. Continuous mixers often consist of a closely fitting agitator element rotating within a stationary housing.

Single-Screw Extruders The use of extruders, like the one shown in Fig. 18-55, is widespread in the plastic industries. The quality and utility of the product often depend on the uniformity of additives, stabilizers, fillers, etc. A typical extruder combines the process functions of melting the base resin, mixing in additives, and developing the pressure required for shaping the product into pellets, sheet, or profiles. Dry ingredients, sometimes premixed in a batch blender, are fed into the feed throat where the channel depth is deepest. As the root diameter of the screw is increased, the plastic is melted by a combination of friction and heat transfer from the barrel. Shear forces can be very high, especially in the melting zone. The mixing is primarily a laminar shear action.

Single-screw extruders can be built with a long length-to-diameter ratio to permit sufficient space and residence time for a sequence of process operations. Capacity is determined by diameter, length, and power. Most extruders are in the 25- to 200-mm-diameter range. Larger units have been made for specific applications, such as polyethylene homogenization. Mixing enhancers (Fig. 18-56) are used to provide both elongation and shearing action to enhance dispersive (axial) and distributive (radial) mixing.

The maximum power (P in kilowatts) supplied for single-screw extruders varies with the screw diameter (D in millimeters) approximately as

$$P = 5.3 \times 10^{-3} D^{2.25} \qquad (18\text{-}24)$$

The energy required for most polymer mixing applications is from 0.15 to 0.30 kWh/kg (230 to 460 Btu/lb).

Twin-Screw Extruders Two screws in a figure-eight barrel have the advantage of interaction between the screws plus action between the screws and the barrel. Twin-screw extruders are used to melt continuously, mix, and homogenize different polymers and additives. Twin-screw extruders can also be used to provide the intimate

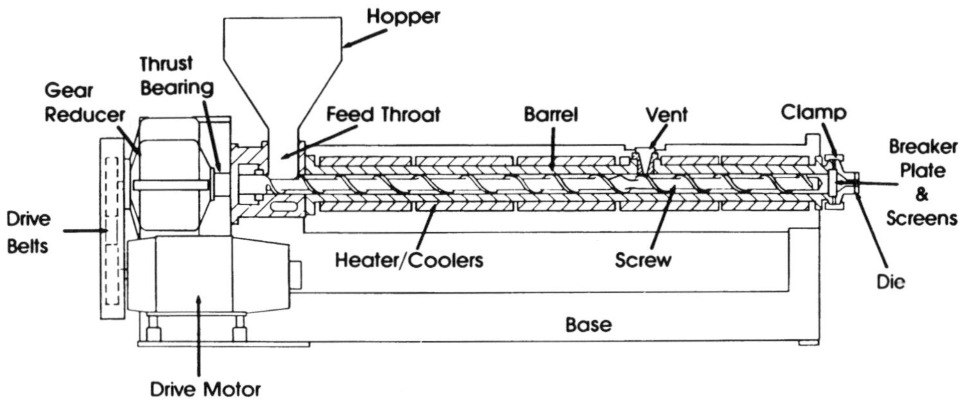

FIG. 18-55 Single-screw extruder. (*Davis Standard.*)

mixing needed to carry out chemical reactions in high-viscosity materials. The screws can be either tangential or intermeshing, with the latter either corotating or counterrotating. Tangential designs allow variability in the channel depth and permit longer lengths.

The most common twin-screw extruder is the counterrotating intermeshing type. The counterrotating intermeshing screws provide a dispersive milling action between the screws and the ability to generate pressure efficiently. The two keyed or splined shafts are fitted with pairs of slip-on kneading or conveying elements, as shown in Fig. 18-57. Each pair of kneading paddles causes an alternating compression and expansion effect that massages the contents and provides a combination of shearing and elongational mixing actions. The arrays of elements can be varied to provide a wide range of mixing effects. The barrel sections are also segmented to allow for optimum positioning of feed ports, vents, barrel valves, etc. The barrels may be heated electrically or with oil or steam and cooled with air or water.

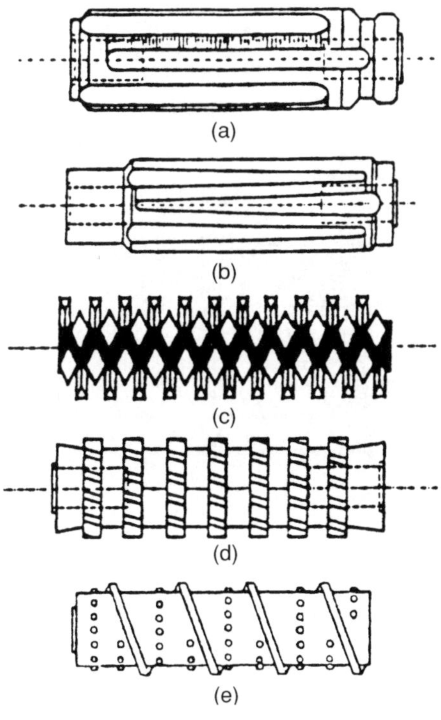

FIG. 18-56 Mixing enhancers for single-screw extruders: (*a*) Maddock, straight; (*b*) Maddock, tapered; (*c*) pineapple; (*d*) gear; (*e*) pin.

Counterrotating twin-screw extruders are available in diameters ranging from 15 to 300 mm (0.5 to 12 in), with length-to-diameter ratios up to 50 and throughput capacities to 7 kg/s (55,000 lb/h). Screw speeds can be as high as 8 r/s (500 rpm) in small production extruders. Residence times for melting are usually less than 120 s (2 min).

Farrel Continuous Mixer The Farrel mixer (Fig. 18-58) consists of rotors similar in cross section to the Banbury batch mixer. The first section of the rotor acts as a screw conveyor, moving the feed ingredients into the mixing section. The mixing action is a combination of intensive shear between the rotor and chamber wall, kneading between the rotors, and a rolling action within the material itself. The amount and quality of mixing are controlled by adjustment of speed, feed rate, and discharge orifice opening. Mixers are available with chamber volumes up to 0.12 m^3 (4.2 ft^3). With speeds to 3.3 r/s (200 rpm), the power range is from 5 to 2200 kW (7.5 to 3000 hp).

Miscellaneous Continuous Mixers Because of the diversity of material properties and process applications involving viscous fluids, pastes, and doughs, the types of mixers are almost as diverse.

Trough-and-screw mixers usually consist of a single rotor or twin rotors that continually turn the feed material over as it progresses toward the discharge end of the mixer. Some mixers have been designed with extensive heat-transfer surface area. The continuous-screw, **Holo-Flite processor** (Fig. 18-59) is used primarily for heat transfer, since the hollow screws provide extended surfaces without creating much shear. Two or four screws may be used.

Another type of trough-and-screw mixer is the **AP Conti paste mixer,** shown in Fig. 18-60. These self-cleaning mixers are particularly appropriate when the product being handled goes through a sticky stage, which could plug the mixer or foul the heat-transfer surfaces.

Pug mills have one or two shafts fitted with short heavy paddles, mounted in a cylinder or trough holding the material to be processed. In the two-shaft mills the shafts are parallel and may be either horizontal or vertical. The paddles may or may not intermesh. Clearances are wide, so considerable mass mixing takes place. Unmixed or partially mixed ingredients are fed at one end of the machine, which is usually totally enclosed. Liquid may be added to the material entering the mixer. The paddles push the material forward as they cut through it. The action of the paddles carries the material toward the discharge end of the mixer. The product may discharge through one or two open ports or through extrusion nozzles. The nozzles create roughly shaped continuous strips of material. Automatic cutters may be used to make blocks or pellets from the strips. Pug mills are most often used for mixing mineral or clay products.

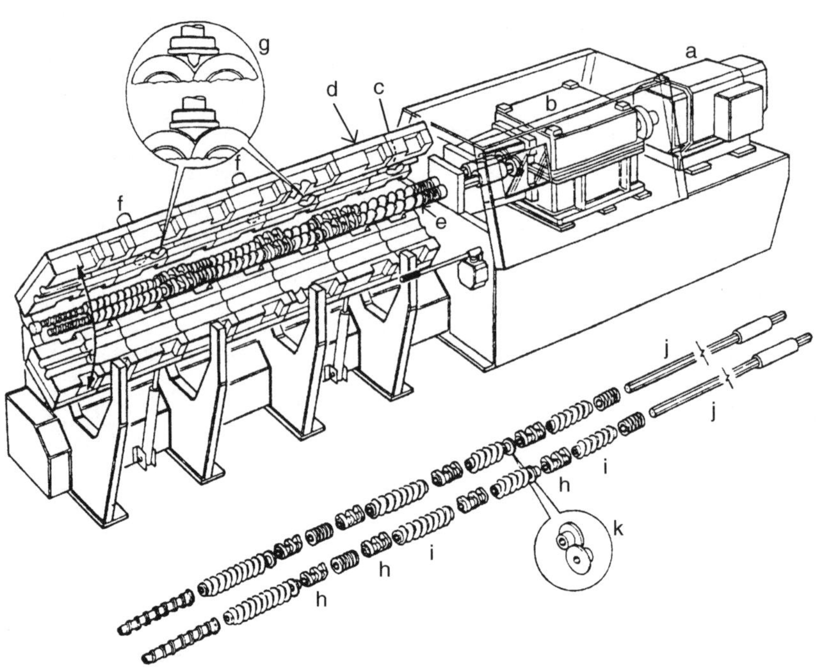

FIG. 18-57 Intermeshing corotating twin-screw extruder: (*a*) drive motor; (*b*) gearbox; (*c*) feed port; (*d*) barrel; (*e*) assembled rotors; (*f*) vent; (*g*) barrel valve; (*h*) kneading paddles; (*I*) conveying screws; (*j*) splined shafts; (*k*) blister rings. (*APV Chemical Machinery, Inc.*)

Motionless mixers are an alternative to rotating impeller mixers. Motionless or static mixers use stationary shaped elements inside pipes or conduits to divide, divert, twist, and recombine flowing material. The dividing, stretching, and recombining processes lead to thinner and thinner striations in viscous materials to achieve uniformity.

The twisted-element mixers, such as the **Kenics static mixer** (Fig. 18-61), create 2^n layers in n divisions. Each element twists the flow, moving material from the center to the wall and from the wall to the center. The twisting also stretches striations having different properties and reorients the material before the next division. The following element

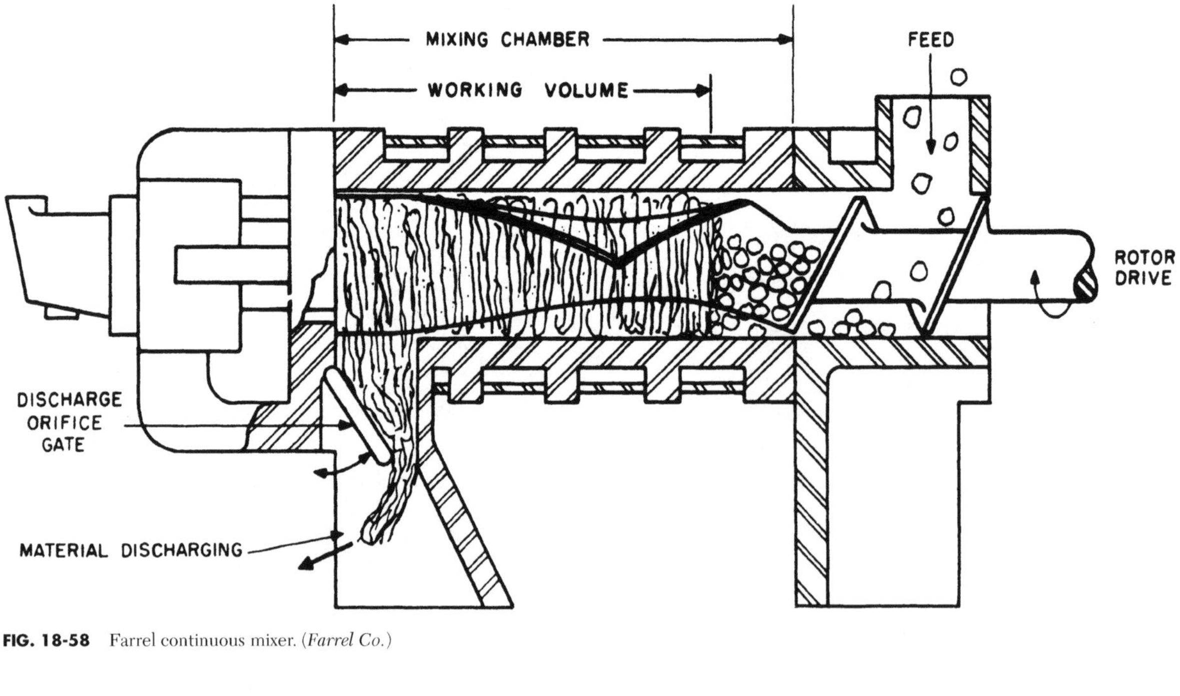

FIG. 18-58 Farrel continuous mixer. (*Farrel Co.*)

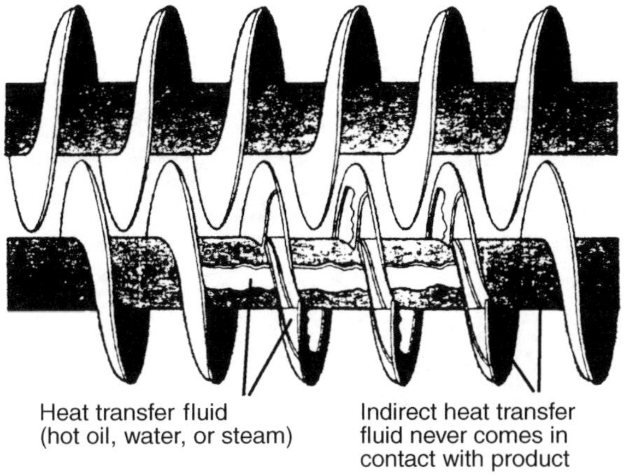

FIG. 18-59 Holo-Flite Processor. (*Metso Minerals.*)

twists the divided material in the opposite direction. The more viscous the material, the more mixing elements are required for uniformity.

Other motionless designs, such as the **Sulzer static mixer** (Fig. 18-62), accomplish mixing by making multiple divisions at each element transition. The flowing material follows a wavy path to stretch and distort the striations. The number of divisions and distorted paths causes more rapid mixing, but at the expense of a greater pressure drop per unit length of the mixer.

The power required to accomplish mixing in a motionless mixer is provided by the pump used to force the fluid through the mixer. The pressure drop through a motionless mixer is usually expressed as a multiplier K of the open pipe loss or as a valve coefficient C_V. The value of the multiplier is strongly dependent on the detail geometry of the mixer, but is usually available through information from the supplier. Fluid properties are taken into account by the value of the Reynolds number for the open pipe. Motionless mixers are usually sized to match the diameter of the connecting pipe. Pumping adjustments are made when necessary to handle the increased pressure drop.

Because motionless mixers continuously interchange fluid between the walls and the center of the conduit, they also provide good heat transfer, especially with the twisted-element style of mixers. Sometimes, high-viscosity heat exchange is best accomplished with a static mixer.

Distributive (radial) mixing is usually excellent; dispersive (axial) mixing is often poor. The result can be a good plug-flow mixer or reactor, with corresponding benefits and limitations.

PROCESS DESIGN CONSIDERATIONS

Scale-up of Batch Mixers While a desirable objective of scale-up might be equal blending uniformity in equal time, practicality dictates that times for blending are longer with larger batches. Scale-up of many processes and applications can be successfully done by holding constant the peripheral speed of the rotating element in the mixer. Equal peripheral speed, often called *equal tip speed,* essentially means that the maximum velocity in the mixer remains constant.

Perhaps one of the most difficult concepts to grasp about viscous mixing is that, unlike in turbulent mixing, greater mixer speed does not always translate to better mixing results. If a rotating mixer blade cuts through a viscous fluid or heavy paste too quickly, the stretching process that reduces striation thickness does not take place throughout the material. At high rotational speeds, rapid shearing between a blade tip and the wall or housing may take place, but flow to create bulk motion may not have time to occur. Thus, slower speeds may actually give better mixing results.

With geometric similarity, equal tip speed means that velocity gradients are reduced and blend times become longer. However, power per volume is also reduced, and viscous heating problems are likely to be more controllable. With any geometric scale-up, the surface-to-volume ratio is reduced, which means that any internal heating, whether by viscous dissipation or chemical reaction, becomes more difficult to remove through the surface of the vessel.

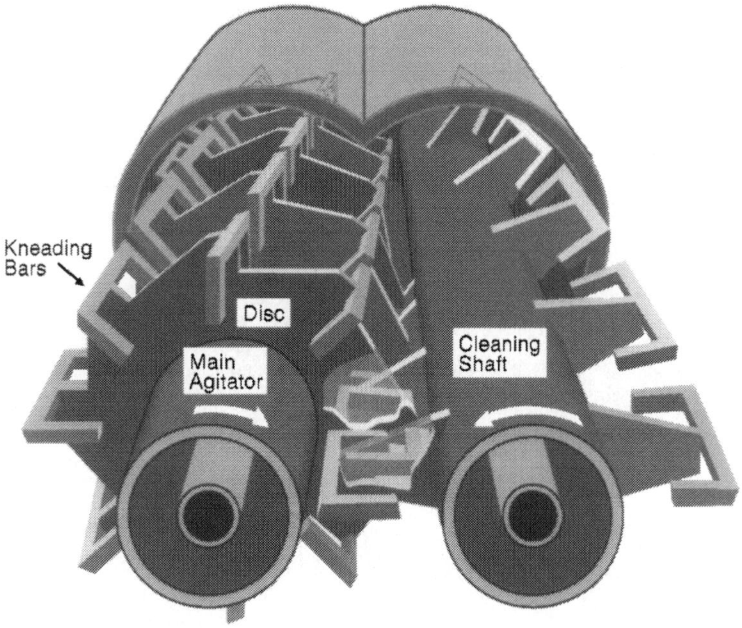

FIG. 18-60 AP Conti paste mixer. (*LIST, Inc.*)

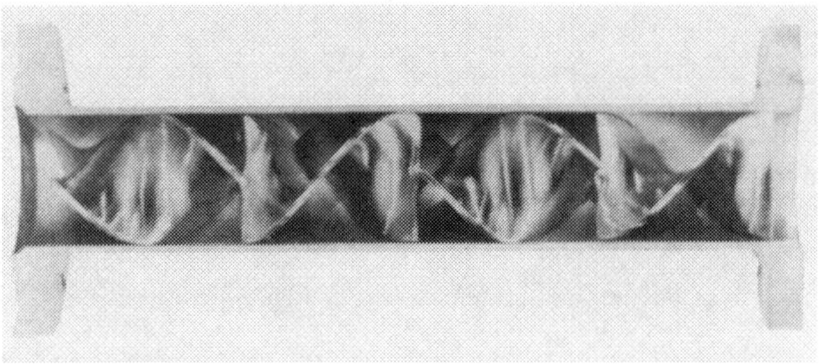

FIG. 18-61 Kenics static mixer. (*Chemineer, Inc.*)

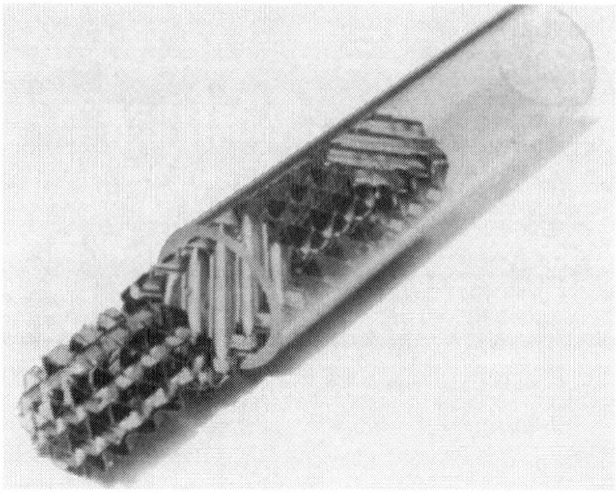

FIG. 18-62 Sulzer static mixer. (*Sulzer Chemtech.*)

In many applications the blend time is closely related to the actual number of revolutions made by the mixing device. Thus if mixing were successfully accomplished in 5 min at 60 rpm in a small mixer, the same uniformity could be achieved in 10 min at 30 rpm in a larger mixer. Other factors, such as the rate of heating, could limit scale-up and mixing times.

The physical properties of a paste are difficult to define because a combination of yield stress, shear dependence, time dependence, and even elasticity may be present. Further, many process applications involve the formation or modification of the physical properties. To relate accurately specific material properties to mixing characteristics or power requirements can be extremely difficult. Actual observation and measurement in small-scale equipment or comparison with similar existing processes may be the only practical way of predicting successful operating conditions. Power measurements in small-scale equipment are often essential to predict large-scale conditions and may form the basis for operating production equipment.

Scale-up of Continuous Mixers While geometric similarity may be practical for most batch mixers, changes in the length-to-diameter ratio or other geometry may be necessary with continuous mixers. The most common problem is heat generation by friction and heat removal by surface transfer.

In single-screw extruders, e.g., Fig. 18-55, channel depth in the flights cannot be increased in proportion to the screw diameter because the distribution of heat generated by friction at the barrel wall requires more time as the channel depth becomes greater. With constant retention time, therefore, a nonhomogeneous product would be discharged from a geometrically similar large-scale extruder.

As the result of the departure from geometric similarity, the throughput rate of single-screw extruders scales up with the diameter to 2.0 to 2.5 power, instead of the diameter cubed, at constant length-to-diameter and screw speed. The throughput rates of twin-screw extruders (Fig. 18-57) and the Farrel continuous mixer (Fig. 18-58) are scaled up with the diameter to about the 2.6 power.

The extent of axial dispersion through a continuous mixer can be characterized either by an axial diffusion coefficient or by analogy to a number of well-mixed stages in series. Retention time can control the performance of a mixing system. As the number of apparent stages increases, the greater is the assurance that all the material will have the required residence time. Under conditions requiring uniform retention time, the feed streams must enter at the correct ratio on a time scale much shorter than the average residence time of the mixer. Otherwise, variations in the feed will appear as changes in the product. Different types of continuous mixers have different degrees of axial dispersion. Thus, appropriate feed conditions must be considered. Single-screw extruders have an equivalent number of stages equal to approximately one-half the length-to-diameter ratio.

HEATING AND COOLING MIXERS

Heat Transfer Pastes and viscous fluids are often heated or cooled by heat transfer through the walls of the mixing container or hollow mixing arms. A uniform temperature throughout the bulk material is almost as important for good heat transfer as a large heat-transfer surface to mixer volume ratio. Bulk temperature uniformity will maximize the temperature-difference driving force for heat transfer. Surface area is a direct factor in overall heat transfer. Effective motion near the surface promotes convection over conduction for better heat transfer. Most mixers for pastes or viscous fluids have some sort of scraper or close-clearance device to move stagnant material away from heat-transfer surfaces.

Typical overall heat-transfer coefficients are between 20 and 200 $J/(m^2 \cdot s \cdot K)$ [4 to 35 $Btu/(h \cdot ft^2 \cdot °F)$].

Heating Methods Steam heating is widely used because it is economical, safe, and easily controlled. The mixer shell must be designed to withstand both the positive pressure of steam and a vacuum caused when the steam condenses. Transfer liquid heating, using water, oil, special organic liquids, or molten salts, permits good temperature control and provides insurance against overheating the process material. Jackets for transfer liquids are usually baffled to provide good circulation. Higher temperatures can be achieved without the heavy vessel construction required by steam pressures.

Electric heating requires that the elements be electrically insulated from the vessel, while still providing good thermal contact. The heaters must be designed for uniform heating to avoid creating hot spots. Temperature control can be precise, maintenance costs low, but utility costs can be very high for large mixers. Electrical heating may be excluded when flammable vapors or dusts are present.

Friction or viscous heating develops rapidly in some mixers, such as a Banbury mixer. The first temperature rise may be beneficial in softening the materials and accelerating chemical reactions. Because energy inputs can be high, higher temperatures detrimental to the products may develop rapidly. So cooling may be required during other portions of a process.

Cooling Methods Air cooling with air blown over external surfaces or external fins may be sufficient for some mixers. Evaporation of excess water or solvent under a vacuum or ambient pressure provides good cooling. A small amount of evaporation produces a large amount of cooling. However, removing too much solvent may damage the product. Some mixers are cooled by circulation of water or refrigerants through jackets or hollow agitators. With viscous fluids, lower temperatures near the cooled surfaces increase viscosity and make heat transfer more difficult.

EQUIPMENT SELECTION

The most common and sometimes the only available approach is by analogy. Many companies manufacture similar products, either of their own or those of competitors. With similar products, both good and bad features of existing or typical mixing equipment need to be considered carefully. Some types of mixing equipment are commonly used throughout certain industries. Sometimes existing equipment can be adapted to a new process. Otherwise, new equipment will be needed.

If new equipment is needed, laboratory or pilot-plant studies are recommended. Often unique product features involve unusual or special fluid properties, which makes prediction of mixer performance almost impossible. The objective is to find potentially suitable equipment and test available mixers. Most equipment vendors have equipment to rent or a demonstration laboratory to test their mixers.

The following list provides some characteristics of a new process that must be considered:

1. List all materials in the process and describe their characteristics.
   a. Method of delivery to the mixer: bags, drums, tote sacks, bulk, pipeline, etc.
   b. Storage and/or weighing requirements at the mixer
   c. Physical form of the material
   d. Specific gravity and bulk characteristics
   e. Particle size or size range
   f. Viscosity
   g. Melting, boiling, or degradation point
   h. Corrosive properties
   i. Abrasive characteristics
   j. Toxicity
   k. Fire or explosion hazards
   l. Irritant characteristics, to skin, eyes, or lungs
   m. Sensitivity of materials when exposed to air, moisture, or heat
2. List pertinent information related to production.
   a. Quantity to be produced per batch
   b. Formulation and order of addition
   c. Analysis required
   d. Cleaning requirements between batches or products
   e. Preceding and/or following process steps
   f. Any changes in physical state during process
   g. Any chemical reactions—exothermic or endothermic
   h. Temperature requirements
   i. Physical form of final product
   j. Removal of product from mixer—pumping or gravity flow through piping, chute, or dumping
3. Describe the controlling features of the finished product.
   a. Degree of uniformity: solution, aggregates, particle size, etc.
   b. Stability of emulsion or dispersion
   c. Ultimate color requirements
   d. Uniformity of active ingredients, as in a pharmaceutical product
   e. Degree of moisture content control

Preparation and Addition of Materials To ensure product quality and productivity, ingredient preparation is important. Order of addition, method and rate of addition, and even preprocessing must be considered.

Some finely powdered materials, such as carbon black or silica, contain a lot of air. If possible, such materials should be compacted, wetted, or agglomerated before addition to the mixture. Air bubbles can be extremely difficult to remove from viscous materials. Holding the product under a vacuum may help release some air or trapped gases. The presence of air in the product may make packaging difficult and may even cause eventual degradation of the product.

Critical ingredients, such as vulcanizers, antioxidants, surfactants, and active agents, are often present in small proportions. If these materials form lumps or aggregates, milling or screening of the materials may be necessary to ensure a uniform product. If small ingredients are soluble in liquid ingredients, adding them as a solution may improve blending. Master batching small quantities of an ingredient into part of a major ingredient often simplifies mixing and makes a more uniform product.

Additional considerations, such as automatic weighing, feed control, liquid metering, and automatic control, may be essential for continuous processes.

CRYSTALLIZATION FROM SOLUTION

General References: *AIChE Testing Procedures: Crystallizers,* American Institute of Chemical Engineers, New York, 1970; *Evaporators,* 1961. Bennett, *Chem. Eng. Prog.,* **58**(9), 76 (1962). Buckley, *Crystal Growth,* Wiley, New York, 1951. Campbell and Smith, *Phase Rule,* Dover, New York, 1951. De Jong and Jancic (eds.), *Industrial Crystallization,* North-Holland Publishing Company, Amsterdam, 1979. "Crystallization from Solution: Factors Influencing Size Distribution," *Chem. Eng. Prog. Symp. Ser.,* **67**(110), (1971). Mullin (ed.), *Industrial Crystallization,* 4th ed., Butterworth-Heinemann, Boston, 2001. Mersmann (ed.), *Crystallization Technology Handbook,* Marcel Dekker, New York, 1995. Jancic and Grootscholten, *Industrial Crystallization,* D. Reidel Publishing, Boston, 1984. Jones, *Crystallization Process Systems,* Butterworth-Heinemann, Boston, 2002. Genck, *Chem. Eng. Prog.* **100** (10), 26 (2004). Newman and Bennett, *Chem. Eng. Prog.,* **55**(3), 65 (1959). Palermo and Larson (eds.), "Crystallization from Solutions and Melts," *Chem. Eng. Prog. Symp. Ser.,* **65**(95), (1969). Randolph (ed.), "Design, Control and Analysis of Crystallization Processes," *Am. Inst. Chem. Eng. Symp. Ser.,* **76**(193), (1980). Randolph and Larson, *Theory of Particulate Processes,* Academic, New York, 2d ed., 1988. Seidell, *Solubilities of Inorganic and Metal Organic Compounds,* American Chemical Society, Washington, 1965.

Crystallization is important as an industrial process because of the number of materials that are and can be marketed in the form of crystals. Its wide use is due to the highly purified and favorable form of a chemical solid which can be obtained from relatively impure solutions in a single processing step. In terms of energy requirements, crystallization requires much less energy for separation than do distillation and other commonly used methods of purification. In addition, it can be performed at relatively low temperatures and on a scale which varies from a few grams up to thousands of tons per day.

Crystallization may be carried out from a vapor, from a melt, or from a solution. Most of the industrial applications of the operation involve crystallization from solutions. Nevertheless, crystal solidification of metals is basically a crystallization process, and much theory has been developed in relation to metal crystallization. This topic is highly specialized, and is outside the scope of this subsection, which is limited to crystallization from solution.

PRINCIPLES OF CRYSTALLIZATION

Crystals A crystal may be defined as a solid composed of atoms or molecules arranged in an orderly, repetitive array. The interatomic distances in a crystal of any definite material are constant and are characteristic of that material. Because the pattern or arrangement of the atoms or molecules is repeated in all directions, there are definite restrictions on the kinds of symmetry that crystals can possess.

There are five main types of crystals, and these types have been arranged into seven crystallographic systems based on the crystal interfacial angles and the relative length of its axes. The treatment of the description and arrangement of the atomic structure of crystals is the science of **crystallography.** The material in this discussion will be limited to a treatment of the growth and production of crystals as a unit operation.

Solubility and Phase Diagrams Equilibrium relations for crystallization systems are expressed in the form of solubility data which are plotted as phase diagrams or solubility curves. Solubility data are ordinarily given as parts by weight of anhydrous material per 100 parts by weight of total solvent. In some cases these data are reported as parts by weight of anhydrous material per 100 parts of solution. If water of crystallization is present in the crystals, this is indicated as a separate phase. The concentration is normally plotted as a function of temperature and has no general shape or slope. It can also be reported as a function of pressure, but for most materials the change in solubility with change in pressure is very small. If there are two components in solution, it is common to plot the concentration of these two components on the X and Y axes and represent the solubility by isotherms. When three or more components are present, there are various techniques for depicting the solubility and phase relations in both three-dimension and two-dimension models. For a description of these techniques, refer to Campbell and Smith (loc. cit.). Shown in Fig. 18-63 is a phase diagram for magnesium sulfate in water. The line *p–a*

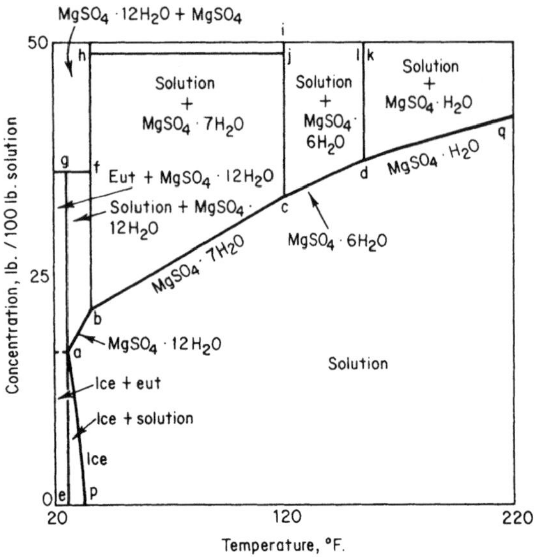

FIG. 18-63 Phase diagram. $MgSO_4 \cdot H_2O$. To convert pounds to kilograms, divide by 2.2; K = (°F + 459.7)/1.8.

represents the freezing points of ice (water) from solutions of magnesium sulfate. Point *a* is the eutectic, and the line *a–b–c–d–q* is the solubility curve of the various hydrates. Line *a–b* is the solubility curve for $MgSO_4 \cdot 12H_2O$, *b–c* is the solubility curve for $MgSO_4 \cdot 7H_2O$, *c–d* is the solubility curve for $MgSO_4 \cdot 6H_2O$, and *d–q* is the portion of the solubility curve for $MgSO_4 \cdot H_2O$.

As shown in Fig. 18-64, the mutual solubility of two salts can be plotted on the *X* and *Y* axes with temperatures as isotherm lines. In the example shown, all the solution compositions corresponding to 100°C with solid-phase sodium chloride present are shown on the line *DE*. All the solution compositions at equilibrium with solid-phase KCl at 100°C are shown by the line *EF*. If both solid-phase KCl and NaCl are present, the solution composition at equilibrium can only be represented by point *E*, which is the invariant point (at constant pressure). Connecting all the invariant points results in the mixed-salt line. The locus of this line is an important consideration in making phase separations.

There are numerous solubility data in the literature; the standard reference is by Seidell (loc. cit.). Valuable as they are, they nevertheless must be used with caution because the solubility of compounds

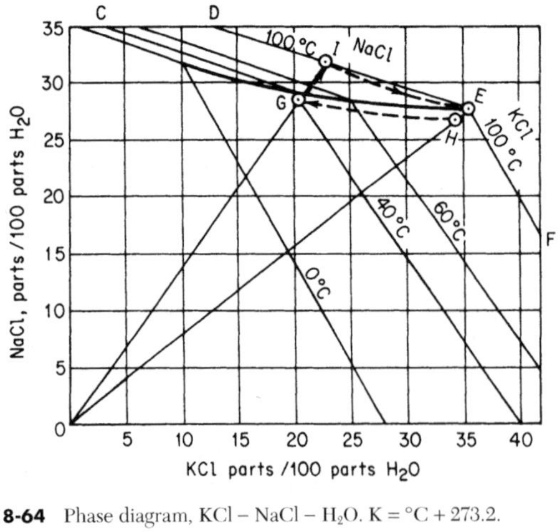

FIG. 18-64 Phase diagram, KCl – NaCl – H_2O. K = °C + 273.2.

is often influenced by pH and/or the presence of other soluble impurities which usually tend to depress the solubility of the major constituents. While exact values for any system are frequently best determined by actual composition measurements, the difficulty of reproducing these solubility diagrams should not be underestimated. To obtain data which are readily reproducible, elaborate pains must be taken to be sure the system sampled is at equilibrium, and often this means holding a sample at constant temperature for a period of from 1 to 100 h. While the published curves may not be exact for actual solutions of interest, they generally will be indicative of the shape of the solubility curve and will show the presence of hydrates or double salts.

Heat Effects in a Crystallization Process The heat effects in a crystallization process can be computed by two methods: (1) a heat balance can be made in which individual heat effects such as sensible heats, latent heats, and the heat of crystallization can be combined into an equation for total heat effects; or (2) an enthalpy balance can be made in which the total enthalpy of all leaving streams minus the total enthalpy of all entering streams is equal to the heat absorbed from external sources by the process. In using the heat-balance method, it is necessary to make a corresponding mass balance, since the heat effects are related to the quantities of solids produced through the heat of crystallization. The advantage of the enthalpy-concentration-diagram method is that both heat and mass effects are taken into account simultaneously. This method has limited use because of the difficulty in obtaining enthalpy-concentration data. This information has been published for only a few systems.

With compounds whose solubility increases with increasing temperature there is an absorption of heat when the compound dissolves. In compounds with decreasing solubility as the temperature increases, there is an evolution of heat when solution occurs. When there is no change in solubility with temperature, there is no heat effect. The solubility curve will be continuous as long as the solid substance of a given phase is in contact with the solution, and any sudden change in the slope of the curve will be accompanied by a change in the heat of solution and a change in the solid phase. Heats of solution are generally reported as the change in enthalpy associated with the dissolution of a large quantity of solute in an excess of pure solvent. Tables showing the heats of solution for various compounds are given in Sec. 2.

At equilibrium the heat of crystallization is equal and opposite in sign to the heat of solution. Using the heat of solution at infinite dilution as equal but opposite in sign to the heat of crystallization is equivalent to neglecting the heat of dilution. With many materials the heat of dilution is small in comparison with the heat of solution and the approximation is justified; however, there are exceptions. Relatively large heat effects are usually found in the crystallization of hydrated salts. In such cases the total heat released by this effect may be a substantial portion of the total heat effects in a cooling-type crystallizer. In evaporative-type crystallizers the heat of crystallization is usually negligible when compared with the heat of vaporizing the solvent.

Yield of a Crystallization Process In most cases the process of crystallization is slow, and the final mother liquor is in contact with a sufficiently large crystal surface so that the concentration of the mother liquor is substantially that of a saturated solution at the final temperature in the process. In such cases it is normal to calculate the yield from the initial solution composition and the solubility of the material at the final temperature. If evaporative crystallization is involved, the solvent removed must be taken into account in determining the final yield. If the crystals removed from solution are hydrated, account must be taken of the water of crystallization in the crystals, since this water is not available for retaining the solute in solution. The yield is also influenced in most plants by the removal of some mother liquor with the crystals being separated from the process. Typically, with a product separated on a centrifuge or filter, the adhering mother liquor would be in the range of 2 to 10 percent of the weight of the crystals.

The actual yield may be obtained from algebraic calculations or trial-and-error calculations when the heat effects in the process and any resultant evaporation are used to correct the initial assumptions on calculated yield. When calculations are made by hand, it is generally preferable to use the trial-and-error system, since it permits easy adjustments for relatively small deviations found in practice, such as the addition of wash water, or instrument and purge water additions. The following calculations are typical of an evaporative crystallizer precipitating a hydrated salt. If SI units are desired, kilograms = pounds × 0.454; K = (°F + 459.7)/1.8.

Example 2: Yield from a Crystallization Process A 10,000-lb batch of a 32.5 percent $MgSO_4$ solution at 120°F is cooled without appreciable evaporation to 70°F. What weight of $MgSO_4 \cdot 7H_2O$ crystals will be formed (if it is assumed that the mother liquor leaving is saturated)?

From the solubility diagram in Fig. 18-56 at 70°F the concentration of solids is 26.3 lb $MgSO_4$ per 100-lb solution.

The mole weight of $MgSO_4$ is 120.38.

The mole weight of $MgSO_4 \cdot 7H_2O$ is 246.49.

For calculations involving hydrated salts, it is convenient to make the calculations based on the hydrated solute and the "free water."

$$0.325 \text{ weight fraction} \times \frac{246.94}{120.38} = 0.6655\ MgSO_4 \cdot 7H_2O \text{ in the feed solution}$$

$$0.263 \times \frac{246.94}{120.38} = 0.5385\ MgSO_4 \cdot 7H_2O \text{ in the mother liquor}$$

Since the free water remains constant (except when there is evaporation), the final amount of soluble $MgSO_4 \cdot 7H_2O$ is calculated by the ratio of

$$\frac{0.538 \text{ lb } MgSO_4 \cdot 7H_2O}{(1 - 0.538) \text{ lb free water}}$$

	Total	$MgSO_4 \cdot 7H_2O$	Free water	$\frac{MgSO_4 \cdot 7H_2O}{\text{Free water}}$
Feed	10,000	6655	3345	1.989
Mother liquor	7249	3904*	3345	1.167
Yield	2751	2751		

*3345 × (0.538/0.462) = 3904

A formula method for calculation is sometimes used where

$$P = R\frac{100W_0 - S(H_0 - E)}{100 - S(R - 1)}$$

where P = weight of crystals in final magma, lb

R = mole weight of hydrate/mole weight of anhydrous = 2.04759

S = solubility at mother-liquor temperature (anhydrous basis) in lb per 100 lb solvent. [0.263/(1 − 0.263)] × 100 = 35.68521

W_0 = weight of anhydrous solute in the original batch. 10,000(0.325) = 3250 lb

H_0 = total weight of solvent at the beginning of the batch. 10,000 − 3250 = 6750 lb

E = evaporation = 0

$$P = 2.04\frac{(100)(3250) - 35.7(6750)}{100 - 35.7(2.04 - 1)} = 2751 \text{ lb}$$

Note that taking the difference between large numbers in this method can increase the chance for error.

Fractional Crystallization When two or more solutes are dissolved in a solvent, it is often possible to (1) separate these into the pure components or (2) separate one and leave the other in the solution. Whether or not this can be done depends on the solubility and phase relations of the system under consideration. Normally alternative 2 is successful only when one of the components has a much more rapid change in solubility with temperature than does the other. A typical example which is practiced on a large scale is the separation of KCl and NaCl from water solution. A phase diagram for this system is shown in Fig. 18-64. In this case the solubility of NaCl is plotted on the *Y* axis in parts per 100 parts of water, and the solubility of KCl is plotted on the *X* axis. The isotherms show a marked decrease in solubility for each component as the amount of the other is increased. This is typical for most inorganic salts. As explained earlier, the mixed-salt line is *CE,* and to make a separation of the solutes into the pure components it is necessary to be on one side of this line or the other. Normally a 95 to 98 percent approach to this line is possible. When evaporation occurs during a cooling or concentration process, this can be represented by movement away from the origin on a straight line through the origin. Dilution by water is represented by movement in the opposite direction.

A typical separation might be represented as follows: Starting at *E* with a saturated brine at 100°C a small amount of water is added to dissolve any traces of solid phase present and to make sure the solids precipitated initially are KCl. Evaporative cooling along line *HG* results in the precipitation of KCl. During this evaporative cooling, part of the water evaporated must be added back to the solution to prevent the coprecipitation of NaCl. The final composition at *G* can be calculated by the NaCl/KCl/H_2O ratios and the known amount of NaCl in the incoming solution at *E.* The solution at point *G* may be concentrated by evaporation at 100°C. During this process the solution will increase in concentration with respect to both components until point *I* is reached. Then NaCl will precipitate, and the solution will become more concentrated in KCl, as indicated by the line *IE,* until the original point *E* is reached. If concentration is carried beyond point *E,* a mixture of KCl and NaCl will precipitate.

Example 3: Yield from Evaporative Cooling Starting with 1000 lb of water in a solution at *H* on the solubility diagram in Fig. 18-64, calculate the yield on evaporative cooling and concentrate the solution back to point *H* so the cycle can be repeated, indicating the amount of NaCl precipitated and the evaporation and dilution required at the different steps in the process.

In solving problems of this type, it is convenient to list the material balance and the solubility ratios. The various points on the material balance are calculated by multiplying the quantity of the component which does not precipitate from solution during the transition from one point to another (normally the NaCl in cooling or the KCl in the evaporative step) by the solubility ratio at the next step, illustrated as follows:

Basis. 1000 lb of water at the initial conditions.

				Solubility ratios		
Solution component	KCl	NaCl	Water	KCl	NaCl	Water
H	343	270	1000	34.3	27.0	100
G(*a*)	194	270	950	20.4	28.4	100
KCl yield	149					
Net evaporation			50			
I(*b*)	194	270	860	22.6	31.4	100
E(*c*)	194	153	554	35.0	27.5	100
NaCl yield		117				
Evaporation			306			
Dilution			11			
H′	194	153	565	34.3	27.0	−100

The calculations for these steps are:

a. 270 lb NaCl (100 lb water/28.4 lb NaCl) = 950 lb water
950 lb water (20.4 lb KCl/100 lb water) = 194 lb KCl

b. 270 lb NaCl (100 lb water/31.4 lb NaCl) = 860 lb water
860 lb water (22.6 lb KCl/100 lb water) = 194 lb KCl

c. 194 lb KCl (100 lb water/35.0 lb KCl) = 554 lb water
554 lb water (27.5 lb NaCl/100 lb water) = 153 lb NaCl

Note that during the cooling step the maximum amount of evaporation which is permitted by the material balance is 50 lb for the step shown. In an evaporative-cooling step, however, the actual evaporation which results from adiabatic cooling is more than this. Therefore, water must be added back to prevent the NaCl concentration from rising too high; otherwise, coprecipitation of NaCl will occur.

Inasmuch as only mass ratios are involved in these calculations, kilograms or any other unit of mass may be substituted for pounds without affecting the validity of the example.

Although the figures given are for a step-by-step process, it is obvious that the same techniques will apply to a continuous system if the fresh feed containing KCl and NaCl is added at an appropriate part of the cycle, such as between steps *G* and *I* for the case of dilute feed solutions.

Another method of fractional crystallization, in which advantage is taken of different crystallization rates, is sometimes used. Thus, a solution saturated with borax and potassium chloride will, in the absence of borax seed crystals, precipitate only potassium chloride on rapid cooling. The borax remains behind as a supersaturated solution, and the potassium chloride crystals can be removed before the slower borax crystallization starts.

Crystal Formation There are two steps involved in the preparation of crystal matter from a solution. The crystals must first form and then grow. The formation of a new solid phase either on an inert particle in the solution or in the solution itself is called **nucleation.** The increase in size of this nucleus with a layer-by-layer addition of solute is called **growth.** The growth process involves two steps, diffusion of the solute to the crystal interface followed by incorporation of the same into the lattice. One of these will control depending on factors such as the degree of agitation and temperature. Nucleation can be classified as primary or secondary. The former usually occurs at high supersaturation and does not involve product crystals. Secondary nucleation involves nuclei generation from product crystals by contact with the agitator, with the crystallizer internals and with one another. Each system has a metastable zone where growth is encouraged in the presence of supersaturation. Secondary nucleation can occur within the zone. Both nucleation and crystal growth have supersaturation as a common driving force. Unless a solution is supersaturated, crystals can neither form nor grow. Supersaturation refers to the quantity of solute present in solution compared with the quantity which would be present if the solution were kept for a very long period of time with solid phase in contact with the solution. The latter value is the equilibrium solubility at the temperature and pressure under consideration. The supersaturation coefficient can be expressed

$$S = \frac{\text{parts solute/100 parts solvent}}{\text{parts solute at equilibrium/100 parts solvent}} \geqq 1.0 \qquad (18\text{-}25)$$

Solutions vary greatly in their ability to sustain measurable amounts of supersaturation. With some materials, such as sucrose, it is possible to develop a supersaturation coefficient of 1.4 to 2.0 with little danger of nucleation. With some common inorganic solutions such as sodium chloride in water, the amount of supersaturation which can be generated stably is so small that it is difficult or impossible to measure.

Certain qualitative facts in connection with supersaturation, growth, and the yield in a crystallization process are readily apparent. If the concentration of the initial solution and the final mother liquor are fixed, the total weight of the crystalline crop is also fixed if equilibrium is obtained. The particle-size distribution of this weight, however, will depend on the relationship between the two processes of nucleation and growth. Considering a given quantity of solution cooled through a fixed range, if there is considerable nucleation initially during the cooling process, the yield will consist of many small crystals. If only a few nuclei form at the start of the crystallization (or seeds are added) and the resulting yield occurs uniformly on these nuclei or seeds without significant secondary nucleation, a crop of large uniform crystals will result. Obviously, many intermediate cases of varying nucleation rates and growth rates can also occur, depending on the nature of the materials being handled, the rate of cooling, agitation, and other factors.

When a process is continuous, nucleation frequently occurs in the presence of a seeded solution by the combined effects of mechanical stimulus and nucleation caused by supersaturation (heterogeneous nucleation). If such a system is completely and uniformly mixed (i.e., the product stream represents the typical magma circulated within the system) and if the system is operating at steady state, the particle-size distribution has definite limits which can be predicted mathematically with a high degree of accuracy, as will be shown later in this section.

Geometry of Crystal Growth Geometrically a crystal is a solid bounded by planes. The shape and size of such a solid are functions of the interfacial angles and of the linear dimension of the faces. As the result of the constancy of its interfacial angles, each face of a growing or dissolving crystal, as it moves away from or toward the center of the crystal, is always parallel to its original position. This concept is known as the "principle of the parallel displacement of faces." The rate at which a face moves in a direction perpendicular to its original position is called the translation velocity of that face or the rate of growth of that face.

From the industrial point of view, the term **crystal habit** or **crystal morphology** refers to the relative sizes of the faces of a crystal. The crystal habit is determined by the internal structure and external influences on the crystal such as the growth rate, solvent used, and impurities present during the crystallization growth period. The crystal habit of commercial products is of very great importance. Long, needlelike crystals tend to be easily broken during centrifugation and drying. Flat, platelike crystals are very difficult to wash during filtration or centrifugation and result in relatively low filtration rates. Complex or twinned crystals tend to be more easily broken in transport than chunky, compact crystal habits. Rounded or spherical crystals (caused generally by attrition during growth and handling) tend to give considerably less difficulty with caking than do cubical or other compact shapes.

Internal structure (unit cell) can be different in crystals that are chemically identical. This is called **polymorphism**. Polymorphs can vary substantially in physical and chemical properties such as bioavailability and solubility. They can be identified by analytical techniques such as X-ray diffraction, infrared, Raman spectro, and microscopic techniques. For the same internal structure, very small amounts of foreign substances will often completely change the crystal habit. The selective adsorption of dyes by different faces of a crystal or the change from an alkaline to an acidic environment will often produce pronounced changes in the crystal habit. The presence of other soluble anions and cations often has a similar influence. In the crystallization of ammonium sulfate, the reduction in soluble iron to below 50 ppm of ferric ion is sufficient to cause significant change in the habit of an ammonium sulfate crystal from a long, narrow form to a relatively chunky and compact form. Additional information is available in the patent literature and Table 18-4 lists some of the better-known additives and their influences.

Since the relative sizes of the individual faces of a crystal vary between wide limits, it follows that different faces must have different translational velocities. A geometric law of crystal growth known as the **overlapping principle** is based on those velocity differences: in growing a crystal, only those faces having the lowest translational velocities survive; and in dissolving a crystal, only those faces having the highest translational velocities survive.

For example, consider the cross sections of a growing crystal as in Fig. 18-65. The polygons shown in the figure represent varying stages in the growth of the crystal. The faces marked A are slow-growing faces (low translational velocities), and the faces marked B are fast-growing (high translational velocities). It is apparent from Fig. 18-65 that the faster B faces tend to disappear as they are overlapped by the slower A faces.

It has been predicted that crystal habit or crystal morphology was related to the internal structure based on energy considerations and speculated that it should be possible to predict the growth shape of crystals from the slice energy of different flat faces. One can predict the calculated attachment energy for various crystal species. Recently computer programs have been developed that predict crystal morphology from attachment energies. These techniques are particularly useful in dealing with organic or molecular crystals and rapid progress in this area is being made by companies such as Molecular Simulations of Cambridge, England.

Purity of the Product If a crystal is produced in a region of the phase diagram where a single-crystal composition precipitates, the crystal itself will normally be pure provided that it is grown at relatively low rates and constant conditions. With many products these purities approach a value of about 99.5 to 99.8 percent. The difference between this and a purity of 100 percent is generally the result of small pockets of mother liquor called inclusions trapped within the crystal. Although frequently large enough to be seen with an ordinary microscope, these inclusions can be submicroscopic and represent dislocations within the structure of the crystal. They can be caused by either attrition or breakage during the growth process or by slip planes within the crystal structure caused by interference between screw-type dislocations and the remainder of the crystal faces. To increase the purity of the crystal beyond the point where such inclusions are normally expected (about 0.1 to 0.5 percent by volume), it is generally necessary to reduce the impurities in the mother liquor itself to an acceptably low level so that the mother liquor contained within these pockets will not contain sufficient impurities to cause an impure product to be formed. It is normally necessary to recrystallize material from a solution which is relatively pure to surmount this type of purity problem.

In addition to the impurities within the crystal structure itself, there is normally an adhering mother-liquid film left on the surface of the

TABLE 18-4 Some Impurities Known to Be Habit Modifiers

Material crystallized	Additive(s)	Effect	Concentration	References
$Ba(NO_2)^2$	Mg, Te^{+4}	Helps growth	—	1
$CaSO_4 \cdot 2H_2O$	Citric, succinic, tartaric acids	Helps growth	Low	
	Sodium citrate	Forms prisms	—	5
$CuSO_4 \cdot 5H_2O$	H_2SO_4	Chunky crystals	0.3%	5
KCl	$K_4Fe(CN)_6$	Inhibits growth, dendrites	1000 ppm	4
	Pb, Bi, Sn^{+2}, Ti, Zr, Th, Cd, Fe, Hg, Mg	Helps growth	Low	1
$KClO_4$	Congo red (dye)	Modifies the 102 face	50 ppm	6
K_2CrO_4	Acid magenta (dye)	Modifies the 010 face	50 ppm	6
KH_2PO_4	$Na_2B_4O_7$	Aids growth	—	1
KNO_2	Fe	Helps growth	Low	1
KNO_3	Acid magenta (dye)	Tabular crystals		7
	Pb, Th, Bi	Helps growth	Low	1
K_2SO_4	Acid magenta (dye)	Forms plates	2000 ppm	6
	Cl, Mn, Mg, Bi, Cu, Al, Fe	Helps growth	Low	1
	Cl_3	Reduces growth rate	1000 ppm	4
	$(NH_4)_3Ce(NO_3)_6$	Reduces growth rate	1000 ppm	4
$LiCl \cdot H_2O$	$Cr \cdot Mn^{+2}$, Sn^{+2}, Co, Ni, Fe^{+3}	Helps growth	Low	1
$MgSO_4 \cdot 7H_2O$	Borax	Aids growth	5%	1
$Na_2B_4O_7 \cdot 10H_2O$	Sodium oleate	Reduces growth & nuc.	5 ppm	
	Casein, gelatin	Promotes flat crystals	—	2, 5
	NaOH, Na_2CO_3	Promotes chunky crystals	—	
$Na_2CO_3 \cdot H_2O$	$SO_4^=$	Reduces *L/D* ratio	0.1–1.0%	Canadian Patent 812,685
	Ca^{+2} and Mg^{+2}	Increase bulk density	400 ppm	U.S. Patent 3,459,497
$NaCO_3 \cdot NaHCO_3 \cdot 2H_2O$	D-40 detergent	Aids growth	20 ppm	U.S. Patent 3,233,983
NaCl	$Na_4Fe(CN)_6$, CdBr	Forms dendrites	100 ppm	4
	Pb, Mn^{+2}, Bi, Sn^{+2}, Ti, Fe, Hg	Helps growth	Low	1
	Urea, formamide	Forms octahedra	Low	2
	Tetraalkyl ammon. salts	Helps growth & hardness	1–100 ppm	U.S. Patent 3,095,281
	Polyethylene-oxy compounds	Helps growth & hardness	—	U.S. Patent 3,000,708
$NaClO_3$	Na_2SO_4, $NaClO_4$	Tetrahedrons	—	3
$NaNO_3$	Acid green (dye)	Flattened rhombahedra		7
Na_2SO_4	NH_4SO_4 @ pH 6.5	Large single crystals	Low	
	$CdCl_2$	Inhibits growth	1000 ppm	4
	Alkyl aryl sulfonates	Aids growth	—	2
	Calgon	Aids growth	100 ppm	
NH_4Cl	Mn, Fe, Cu, Co, Ni, Cr	Aid growth	Low	1
	Urea	Forms octahedra		5
NH_4ClO_4	Azurine (dye)	Modifies the 102 face	22 ppm	6
NF_4F	Ca	Helps growth	Low	1
$(NH_4)NO_3$	Acid magenta (dye)	Forms 010 face plates	1%	6
$(NH_4)_2HPO_4$	H_2SO_4	Reduces *L/D* ratio	7%	
$NH_4H_2PO_4$	Fe^{+3}, Cr, Al, Sn	Helps growth	Traces	1
$(NH_4)_2SO_4$	Cr^{+3}, Fe^{+3}, Al^{+3}	Promotes needles	50 ppm	
	H_2SO_4	Promotes needles	2–6%	U.S. Patent 2,092,073
	Oxalic acid, citric acid	Promotes chunky crystals	1000 ppm	U.S. Patent 2,228,742
	H_3PO_4, SO_2	Promotes chunky crystals	1000 ppm	
$ZnSO_4 \cdot 7H_2O$	Borax	Aids growth	—	1
Adipic acid	Surfactant-SDBS	Aids growth	50–100 ppm	2
Fructose	Glucose, difructose	Affects growth		8
L-asparagine	L-glutamic acid	Affects growth		8
Naphthalene	Cyclohexane (solvent)	Forms needles	—	2
	Methanol (solvent)	Forms plates		
Pentaerythritol	Sucrose	Aids growth	—	1
	Acetone (solvent)	Forms plates	—	2
Sodium glutamate	Lysine, CaO	Affects growth		8
Sucrose	Raffinose, KCl, NaBr	Modify growth rate		
Urea	Biuret	Reduces *L/D* & aids growth	2–7%	
	NH_4Cl	Reduces *L/D* & aids growth	5–10%	

1. Gillman, *The Art and Science of Growing Crystals,* Wiley, New York, 1963.
2. Mullin, *Crystallization,* Butterworth, London, 1961.
3. Buckley, *Crystal Growth,* Wiley, New York, 1961.
4. Phoenix, L., *British Chemical Engineering,* vol. II, no. 1 (Jan. 1966), pp. 34–38.
5. Garrett, D. E., *British Chemical Engineering,* vol. I, no. 12 (Dec. 1959), pp. 673–677.
6. Buckley, *Crystal Growth,* (Faraday Soc.) Butterworths, 1949, p. 249.
7. Butchart and Whetstone, *Crystal Growth,* (Faraday Soc.) Butterworths, 1949, p. 259.
8. Nyvlt, J., *Industrial Crystallization,* Verlag Chemie Publishers, New York, 1978, pp. 26–31.

crystal after separation in a centrifuge or on a filter. Typically a centrifuge may leave about 2 to 10 percent of the weight of the crystals as adhering mother liquor on the surface. This varies greatly with the size and shape or habit of the crystals. Large, uniform crystals from low-viscosity mother liquors will retain small quantities of mother liquor, while nonuniform or small crystals crystallized from viscous solutions will retain a considerably larger proportion. Comparable statements apply to the filtration of crystals, although normally the amounts of mother liquor adhering to the crystals are considerably larger. It is common practice when crystallizing materials from solutions which contain appreciable quantities of impurities to wash the crystals on the centrifuge or filter with either fresh solvent or feed solution. In principle, such washing can reduce the impurities quite substantially. It is also possible in many cases to reslurry the crystals in

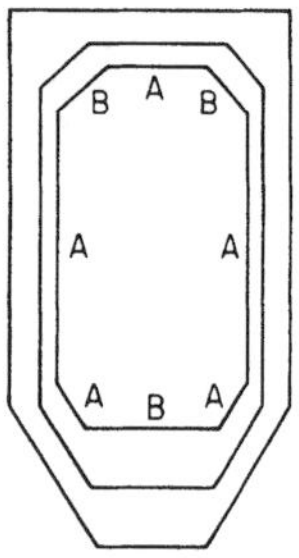

FIG. 18-65 Overlapping principle.

fresh solvent and recentrifuge the product in an effort to obtain a longer residence time during the washing operation and better mixing of the wash liquors with the crystals. Mother liquor inclusions and residual moisture after drying can present caking problems.

Coefficient of Variation One of the problems confronting any user or designer of crystallization equipment is the expected particle-size distribution of the solids leaving the system and how this distribution may be adequately described. Most crystalline-product distributions plotted on arithmetic-probability paper will exhibit a straight line for a considerable portion of the plotted distribution. In this type of plot the particle diameter should be plotted as the ordinate and the cumulative percent on the log-probability scale as the abscissa.

It is common practice to use a parameter characterizing crystal-size distribution called the coefficient of variation. This is defined as follows:

$$\mathrm{CV} = 100\,\frac{\mathrm{PD}_{16\%} - \mathrm{PD}_{84\%}}{2\mathrm{PD}_{50\%}} \tag{18-26}$$

where CV = coefficient of variation, as a percentage
PD = particle diameter from intercept on ordinate axis at percent indicated

In order to be consistent with normal usage, the particle-size distribution when this parameter is used should be a straight line between approximately 10 percent cumulative weight and 90 percent cumulative weight. By giving the coefficient of variation and the mean particle diameter, a description of the particle-size distribution is obtained which is normally satisfactory for most industrial purposes. If the product is removed from a mixed-suspension crystallizer, this coefficient of variation should have a value of approximately 50 percent (Randolph and Larson, op. cit., chap. 2).

Crystal Nucleation and Growth

Rate of Growth Crystal growth is a layer-by-layer process, and since growth can occur only at the face of the crystal, material must be transported to that face from the bulk of the solution. Diffusional resistance to the movement of molecules (or ions) to the growing crystal face, as well as the resistance to integration of those molecules into the face, must be considered. As discussed earlier, different faces can have different rates of growth, and these can be selectively altered by the addition or elimination of impurities.

If L is a characteristic dimension of a crystal of selected material and shape, the rate of growth of a crystal face that is perpendicular to L is, by definition,

$$G \equiv \lim_{\Delta L \to 0} \frac{\Delta L}{\Delta t} = \frac{dL}{dt} \tag{18-27}$$

where G is the growth rate over time interval t. It is customary to measure G in the practical units of millimeters per hour. It should be noted that growth rates so measured are actually twice the facial growth rate.

The delta L *law.* It has been shown by McCabe [*Ind. Eng. Chem.*, **21,** 30, 112 (1929)] that all geometrically similar crystals of the same material suspended in the same solution grow at the same rate if growth rate is defined as in Eq. (18-27). The rate is independent of crystal size, provided that all crystals in the suspension are treated alike. This generalization is known as the delta L law. Although there are some well-known exceptions, they usually occur when the crystals are very large or when movement of the crystals in the solution is so rapid that substantial changes occur in diffusion-limited growth of the faces.

It is emphasized that the delta L law does not apply when similar crystals are given preferential treatment based on size. It fails also when surface defects or dislocations significantly alter the growth rate of a crystal face. Nevertheless, it is a reasonably accurate generalization for a surprising number of industrial cases. When it is, it is important because it simplifies the mathematical treatment in modeling real crystallizers and is useful in predicting crystal-size distribution in many types of industrial crystallization equipment.

Important exceptions to McCabe's growth-rate model have been noted by Bramson, by Randolph, and by Abegg. These are discussed by Canning and Randolph, *Am. Inst. Chem. Eng. J.*, **13,** 5 (1967).

Nucleation The mechanism of crystal nucleation from solution has been studied by many scientists, and their work suggests that—in commercial crystallization equipment, at least—the nucleation rate is the sum of contributions by (1) primary nucleation and (2) nucleation due to contact between crystals and (*a*) other crystals, (*b*) the walls of the container, and (*c*) the impeller. If B^0 is the net number of new crystals formed in a unit volume of solution per unit of time,

$$B^0 = B_{ss} + B_{cc} + B_{ci} \tag{18-28}$$

where B_{ci} is the rate of nucleation due to crystal-impeller contacts, B_{cc} is that due to crystal-crystal contacts, and B_{ss} is the primary nucleation rate due to the supersaturation driving force. The mechanism of the last-named is not precisely known, although it is obvious that molecules forming a nucleus not only have to coagulate, resisting the tendency to redissolve, but also must become oriented into a fixed lattice. The number of atoms or molecules required to form a stable crystal nucleus has been variously estimated at from 80 to 100 (with ice), and the probability that a stable nucleus will result depends on many factors such as activation energies and supersaturation. In commercial crystallization equipment, in which supersaturation is low and agitation is employed to keep the growing crystals suspended, the predominant mechanism is contact nucleation or, in extreme cases, attrition.

In order to treat crystallization systems both dynamically and continuously, a mathematical model has been developed which can correlate the nucleation rate to the level of supersaturation and/or the growth rate. Because the growth rate is more easily determined and because nucleation is sharply nonlinear in the regions normally encountered in industrial crystallization, it has been common to assume

$$B^0 = ks^b \tag{18-29}$$

where s, the supersaturation, is defined as $(C - C_s)$, C being the concentration of the solute and C_s its saturation concentration; and the exponent b and dimensional coefficient k are values characteristic of the material.

While Eq. (18-29) has been popular among those attempting correlations between nucleation rate and supersaturation, it has become common to use a derived relationship between nucleation rate and growth rate by assuming that

$$G = k's^g \tag{18-30}$$

whence, in consideration of Eq. (18-29),

$$B^0 = k''G^i \tag{18-31}$$

where the dimensional coefficient k' and exponent g are characteristic of the material and the conditions of crystallization and $k'' = k/(k')^i$ with $i = b/g$, a measure of the relative dependence of B^0 and G on supersaturation. Feeling that a model in which nucleation depends only on supersaturation or growth rate is simplistically deficient, some have proposed that contact nucleation rate is also a power function of slurry density and that

$$B^0 = k_n G^i M_T^j \tag{18-32}$$

where M_T is the density of the crystal slurry, g/L.

Although Eqs. (18-31) and (18-32) have been adopted by many as a matter of convenience, they are oversimplifications of the very complex relationship that is suggested by Eq. (18-28); Eq. (18-32) implicitly and quite arbitrarily combines the effects of homogeneous nucleation and those due to contact nucleation. They should be used only with caution.

In work pioneered by Clontz and McCabe [*Chem. Eng. Prog. Symp. Ser.*, **67**(110), 6 (1971)] and subsequently extended by others, contact nucleation rate was found to be proportional to the input of energy of contact, frequency of contact, as well as being a function of contact area and supersaturation. This observation is important to the scaling up of crystallizers. At the laboratory or bench scale, particle contact frequency with the agitator is high while in commercial equipment the contact energy input is higher at the impeller but the contact frequency is less. Scale-up modeling of a crystallizer, therefore, must include its mechanical characteristics as well as the physiochemical driving force.

Nucleation and Growth From the preceding, it is clear that no analysis of a crystallizing system can be truly meaningful unless the simultaneous effects of nucleation rate, growth rate, heat balance, and material balance are considered. The most comprehensive treatment of this subject is by Randolph and Larson (op. cit), who developed a mathematical model for continuous crystallizers of the mixed-suspension or circulating-magma type [*Am. Inst. Chem. Eng. J.*, **8**, 639 (1962)] and subsequently examined variations of this model that include most of the aberrations found in commercial equipment. Randolph and Larson showed that when the total number of crystals in a given volume of suspension from a crystallizer is plotted as a function of the characteristic length as in Fig. 18-66, the slope of the line is usefully identified as the crystal population density, n:

$$n = \lim_{\Delta L \to 0} \frac{\Delta N}{\Delta L} = \frac{dN}{dL} \tag{18-33}$$

where N = total number of crystals up to size L per unit volume of magma. The population density thus defined is useful because it characterizes the nucleation-growth performance of a particular crystallization process or crystallizer.

The data for a plot like Fig. 18-67 are easily obtained from a screen analysis of the total crystal content of a known volume (e.g., a liter) of magma. The analysis is made with a closely spaced set of testing sieves (or intervals for a particle counter), the cumulative number of particles smaller than each sieve in the nest being plotted against the aperture dimension of that sieve. The fraction retained on each sieve is weighed, and the mass is converted to the equivalent number of particles by dividing by the calculated mass of a particle whose dimension is the arithmetic mean of the mesh sizes of the sieve on which it is retained and the sieve immediately above it.

In industrial practice, the size-distribution curve usually is not actually constructed. Instead, a mean value of the population density for any sieve fraction of interest (in essence, the population density of the particle of average dimension in that fraction) is determined directly as $\Delta N/\Delta L$, ΔN being the number of particles retained on the sieve and ΔL being the difference between the mesh sizes of the retaining sieve and its immediate predecessor. It is common to employ the units of $(\text{mm}\cdot\text{L})^{-1}$ for n.

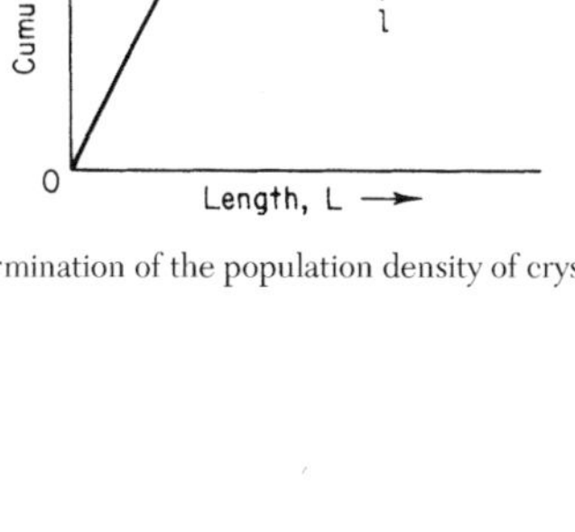

FIG. 18-66 Determination of the population density of crystals.

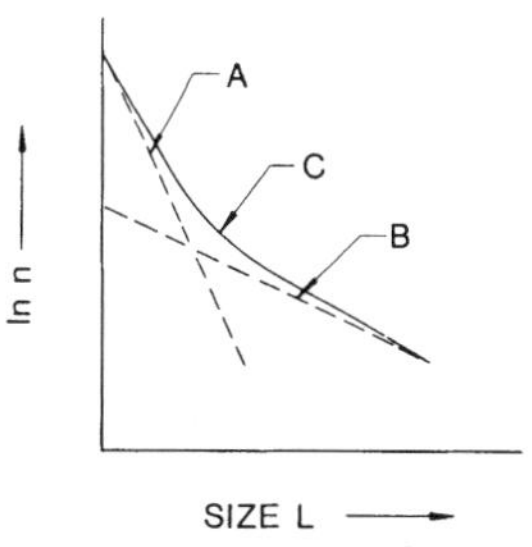

FIG. 18-67 Population density of crystals resulting from Bujacian behavior.

For a steady-state crystallizer receiving solids-free feed and containing a well-mixed suspension of crystals experiencing negligible breakage, a material-balance statement yields negligible agglomeration and breakage to a particle balance (the Randolph-Larson general-population balance); in turn, it simplifies to

$$\frac{dn}{dL} + \frac{n}{Gt} = 0 \tag{18-34}$$

if the delta L law applies (i.e., G is independent of L) and the draw-down (or retention) time is assumed to be invariant and calculated as $t = V/Q$. Integrated between the limits n^0, the population density of nuclei (for which L is assumed to be zero), and n, that of any chosen crystal size L, Eq. (18-34) becomes

$$\int_{n^0}^{n} \frac{dn}{n} = -\int_{0}^{L} \frac{dL}{Gt} \tag{18-35}$$

$$\ln n = \frac{-L}{Gt} + \ln n^0 \tag{18-36a}$$

or

$$n = n^0 e^{-L/Gt} \tag{18-36b}$$

It can be shown that

$$B^0 = n^0 G \tag{18-36c}$$

A plot of ln n versus L is a straight line whose intercept is ln n^0 and whose slope is $-1/Gt$. (For plots on base-10 log paper, the appropriate slope correction must be made.) Thus, from a given product sample of known slurry density and retention time it is possible to obtain the nucleation rate and growth rate for the conditions tested if the sample satisfies the assumptions of the derivation and yields a straight line. A number of derived relations which describe the nucleation rate, size distribution, and average properties are summarized in Table 18-5.

If a straight line does *not* result (Fig. 18-67), at least part of the explanation may be violation of the delta L law (Canning and Randolph, loc. cit.). The best current theory about what causes size-dependent growth suggests what has been called growth dispersion or "Bujacian behavior" [Mullen (ed.), op. cit., p. 254]. In the same environment different crystals of the same size can grow at different rates owing to differences in dislocations or other surface effects. The graphs of "slow" growers (Fig. 18-67, curve A) and "fast" growers (curve B) sum to a resultant line (curve C), concave upward, that is described by Eq. (18-37) (Randolph, in deJong and Jancic, op. cit., p. 254):

$$n = \sum \frac{B^0{}_i}{G_i} e^{(-L/G_i t)} \tag{18-37}$$

Equation (18-34) contains no information about the crystallizer's influence on the nucleation rate. If the crystallizer is of a mixed-suspension, mixed-product-removal (MSMPR) type, satisfying the criteria for Eq. (18-34), and if the model of Clontz and McCabe is valid, the contribution to the nucleation rate by the circulating pump

TABLE 18-5 Common Equations for Population-Balance Calculations

Name	Symbol	Units	Systems without fines removal	Systems with fines removal: Fines stream	Systems with fines removal: Product stream	References
Drawdown time (retention time)	t	h	$t = V/Q$	$t_F = V_{\text{liquid}}/Q_F$	$t = V/Q$	
Growth rate	G	mm/h	$G = dL/dt$	$G = dL/dt$	$G = dL/dt$	
Volume coefficient	K_v	1/no. (crystals)	$K_v = \dfrac{\text{volume of one crystal}}{L^3}$	$K_v = \dfrac{\text{volume of one crystal}}{L^3}$	$K_v = \dfrac{\text{volume of one crystal}}{L^3}$	
Population density	n	No. (crystals)/mm	$n = dN/dL$	$n = dN/dL$	$n = dN/dL$	1
Nuclei population density	n^o	No. (crystals)/mm	$n^o = K_M M^j G^{i-1}$			2
Population density	n	No. (crystals)/mm	$n = n^o e^{-L/Gt}$	$n_F = n^o e^{-L/Gt_F}$	$n = n^{o-L_e/Gt_F} e^{-L/Gt}$	1, 3
Nucleation rate	B_0	No. (crystals)/h	$B_0 = Gn^o = K_M M^j G^i$	$B_0 = G^o_n$		4
Dimensionless length	x	None	$x = \dfrac{L}{Gt}$	$x_F = \dfrac{L}{Gt_F}, L_0 \rightarrow L_f$	$x = \dfrac{L}{Gt}, L_f \rightarrow L$	1
Mass/unit volume (slurry density)	M_T	g/L	$M_t = K_v \rho \int_0^\infty nL^3\,dL$ $M_t = K_v \rho 6\, n^o (Gt)^4$	$M_{T_F} = K_v \rho \int_0^{L_f} n^o e^{-L/Gt_F} L^3\,dL$	$M_T = K_v \rho \int_{L_f}^\infty n^o e^{-L/Gt_F} e^{-L/Gt} L^3\,dL$	1
Cumulative mass to x / Total mass	W_x	None	$W_x = 1 - e^{-x}\left(\dfrac{x^3}{6} + \dfrac{x^2}{2} + x + 1\right)$	$W_F = \dfrac{e^{-x}(x^3 + 3x^2 + 6x + 6) - 6}{e^{-x_c}(x_c^3 + 3x_c^2 + 6x_c + 6) - 6}$	$W = \dfrac{6K_v \rho n^o e^{-L_e/Gt_F} (Gt)^4 \left[1 - e^{-x}\left(\dfrac{x^3}{6} + \dfrac{x^2}{2} + x + 1\right)\right]}{\text{Slurry density } M, g/L}$ when $L_c \approx 0$, compared with L_a	5
Dominant particle	L_d	mm	$L_d = 3Gt$			
Average particle, weight	L_a	mm	$L_a = 3.67Gt$			6
Total number of crystals	N_T	No./L	$N_T = \int_0^\infty n\,dl$	$N_F = \int_0^{L_f} n_F\,dL$	$N_T = \int_{L_f}^\infty n\,dL$	1, 3

1. Randolph and Larson, *Am. Inst. Chem. Eng. J.*, **8,** 639 (1962).
2. Timm and Larson, *Am. Inst. Chem. Eng. J.*, **14,** 452 (1968).
3. Larson, private communication.
4. Larson, Timm, and Wolff, *Am. Inst. Chem. Eng. J.*, **14,** 448 (1968).
5. Larson and Randolph, *Chem. Eng. Prog. Symp. Ser.*, **65**(95), 1 (1969).
6. Schoen, *Ind. Eng. Chem.*, **53,** 607 (1961).

can be calculated [Bennett, Fiedelman, and Randolph, *Chem. Eng. Prog.*, **69**(7), 86 (1973)]:

$$B_e = K_e\left(\frac{I^2}{P}\right)\rho G \int_0^\infty nL^4\,dL \qquad (18\text{-}38)$$

where I = tip speed of the propeller or impeller, m/s
ρ = crystal density, g/cm³
P = volume of crystallizer/circulation rate (turnover), $m^3/(m^3/s) = s$

Since the integral term is the fourth moment of the distribution (m_4), Eq. (18-38) becomes

$$B_e = K_e\rho G\left(\frac{I^2}{P}\right)m_4 \qquad (18\text{-}39)$$

Equation (18-39) is the general expression for impeller-induced nucleation. In a fixed-geometry system in which only the speed of the circulating pump is changed and in which the flow is roughly proportional to the pump speed, Eq. (18-39) may be satisfactorily replaced with

$$B_e = K_e''\rho G(S_R)^3 m_4 \qquad (18\text{-}40)$$

where S_R = rotation rate of impeller, r/min. If the maximum crystal-impeller impact stress is a nonlinear function of the kinetic energy, shown to be the case in at least some systems, Eq. (18-40) no longer applies.

In the specific case of an MSMPR exponential distribution, the fourth moment of the distribution may be calculated as

$$m_4 = 4!n^0(Gt)^5 \qquad (18\text{-}41)$$

Substitution of this expression into Eq. (18-39) gives

$$B_e = kn^0\,G(S_R)^3L_D^5 \qquad (18\text{-}42)$$

where $L_D = 3Gt$, the dominant crystal (mode) size.

Equation (18-42) displays the competing factors that stabilize secondary nucleation in an operating crystallizer when nucleation is due mostly to impeller-crystal contact. Any increase in particle size produces a fifth-power increase in nucleation rate, tending to counteract the direction of the change and thereby stabilizing the crystal-size distribution. From dimensional argument alone the size produced in a mixed crystallizer for a (fixed) nucleation rate varies as $(B^0)^{1/3}$. Thus, this fifth-order response of contact nucleation does not wildly upset the crystal size distribution but instead acts as a stabilizing feedback effect.

Nucleation due to crystal-to-crystal contact is greater for equal striking energies than crystal-to-metal contact. However, the viscous drag of the liquid on particle sizes normally encountered limits the velocity of impact to extremely low values. The assumption that only the largest crystal sizes contribute significantly to the nucleation rate by crystal-to-crystal contact permits a simple computation of the rate:

$$B_c = K_c\,\rho G m_j^2 \qquad (18\text{-}43)$$

where m_j = the fourth, fifth, sixth, or higher moments of the distribution.

A number of different crystallizing systems have been investigated by using the Randolph-Larson technique, and some of the published growth rates and nucleation rates are included in Table 18-6. Although the usefulness of these data is limited to the conditions tested, the table gives a range of values which may be expected, and it permits resolution of the information gained from a simple screen analysis into the fundamental factors of growth rate and nucleation rate. Experiments may then be conducted to determine the independent effects of operation and equipment design on these parameters.

Although this procedure requires laborious calculations because of the number of samples normally needed, these computations and the determination of the best straight-line fit to the data are readily programmed for digital computers.

Example 4: Population Density, Growth, and Nucleation Rate Calculate the population density, growth, and nucleation rates for a crystal sample of urea for which there is the following information. These data are from Bennett and Van Buren [*Chem. Eng. Prog. Symp. Ser.*, **65**(95), 44 (1969)].

Slurry density = 450 g/L
Crystal density = 1.335 g/cm³
Drawdown time t = 3.38 h
Shape factor k_v = 1.00
Product size:

−14 mesh, +20 mesh	4.4 percent
−20 mesh, +28 mesh	14.4 percent
−28 mesh, +35 mesh	24.2 percent
−35 mesh, +48 mesh	31.6 percent
−48 mesh, +65 mesh	15.5 percent
−65 mesh, +100 mesh	7.4 percent
−100 mesh	2.5 percent

n = number of particles per liter of volume
14 mesh = 1.168 mm, 20 mesh = 0.833 mm, average opening 1.00 mm
Size span = 0.335 mm = ΔL

$$n_{20} = \frac{(450\text{ g/L})(0.044)}{(1.335/1000)\text{ g/mm}^3(1.00^3\text{ mm}^3\text{/particle})(0.335\text{ mm})(1.0)}$$

$$n_{20} = 44{,}270$$

$$\ln n_{20} = 10{,}698$$

Repeating for each screen increment:

Screen size	Weight, %	k_v	ln n	L, average diameter, mm
100	7.4	1.0	18.099	0.178
65	15.5	1.0	17.452	0.251
48	31.6	1.0	16.778	0.356
35	24.2	1.0	15.131	0.503
28	14.4	1.0	13.224	0.711
20	4.4	1.0	10.698	1.000

Plotting ln n versus L as shown in Fig. 18-68, a straight line having an intercept at zero length of 19.781 and a slope of −9.127 results. As mentioned in discussing Eq. (18-27), the growth rate can then be found.

$$\text{Slope} = -1/Gt \text{ or } -9.127 = -1/[G(3.38)]$$

or $$G = 0.0324 \text{ mm/h}$$

and $$B_0 = Gn^0 = (0.0324)(e^{19.781}) = 12.65\times10^6\,\frac{n^0}{\text{L·h}}$$

and $$L_a = 3.67(0.0324)(3.38) = 0.40 \text{ mm}$$

An additional check can be made of the accuracy of the data by the relation

$$M_T = 6k_v\rho n^0(Gt)^4 = 450 \text{ g/L}$$

$$M_T = (6)(1.0)\frac{1.335\text{ g/cm}^3}{1000\text{ mm}^3\text{/cm}^3}e^{19.78}[(0.0324)(3.38)]^4$$

$$M_T = 455 \text{ g/L} \approx 450 \text{ g/L}$$

Had only the growth rate been known, the size distribution of the solids could have been calculated from the equation

$$W_f = 1 - e^{-x}\left(\frac{x^3}{6}+\frac{x^2}{2}+x+1\right)$$

where W_f is the weight fraction *up to* size L and $x = L/Gt$.

$$x = \frac{L}{(0.0324)(3.38)} = \frac{L}{0.1095}$$

Screen size	L, mm	x	W_f*	Cumulative % retained 100 $(1-W_f)$	Measured cumulative % retained
20	0.833	7.70	0.944	5.6	4.4
28	0.589	5.38	0.784	21.6	18.8
35	0.417	3.80	0.526	47.4	43.0
48	0.295	2.70	0.286	71.4	74.6
65	0.208	1.90	0.125	87.5	90.1
100	0.147	1.34	0.048	95.2	97.5

*Values of W_f as a function of x may be obtained from a table of Wick's functions.

Note that the calculated distribution shows some deviation from the measured values because of the small departure of the actual sample from the theoretical coefficient of variation (i.e., 47.5 versus 50 percent).

The critical value of i, which is defined in Eq. (18-31) as the ratio of b/g or the relative dependence of nucleation and growth on supersaturation, can be determined by a few extra experiments. This is done by varying the residence time of the crystals (changing feed rate) while maintaining everything else constant. The

TABLE 18-6 Growth Rates and Kinetic Equations for Some Industrial Crystallized Products

Material crystallized	G, m/s × 10^8	Range t, h	Range M_T, g/L	Temp., °C	Scale*	Kinetic equation for B_0 no./(L·s)	References†
$(NH_4)_2SO_4$	1.67	3.83	150	70	P	$B_0 = 6.62 \times 10^{-25} G^{0.82} p^{-0.92} m_2^{2.05}$	Bennett and Wolf, *AIChE*, SFC, 1979.
$(NH_4)_2SO_4$	0.20	0.25	38	18	B	$B_0 = 2.94(10^{10})G^{1.03}$	Larsen and Mullen, *J. Crystal Growth* **20:** 183 (1973).
$(NH_4)_2SO_4$	—	0.20	—	34	B	$B_0 = 6.14(10^{-11})S_R^{7.84} M_T^{0.98} G^{1.22}$	Youngquist and Randolph, *AIChE J.* **18:** 421 (1972).
$MgSO_4 \cdot 7H_2O$	3.0–7.0	—	—	25	B	$B_0 = 9.65(10^{12})M_T^{0.67} G^{1.24}$	Sikdar and Randolph, *AIChE J.* **22:** 110 (1976).
$MgSO_4 \cdot 7H_2O$	—	—	Low	29	B	$B_0 = f\,(N, L^4, N^{4.2}, S^{2.5})$	Ness and White, *AIChE Symposium Series 153,* vol. 72, p. 64.
KCl	2–12	—	200	32	P	$B_0 = 7.12(10^{39})M_T^{0.14} G^{4.99}$	Randolph et al., *AIChE J.* **23:** 500 (1977).
KCl	3.3	1–2	100	37	B	$B_0 = 5.16(10^{22})M_T^{0.91} G^{2.77}$	Randolph et al., *Ind. Eng. Chem. Proc. Design Dev.* **20:** 496 (1981).
KCl	0.3–0.45	—	50–147	25–68	B	$B_0 = 5 \times 10^{-3} G^{2.78}(M_T TIP^2)^{1.2}$	Qian et al., *AIChE J.* **33**(10): 1690 (1987).
KCr_2O_7	1.2–9.1	0.25–1	14–42	—	B	$B_0 = 7.33(10^4)M_T^{0.6} G^{0.5}$	Desari et al., *AIChE J.* **20:** 43 (1974).
KCr_2O_7	2.6–10	0.15–0.5	20–100	26–40	B	$B_0 = 1.59(10^{-3})S_R^3 M_T G^{0.48}$	Janse, Ph.D. thesis, Delft Technical University, 1977.
KNO_3	8.13	0.25–0.050	10–40	20	B	$B_0 = 3.85(10^{16})M_T^{0.5} G^{2.06}$	Juraszek and Larson, *AIChE J.* **23:** 460 (1977).
K_2SO_4	—	0.03–0.17	1–7	30	B	$B_0 = 2.62(10^3)S_R^{2.5} M_T^{0.5} G^{0.54}$	Randolph and Sikdar, *Ind. Eng. Chem. Fund.* **15:** 64 (1976).
K_2SO_4	2–6	0.25–1	2–20	10–50	B	$B_0 = 4.09(10^6) \exp\left(\frac{10900}{RT}\right) M_T G^{0.5}$	Jones, Budz, and Mullin, *AIChE J.* **33:** 12 (1986).
K_2SO_4	0.8–1.6	—	—	—	B	$\frac{G}{G_0} = 1 + 2L^{2/3}$ (L in μm)	White, Bendig, and Larson, *AIChE Mtg.*, Washington, D.C., Dec. 1974.
NaCl	4–13	0.2–1	25–200	50	B	$B_0 = 1.92(10^{10})S_R^2 M_T G^2$	Asselbergs, Ph.D. thesis, Delft Technical University, 1978.
NaCl	—	0.6	35–70	55	P	$B_0 = 8 \times 10^{10} N^2 G^2 M_T$	Grootscholten et al., *Chem. Eng. Design* **62:** 179 (1984).
NaCl	0.5	1–2.5	70–190	72	P	$B_0 = 1.47(10^2)\left(\frac{I^2}{P}\right) m_4^{0.84} G^{0.98}$	Bennett et al., *Chem. Eng. Prog.* **69**(7): 86 (1973).
Citric acid	1.1–3.7	—	—	16–24	B	$B_0 = 1.09(10^{10})m_4^{0.084} G^{0.84}$	Sikdar and Randolph, *AIChE J.* **22:** 110 (1976).
Fructose	0.1–0.25	—	—	50	B	—	Shiau and Berglund, *AIChE J.* **33:** 6 (1987).
Sucrose	—	—	—	80	B	$B_0 = 5 \times 10^6 N^{0.7} M_T^{0.3} G^{0.4}$	Berglund and deJong, *Separations Technology* **1:** 38 (1990).
Sugar	2.5–5	0.375	50	45	B	$B_0 = 4.38(10^6)M_T^{1.01}(\Delta C - 0.5)^{1.42}$	Hart et al., *AIChE Symposium Series 193,* vol. 76, 1980.
Urea	0.4–4.2	2.5–6.8	350–510	55	P	$B_0 = 5.48(10^{-1})M_T^{-3.87} G^{1.66}$	Bennett and Van Buren, *Chem. Eng. Prog. Symposium Series* **95**(7): 65 (1973).
Urea	—	—	—	3–16	B	$B_0 = 1.49(10^{-31})S_R^{2.3} M_T^{1.07} G^{-3.54}$	Lodaya et al., *Ind. Eng. Chem. Proc. Design Dev.* **16:** 294 (1977).

*B = bench scale; P = pilot plant.
†Additional data on many components are in Garside and Shah, *Ind. Eng. Chem. Proc. Design Dev.*, **19,** 509 (1980).

B^0 and G values are determined at each residence time, and a plot of ln B^0 versus ln G should yield a straight line of slope i. High values of i indicate a propensity to nucleate versus grow and dictate the need to ensure low values of supersaturation.

Had sufficient data indicating a change in n^0 for various values of M_T at constant G been available, a plot of ln n^0 versus ln M_T at corresponding G's would permit determination of the power j.

Crystallizers with Fines Removal In Example 4, the product was from a forced-circulation crystallizer of the MSMPR type. In many cases, the product produced by such machines is too small for commercial use; therefore, a separation baffle is added within the crystallizer to permit the removal of unwanted fine crystalline material from the magma, thereby controlling the population density in the machine so as to produce a coarser crystal product. When this is done, the product sample plots on a graph of ln n versus L as shown in line P, Fig. 18-69. The line of steepest slope, line F, represents the particle-size distribution of the fine material, and samples which show this distribution can be taken from the liquid leaving the fines-separation baffle. The product crystals have a slope of lower value, and typically there should be little material present smaller than L_f, the size which the baffle is designed to separate. However, this is not to imply that there are no fines in the product stream. The effective nucleation rate for the product material is the intersection of the extension of line P to zero size.

As long as the largest particle separated by the fines-destruction baffle is small compared with the mean particle size of the product, the seed for the product may be thought of as the particle-size distribution corresponding to the fine material which ranges in length from zero to L_f, the largest size separated by the baffle.

The product discharged from the crystallizer is characterized by the integral of the distribution from size L_f to infinity:

$$M_T = k_v \rho \int_{L_f}^{\infty} n^0 \exp(-L_f/Gt_f) \exp(L/GT) L^3 \, dL \qquad (18\text{-}44)$$

The integrated form of this equation is shown in Table 18-5.

For a given set of assumptions it is possible to calculate the characteristic curves for the product from the crystallizer when it is operated

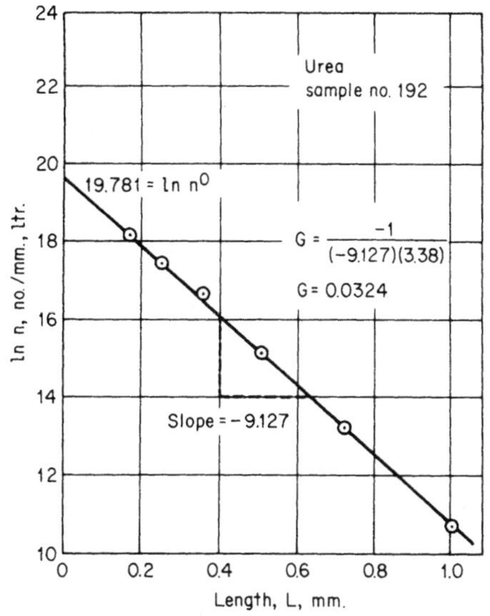

FIG. 18-68 Population density plot for Example 4.

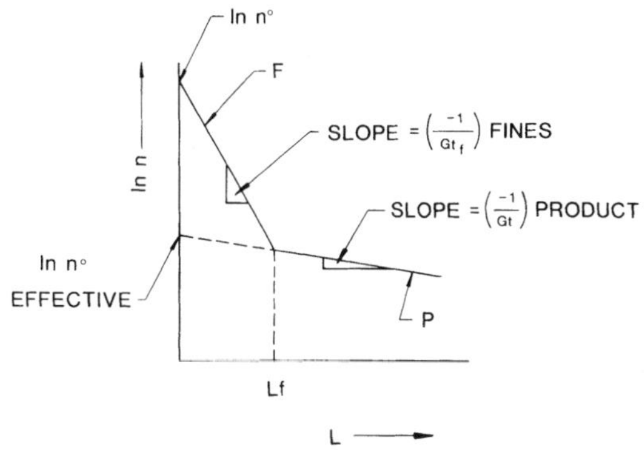

FIG. 18-69 Plot of Log N against L for a crystallizer with fines removal.

at various levels of fines removal as characterized by L_f. This has been done for an ammonium sulfate crystallizer in Fig. 18-70. Also shown in that figure is the actual size distribution obtained. In calculating theoretical size distributions in accordance with the Eq. (18-44), it is assumed that the growth rate is a constant, whereas in fact larger values of L_f will interact with the system driving force to raise the growth rate and the nucleation rate. Nevertheless, Fig. 18-70 illustrates clearly the empirical result of the operation of such equipment, demonstrating that the most significant variable in changing the particle-size distribution of the product is the size removed by the baffle. Conversely, changes in retention time for a given particle-removal size L_f make a relatively small change in the product-size distribution. Jancic and Grootscholten (op. cit., p. 318) have found that the size enlargement is dependent on the fines size, the relative kinetic order i, and the rate of flow to the fines circuit versus product flow.

It is implicit that increasing the value of L_f will raise the supersaturation and growth rate to levels at which mass nucleation can occur, thereby leading to periodic upsets of the system or cycling [Randolph, Beer, and Keener, *Am. Inst. Chem. Eng. J.*, **19,** 1140 (1973)]. That this could actually happen was demonstrated experimentally by Randolph, Beckman, and Kraljevich [*Am. Inst. Chem. Eng. J.*, **23,** 500 (1977)], and that it could be controlled dynamically by regulating the

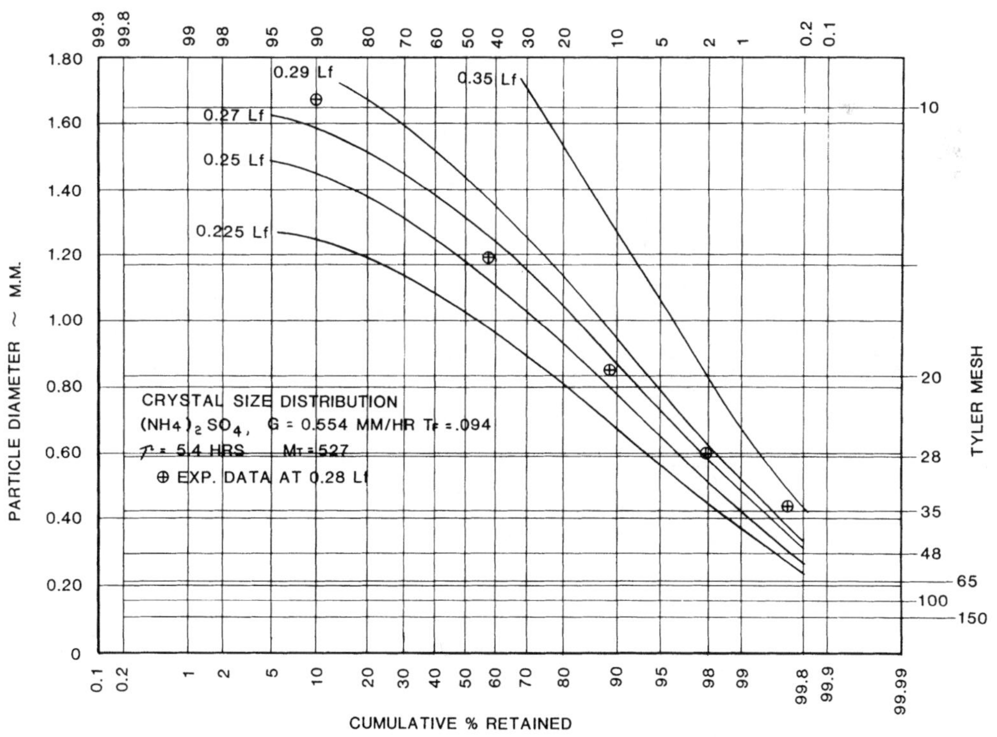

FIG. 18-70 Calculated product-size distribution for a crystallizer operation at different fine-crystal-separation sizes.

fines-destruction system was shown by Beckman and Randolph [ibid., (1977)]. Dynamic control of a crystallizer with a fines-destruction baffle and fine-particle-detection equipment employing a light-scattering (laser) particle-size-measurement instrument is described in U.S. Patent, 4,263,010 and 5,124,265.

CRYSTALLIZATION EQUIPMENT

Whether a vessel is called an evaporator or a crystallizer depends primarily on the criteria used in arriving at its sizing. In an evaporator of the salting-out type, sizing is done on the basis of vapor release. In a crystallizer, sizing is normally done on the basis of the volume required for crystallization or for special features required to obtain the proper product size. In external appearance, the vessels could be identical. Evaporators are discussed in Sec. 11. Genck (loc. cit., 2004) provides a detailed discussion of guidelines for crystallizer selection and operation.

In the discussion which follows, crystallization equipment has been classified according to the means of suspending the growing product. This technique reduces the number of major classifications and segregates those to which Eq. (18-34) applies.

Mixed-Suspension, Mixed-Product-Removal Crystallizers This type of equipment, sometimes called the circulating-magma crystallizer, is by far the most important in use today. In most commercial equipment of this type, the uniformity of suspension of product solids within the crystallizer body is sufficient for the theory [Eqs. (18-34) to (18-36*c*)] to apply. Although a number of different varieties and features are included within this classification, the equipment operating with the highest capacity is the kind in which the vaporization of a solvent, usually water, occurs.

Although surface-cooled types of MSMPR crystallizers are available, most users prefer crystallizers employing vaporization of solvents or of refrigerants. The primary reason for this preference is that heat transferred through the critical supersaturating step is through a boiling-liquid-gas surface, avoiding the troublesome solid deposits that can form on a metal heat-transfer surface. In this case very low LMTDs are required to stay within the metastable zone to promote growth and reduce scaling. The result is multipass, large-surface-area heat exchangers.

Forced-Circulation Evaporator-Crystallizer This crystallizer is shown in Fig. 18-71. Slurry leaving the body is pumped through a circulating pipe and through a tube-and-shell heat exchanger, where its temperature increases by about 2 to 6°C (3 to 10°F). Since this heating is done without vaporization, materials of normal solubility should produce no deposition on the tubes. The heated slurry, returned to the body by a recirculation line, mixes with the body slurry and raises its temperature locally near the point of entry, which causes boiling at the liquid surface. During the consequent cooling and vaporization to achieve equilibrium between liquid and vapor, the supersaturation which is created causes growth on the swirling body of suspended crystals until they again leave via the circulating pipe. Severe vortexing must be eliminated to ensure that the supersaturation is relieved. The quantity and the velocity of the recirculation, the size of the body, and the type and speed of the circulating pump are critical design items if predictable results are to be achieved. A further discussion of the parameters affecting this type of equipment is given by Bennett, Newman, and Van Buren [*Chem. Eng. Prog.*, **55**(3), 65 (1959); *Chem. Eng. Prog. Symp. Ser.*, **65**(95), 34, 44 (1969)].

If the crystallizer is not of the evaporative type but relies only on *adiabatic evaporative cooling* to achieve the yield, the heating element is omitted. The feed is admitted into the circulating line after withdrawal of the slurry, at a point sufficiently below the free-liquid surface to prevent flashing during the mixing process.

FC units typically range from 2 to 20 ft in diameter. They are especially useful for high evaporation loads. For example, a unit used to evaporate water at 380 mmHg can typically be designed to handle 250 to 300 lb/(h·ft^2). Other than allowing one to adjust the residence time or slurry density, the FC affords little opportunity to change the size distribution.

Draft-Tube-Baffle (DTB) Evaporator-Crystallizer Because mechanical circulation greatly influences the level of nucleation within the crystallizer, a number of designs have been developed that use circulators located within the body of the crystallizer, thereby reducing the head against which the circulator must pump. This technique reduces the power input and circulator tip speed and therefore the rate of nucleation. A typical example is the draft-tube-baffle (DTB) evaporator-crystallizer (Swenson Process Equipment, Inc.) shown in Fig. 18-72. The suspension of product crystals is maintained by a large, slow-moving propeller surrounded by a draft tube within the body. The propeller directs the slurry to the liquid surface so as to prevent solids from short-circuiting the zone of the most intense supersaturation. Slurry which has been cooled is returned to the bottom of the vessel and recirculated through the propeller. At the propeller, heated solution is mixed with the recirculating slurry.

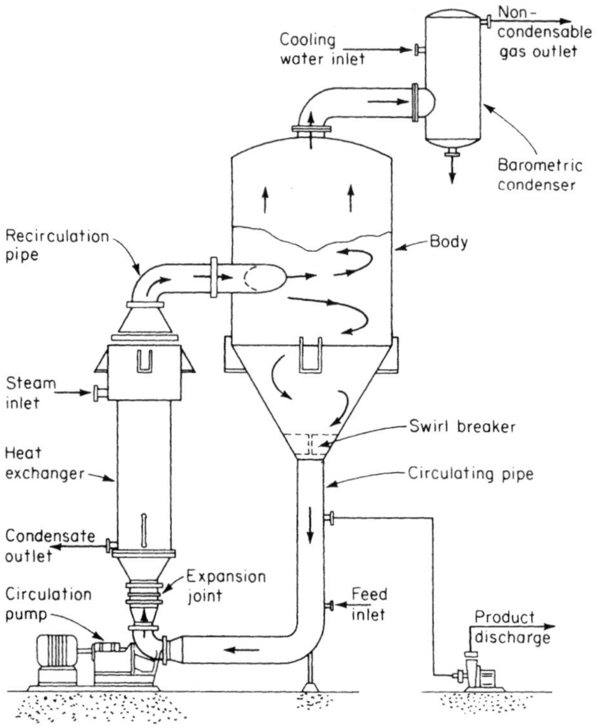

FIG. 18-71 Forced-circulation (evaporative) crystallizer. *(Swenson Process Equipment, Inc.)*

The design of Fig. 18-72 contains a fines-destruction feature comprising the settling zone surrounding the crystallizer body, the circulating pump, and the heating element. The heating element supplies sufficient heat to meet the evaporation requirements and to raise the temperature of the solution removed from the settler so as to destroy any small crystalline particles withdrawn. Coarse crystals are separated from the fines in the settling zone by gravitational sedimentation, and therefore this fines-destruction feature is applicable only to systems in which there is a substantial density difference between crystals and mother liquor.

This type of equipment can also be used for applications in which the only heat removed is that required for adiabatic cooling of the incoming feed solution. When this is done and the fines-destruction feature is to be employed, a stream of liquid must be withdrawn from the settling zone of the crystallizer and the fine crystals must be separated or destroyed by some means other than heat addition—for example, either dilution or thickening and physical separation.

In some crystallization applications it is desirable to increase the solids content of the slurry within the body above the natural make, which is that developed by equilibrium cooling of the incoming feed solution to the final temperature. This can be done by withdrawing a stream of mother liquor from the baffle zone, thereby thickening the

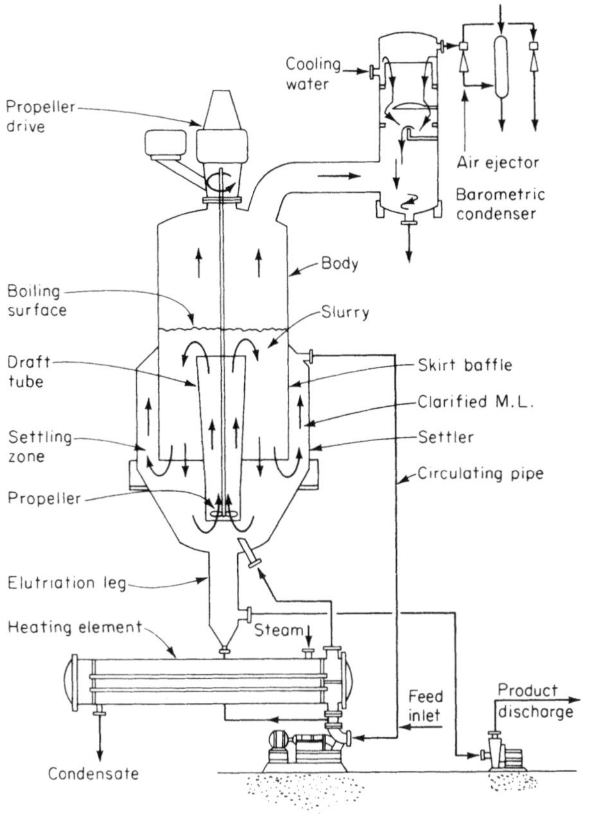

FIG. 18-72 Draft-tube-baffle (DTB) crystallizer. (*Swenson Process Equipment, Inc.*)

slurry within the growing zone of the crystallizer. This mother liquor is also available for removal of fine crystals for size control of the product.

Draft-Tube (DT) Crystallizer This crystallizer may be employed in systems in which fines destruction is not needed or wanted. In such cases the baffle is omitted, and the internal circulator is sized to have the minimum nucleating influence on the suspension.

In DTB and DT crystallizers the circulation rate achieved is generally much greater than that available in a similar forced-circulation crystallizer. The equipment therefore finds application when it is necessary to circulate large quantities of slurry to minimize supersaturation levels within the equipment. In general, this approach is required to obtain long operating cycles with material capable of growing on the walls of the crystallizer. The draft-tube and draft-tube-baffle designs are commonly used in the production of granular materials such as ammonium sulfate, potassium chloride, photographic hypo, and other inorganic and organic crystals for which product in the range 8 to 30 mesh is required.

Surface-Cooled Crystallizer For some materials, such as sodium chlorate, it is possible to use a forced-circulation tube-and-shell exchanger in direct combination with a draft-tube-crystallizer body, as shown in Fig. 18-73. Careful attention must be paid to the temperature difference between the cooling medium and the slurry circulated through the exchanger tubes. In addition, the path and rate of slurry flow within the crystallizer body must be such that the volume contained in the body is "active." That is to say, crystals must be so suspended within the body by the turbulence that they are effective in relieving supersaturation created by the reduction in temperature of the slurry as it passes through the exchanger. Obviously, the circulating pump is part of the crystallizing system, and careful attention must be paid to its type and its operating parameters to avoid undue nucleating influences.

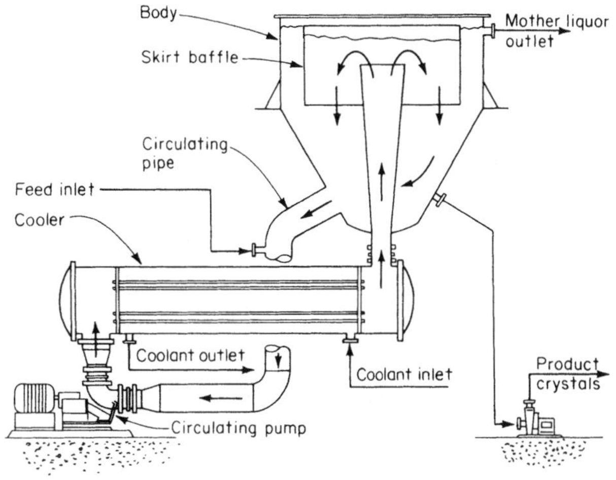

FIG. 18-73 Forced-circulation baffle surface-cooled crystallizer. (*Swenson Process Equipment, Inc.*)

The use of the internal baffle permits operation of the crystallizer at a slurry consistency other than that naturally obtained by the cooling of the feed from the initial temperature to the final mother-liquor temperature. The baffle also permits fines removal and destruction.

With most inorganic materials this type of equipment produces crystals in the range 30 to 100 mesh. The design is based on the allowable rates of heat exchange and the retention required to grow the product crystals.

Direct-Contact-Refrigeration Crystallizer For some applications, such as the freezing of ice from seawater, it is necessary to go to such low temperatures that cooling by the use of refrigerants is the only economical solution. In such systems it is sometimes impractical to employ surface-cooled equipment because the allowable temperature difference is so small (under 3°C) that the heat-exchanger surface becomes excessive or because the viscosity is so high that the mechanical energy put in by the circulation system requires a heat-removal rate greater than can be obtained at reasonable temperature differences. In such systems, it is convenient to admix the refrigerant with the slurry being cooled in the crystallizer, as shown in Fig. 18-74, so that the heat of vaporization of the refrigerant cools the slurry by direct contact. The successful application of such systems requires that the refrigerant be relatively immiscible with the mother liquor and be capable of separation, compression, condensation, and subsequent recycle into the crystallizing system. The operating pressures and temperatures chosen have a large bearing on power consumption.

This technique has been very successful in reducing the problems associated with buildup of solids on a cooling surface. The use of direct-contact refrigeration also reduces overall process-energy requirements, since in a refrigeration process involving two fluids a greater temperature difference is required on an overall basis when the refrigerant must first cool some intermediate solution, such as calcium chloride brine, and that solution in turn cools the mother liquor in the crystallizer.

Equipment of this type has been successfully operated at temperatures as low as −59°C (−75°F).

Reaction-Type Crystallizers In chemical reactions in which the end product is a solid-phase material such as a crystal or an amorphous solid the type of equipment described in the preceding subsections or shown in Fig. 18-75 may be used. By mixing the reactants in a large circulated stream of mother liquor containing suspended solids of the equilibrium phase, it is possible to minimize the driving force created during their reaction and remove the heat of reaction through the vaporization of a solvent, normally water. Depending on the final particle size required, it is possible to incorporate a fines-destruction baffle as shown in Fig. 18-75 and take advantage of the control over particle size afforded by this technique. In the case of ammonium sulfate crystallization from ammonia gas and concentrated sulfuric acid, it is necessary to vaporize water to remove the heat of reaction, and this water so

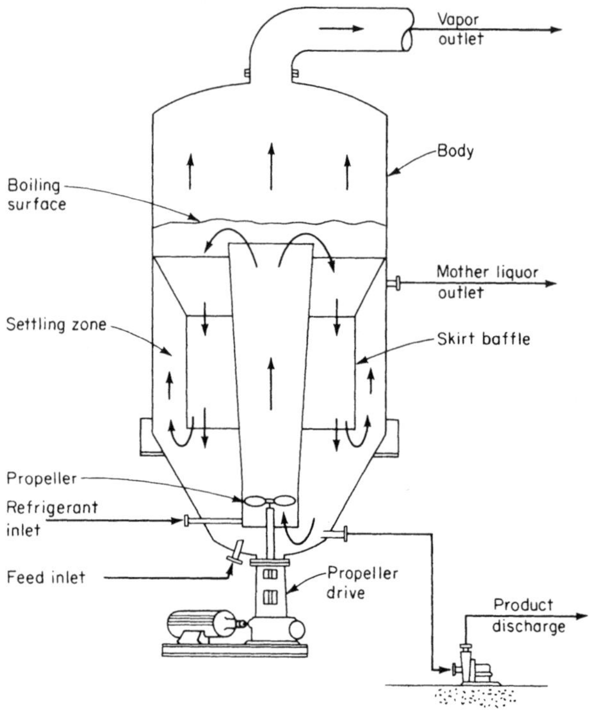

FIG. 18-74 Direct-contact-refrigeration crystallizer (DTB type). (*Swenson Process Equipment, Inc.*)

removed can be reinjected after condensation into the fines-destruction stream to afford a very large amount of dissolving capability.

Other examples of this technique are where a solid material is to be decomposed by mixing it with a mother liquor of a different composition, as shown in Fig. 18-76. Carnallite ore ($KCl{\cdot}MgCl_2{\cdot}4H_2O$) can be added to a mother liquor into which water is also added so that decomposition of the ore into potassium chloride (KCl) crystals and magnesium chloride–rich mother liquor takes place. Circulated slurry in the draft tube suspends the product crystals as well as the incoming ore particles until the ore can decompose into potassium chloride crystals and mother liquor. By taking advantage of the fact that water must be added to the process, the fines-bearing mother liquor can be removed behind the baffle and then water added so that the finest particles are dissolved before being returned to the crystallizer body.

Other examples of this technique involve neutralization reactions such as the neutralization of sulfuric acid with calcium chloride to result in the precipitation of gypsum.

Mixed-Suspension, Classified-Product-Removal Crystallizers Many of the crystallizers just described can be designed for classified-product discharge. Classification of the product is normally done by means of an elutriation leg suspended beneath the crystallizing body as shown in Fig. 18-72. Introduction of clarified mother liquor to the lower portion of the leg fluidizes the particles prior to discharge and selectively returns the finest crystals to the body for further growth. A relatively wide distribution of material is usually produced unless the elutriation leg is extremely long. Inlet conditions at the leg are critical if good classifying action or washing action is to be achieved.

If an elutriation leg or other product-classifying device is added to a crystallizer of the MSMPR type, the plot of the population density versus L is changed in the region of largest sizes. Also the incorporation of an elutriation leg destabilizes the crystal-size distribution and under some conditions can lead to cycling. To reduce cycling, fines destruction is usually coupled with classified product removal. The theoretical treatment of both the crystallizer model and the cycling relations is discussed by Randolph, Beer, and Keener (loc. cit.). Although such a feature can be included on many types of classified-suspension or mixed-suspension crystallizers, it is most common to use this feature with the forced-circulation evaporative-crystallizer and the DTB crystallizer.

Classified-Suspension Crystallizer This equipment is also known as the **growth** or **Oslo crystallizer** and is characterized by the production of supersaturation in a circulating stream of liquor. Supersaturation is developed in one part of the system by evaporative cooling or by cooling in a heat exchanger, and it is relieved by passing the liquor through a fluidized bed of crystals. The fluidized bed may be contained in a simple tank or in a more sophisticated vessel arranged for a pronounced classification of the crystal sizes. Ideally this equipment operates within the metastable supersaturation field described by Miers and Isaac, *J. Chem. Soc.,* **1906,** 413.

In the **evaporative crystallizer** of Fig. 18-77, solution leaving the vaporization chamber at B is supersaturated slightly within the metastable zone so that new nuclei will not form. The liquor contacting the bed at E relieves its supersaturation on the growing crystals and leaves through the circulating pipe F. In a cooling-type crystallization hot feed is introduced at G, and the mixed liquor flashes when it reaches the vaporization chamber at A. If further evaporation is required to produce the driving force, a heat exchanger is installed between the circulating pump and the vaporization changer to supply the heat for the required rate of vaporization.

The transfer of supersaturated liquor from the vaporizer (point B, Fig. 18-77) can cause salt buildup in the piping and reduction of the operating cycle in equipment of this type. The rate of buildup can be reduced by circulating a thin suspension of solids through the vaporizing chamber; however, the presence of such small seed crystals tends to rob the supersaturation developed in the vaporizer, thereby lowering the efficiency of the recirculation system.

The decrease in temperature due to flashing is typically less than 4 to 6°F, and the increase in solute concentration in the circulating liquor is often around 1 to 3 g/L solvent. Care must be taken to ensure that the liquid velocities in the tapered cross-section of the lower body allow classification of the solids. One must know the settling rates and morphologies of the crystals for proper design and operation. An unclassified operation will perform as an FC unit.

The Oslo crystallizer is best suited for use with compounds with high settling velocities such as greater than 20 to 40 mm/s. If the crystals have high settling rates, the larger particles will settle out quickly. Crystals with low settling velocities require large cross-sectional areas which implies large crystallizers and low crystal production rates.

The suggested productivities for concentration driving forces depend on the settling velocities of the crystals. For a given ΔC, the higher the settling velocity, the higher the allowable crystallizer productivity. For example, for a change in concentration of 2 g/L, the recommended productivity increases from 125 to 250 kg/(h·m^3) as the settling rate increases from 20 to 30 mm/s.

An **Oslo surface-cooled crystallizer** is illustrated in Fig. 18-78. Supersaturation is developed in the circulated liquor by chilling in the cooler H. This supersaturated liquor is contacted with the suspension of crystals in the suspension chamber at E. At the top of the suspension chamber a stream of mother liquor D can be removed to be used for fines removal and destruction. This feature can be added on either type of equipment. Fine crystals withdrawn from the top of the suspension are destroyed, thereby reducing the overall number of crystals in the system and increasing the particle size of the remaining product crystals.

Scraped-Surface Crystallizer A number of crystallizer designs employing direct heat exchange between the slurry and a jacket or double wall containing a cooling medium have been developed. The heat-transfer surface is scraped or agitated in such a way that the deposits cannot build up. The scraped-surface crystallizer provides an effective and inexpensive method of producing slurry in equipment which does not require expensive installation or supporting structures. At times these units are employed to provide auxiliary cooling capacity for existing units.

Double-Pipe Scraped-Surface Crystallizer This type of equipment consists of a double-pipe heat exchanger with an internal agitator fitted with spring-loaded scrapers that wipe the wall of the inner pipe.

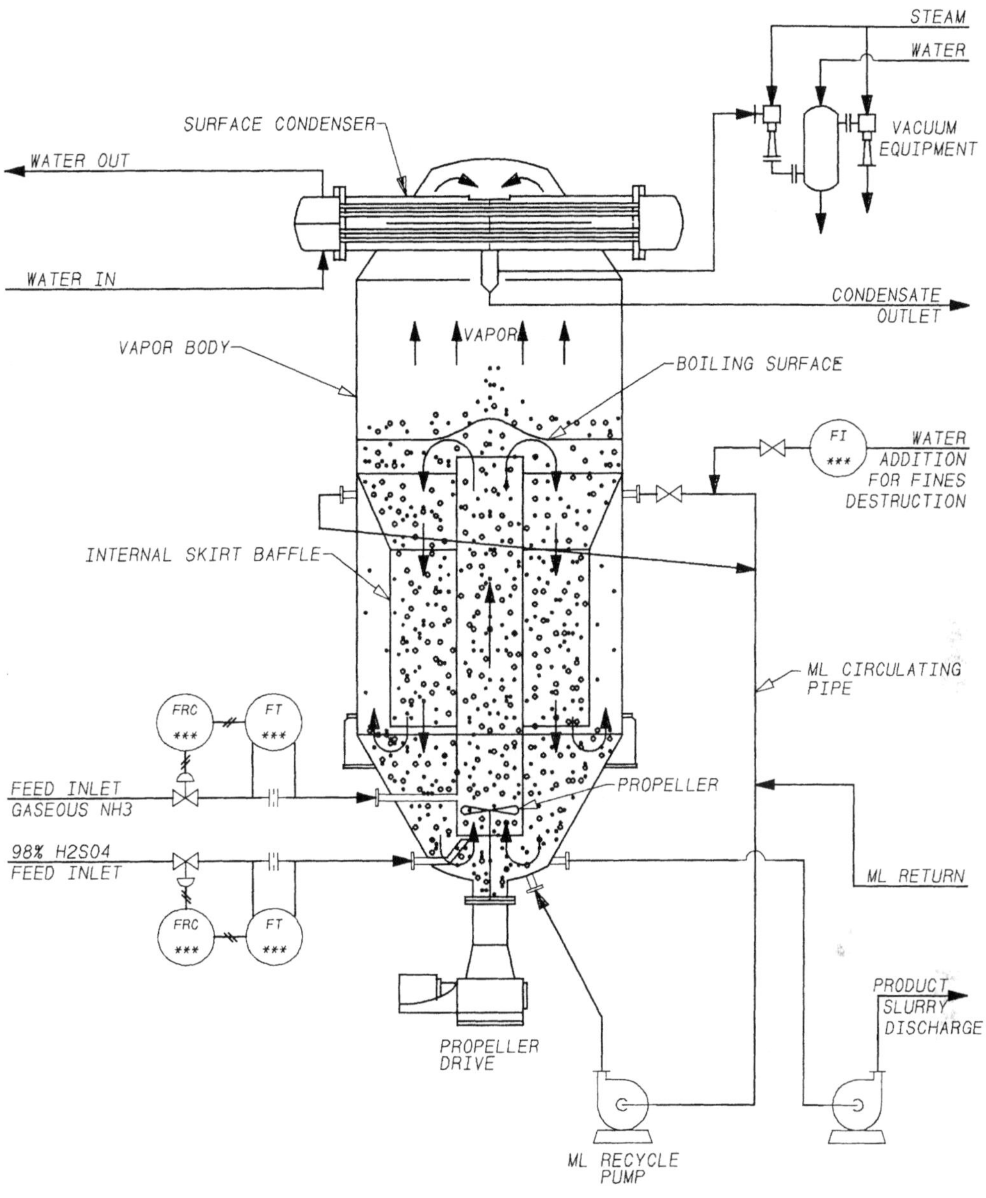

FIG. 18-75 Swenson reaction type DTB crystallizer. (*Swenson Process Equipment, Inc.*)

The cooling liquid passes between the pipes, this annulus being dimensioned to permit reasonable shell-side velocities. The scrapers prevent the buildup of solids and maintain a good film coefficient of heat transfer. Crystal growth is in the bulk of the liquid. The equipment can be operated in a continuous or in a recirculating batch manner.

Such units are generally built in lengths to above 12 m (40 ft). They can be arranged in parallel or in series to give the necessary liquid velocities for various capacities. Heat-transfer coefficients have been reported in the range of 170 to 850 W/(m^2·K) [30 to 150 Btu/(h·ft^2·°F)] at temperature differentials of 17°C (30°F) and higher [Garrett and Rosenbaum, *Chem. Eng.*, **65**(16), 127 (1958)]. Equipment of this type is marketed as the **Votator** and the **Armstrong crystallizer.**

Batch Crystallization Batch crystallization has been practiced longer than any other form of crystallization in both atmospheric tanks, which are either static or agitated, as well as in vacuum or pressure vessels. It is widely practiced in the pharmaceutical and fine chemical industry or in those applications where the capacity is very small. This supersaturation can be generated by a number of modes including antisolvent addition, cooling, evaporation, pH adjustment, and chemical reaction.

A typical batch process involves charging the crystallizer with concentrated or near-saturated solution, producing supersaturation by means of a cooling temperature profile or evaporation profile and seeding the batch in the metastable zone or by allowing spontaneous nucleation to occur. The final mother liquor temperature and concentration is achieved by a time-dependent profile and the batch is then held for ripening followed by transferring the same to downstream processing such as centrifuging, filtration, and drying.

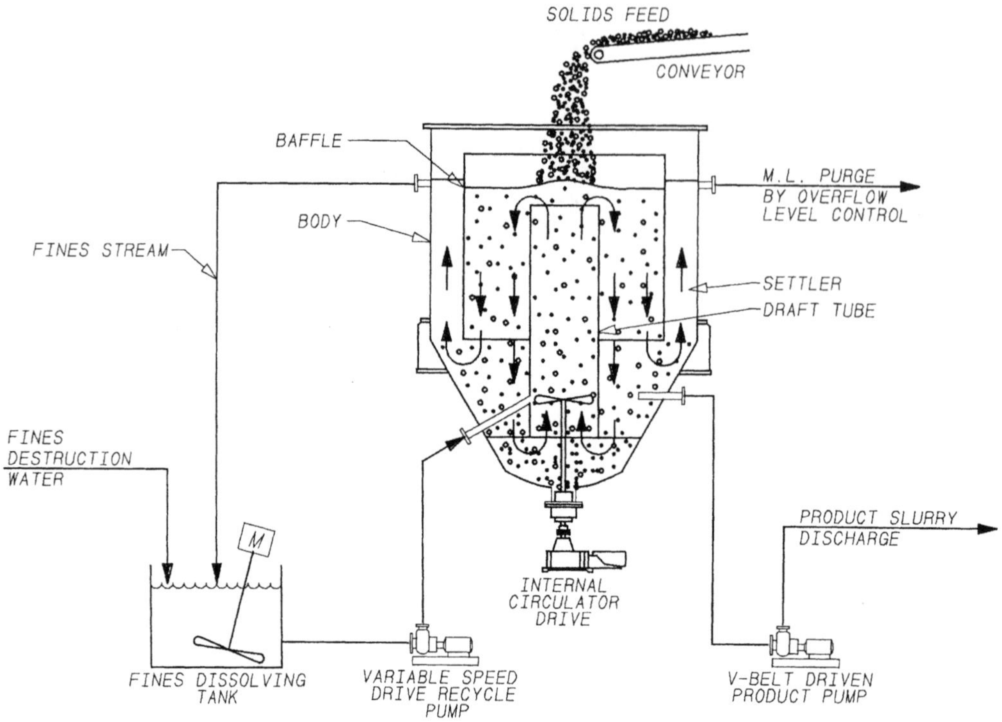

FIG. 18-76 Swenson atmospheric reaction–type DTB crystallizer. (Swenson Process Equipment, Inc.)

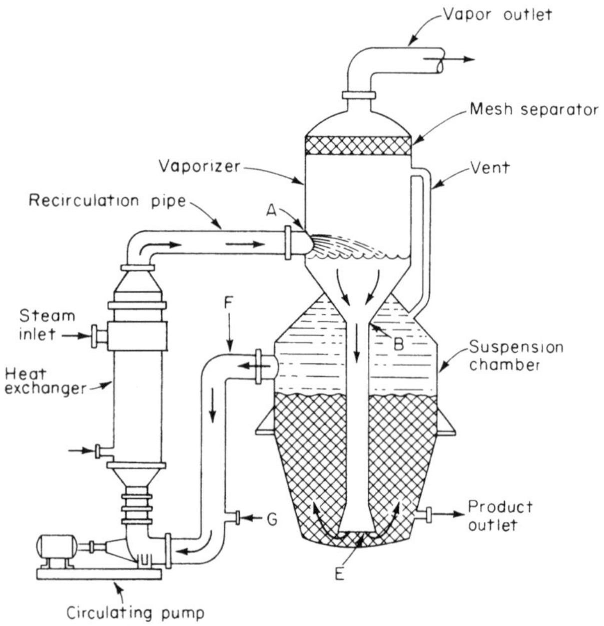

FIG. 18-77 OSLO evaporative crystallizer.

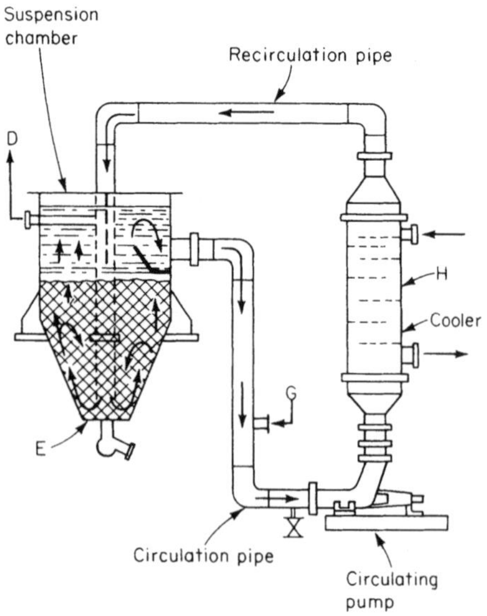

FIG. 18-78 OSLO surface-cooled crystallizer.

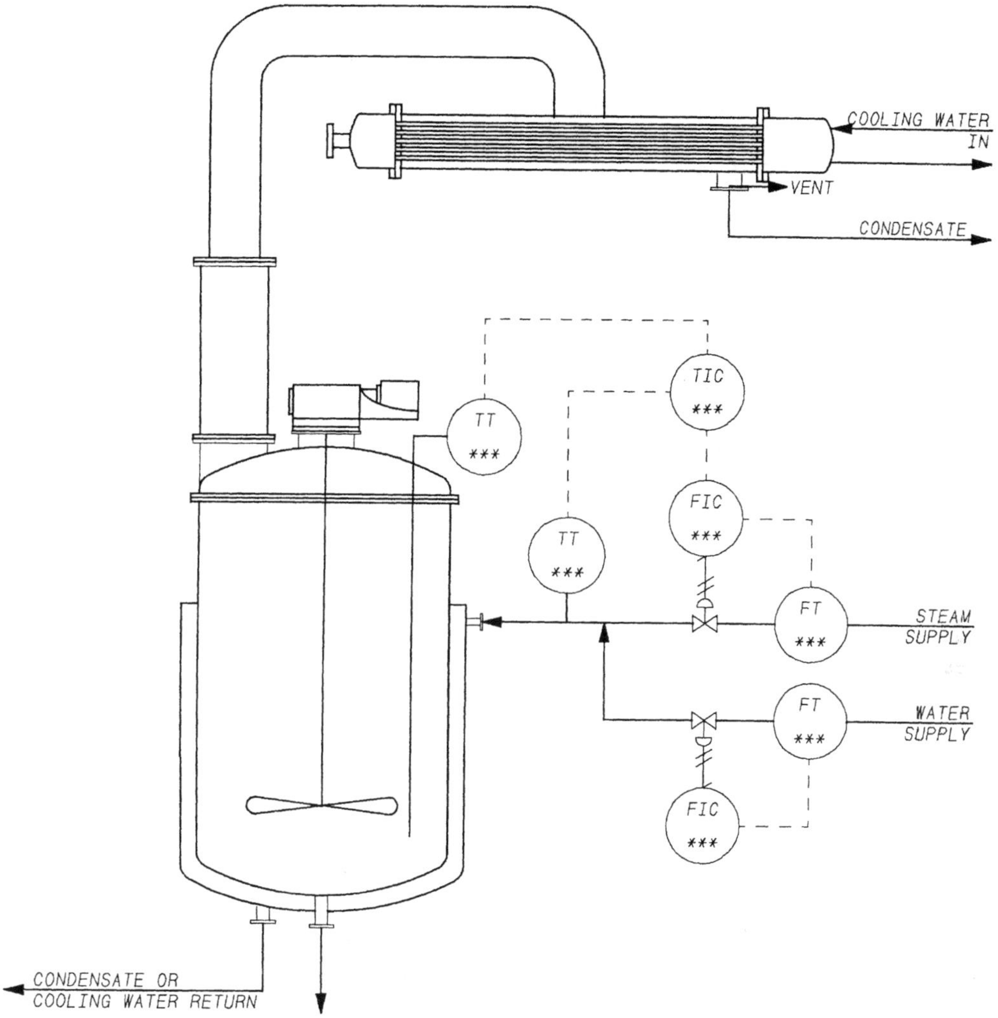

FIG. 18-79 Typical agitated batch crystallizer. (*Swenson Process Equipment, Inc.*)

Control of a batch crystallizer is critical to achieve the desired size distribution. It is necessary to have some means for determining when the initial solution is supersaturated so that seed of the appropriate size, quantity, and habit may be introduced into the batch. Following seeding, it is necessary to limit the cooling or evaporation in the batch to that which permits the generated supersaturation to be relieved on the seed crystals. This means that the first cooling or evaporation following seeding must be at a very slow rate, which is increased nonlinearly in order to achieve the optimum batch cycle and product properties. Frequently, such controls are operated by cycle timers or computers so as to achieve the required conditions. Shown in Fig. 18-79 is a typical batch crystallizer comprising a jacketed closed tank with top-mounted agitator and feed connections. The tank is equipped with a short distillation column and surface condenser so that volatile materials may be retained in the tank and solvent recycled to maintain the batch integrity. Provisions are included so that the vessel may be heated with steam addition to the shell or cooling solution circulated through the jacket so as to control the temperature. Tanks of this type are intended to be operated with a wide variety of chemicals under both cooling and solvent evaporation conditions.

A detailed discussion of crystallization practice is provided by Genck in the following articles: Genck, *Chem. Eng.,* **104**(11), 94 (1997); Genck, *Chem. Eng.,* **107**(8), 90 (2000), and Genck, *Chem. Eng. Progress,* **99**(6), 36 (2003).

Recompression Evaporation-Crystallization In all types of crystallization equipment wherein water or some other solvent is vaporized to produce supersaturation and/or cooling, attention should be given to the use of mechanical vapor recompression, which by its nature permits substitution of electrical energy for evaporation and solvent removal rather than requiring the direct utilization of heat energy in the form of steam or electricity. A typical recompression crystallizer flowsheet is shown in Fig. 18-80, which shows a single-stage evaporative crystallizer operating at approximately atmospheric pressure. The amount of heat energy necessary to remove 1 kg of water to produce the equivalent in crystal product is approximately 550 kilocalories. If the water evaporated is compressed by a mechanical compressor of high efficiency

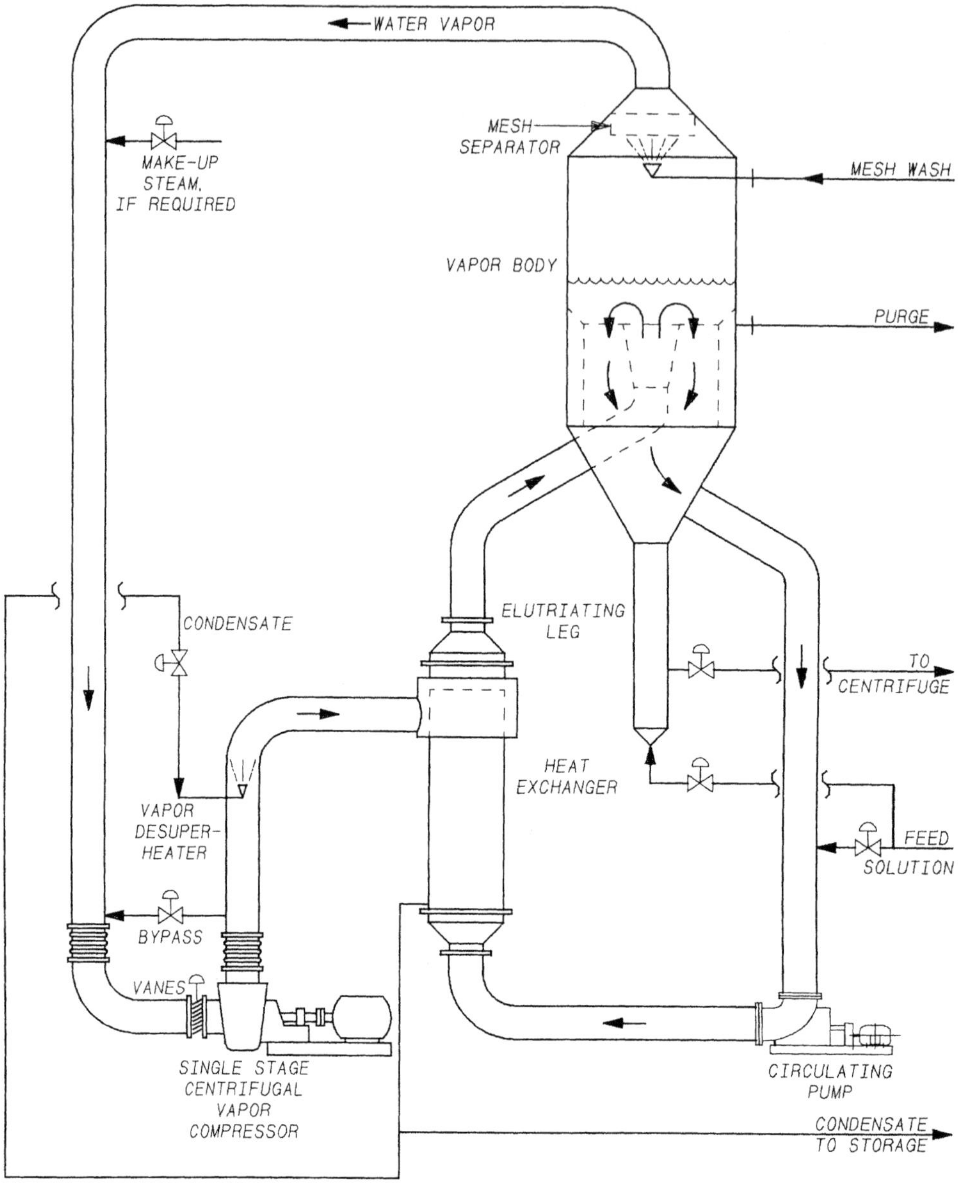

FIG. 18-80 Swenson single-stage recompression evaporator. (*Swenson Process Equipment, Inc.*)

to a pressure where it can be condensed in the heat exchanger of the crystallizer, it can thereby supply the energy needed to sustain the process. Then the equivalent power for this compression is about 44 kilocalories (Bennett, *Chem. Eng. Progress,* 1978, pp. 67–70).

Although this technique is limited economically to those large-scale cases where the materials handled have a relatively low boiling point elevation and in those cases where a significant amount of heat is required to produce the evaporation for the crystallization step, it nevertheless offers an attractive technique for reducing the use of heat energy and substituting mechanical energy or electrical energy in those cases where there is a cost advantage for doing so. This technique finds many applications in the crystallization of sodium sulfate, sodium carbonate monohydrate, and sodium chloride. Shown in Fig. 18-81 is the amount of vapor compressed per kilowatt-hour for water vapor at 100°C and various ΔTs. The amount of water vapor compressed per horsepower decreases rapidly with increasing ΔT and, therefore, normal design considerations dictate that the recompression evaporators have a relatively large amount of heat-transfer surface so as to minimize the power cost. Often this technique is utilized only with the initial stages of evaporation where concentration of the solids is relatively low and, therefore, the boiling-point elevation is negligible. In order to maintain adequate tube velocity for heat transfer and suspension of crystals, the increased surface requires a large internal recirculation within the crystallizer body, which consequently lowers the supersaturation in the fluid pumped through the tubes.

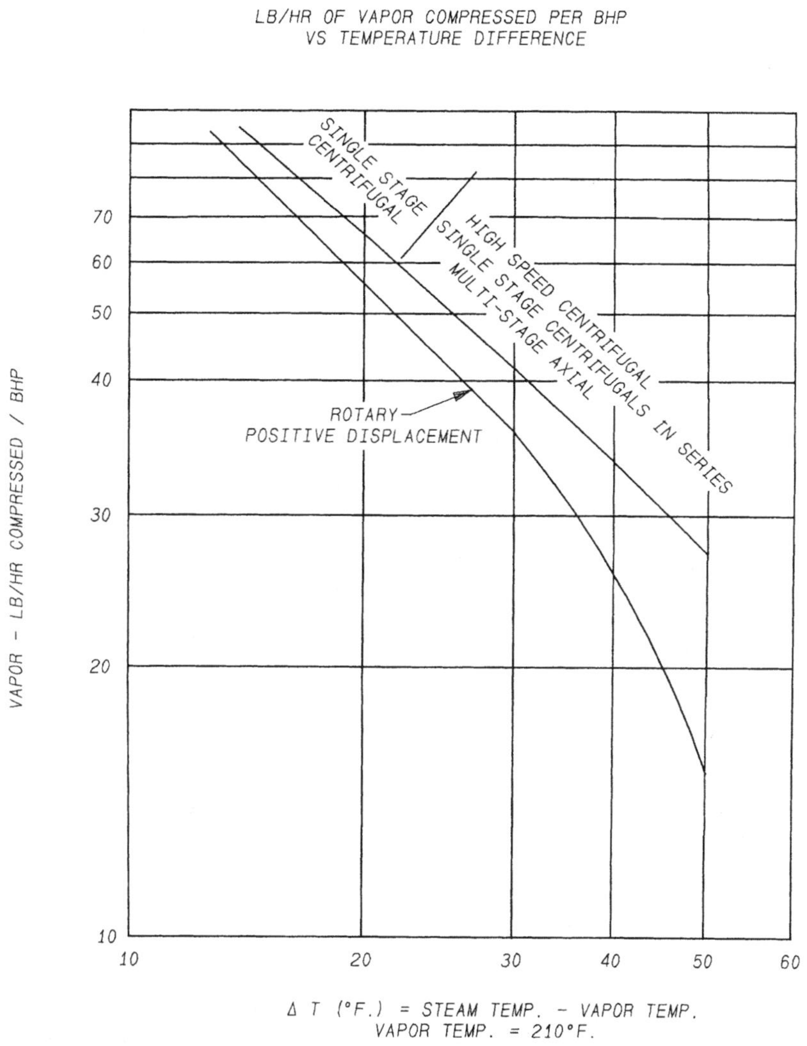

FIG. 18-81 Recompression evaporator horsepower as a function of overall ΔT.

One benefit of this design is that with materials of flat or inverted solubility, the use of recompression complements the need to maintain low ΔTs to prevent fouling of the heat-transfer surface.

INFORMATION REQUIRED TO SPECIFY A CRYSTALLIZER

The following information regarding the product, properties of the feed solution, and required materials of construction must be available before a crystallizer application can be properly evaluated and the appropriate equipment options identified. Is the crystalline material being produced a hydrated or an anhydrous material? What is the solubility of the compound in water or in other solvents under consideration, and how does this change with temperature? Are other compounds in solution which coprecipitate with the product being crystallized, or do these remain in solution, increasing in concentration until some change in product phase occurs? What will be the influence of impurities in the solution on the crystal habit, growth, and nucleation rates? What are the physical properties of the solution and its tendency to foam? What is the heat of crystallization of the product crystal? What is the production rate, and what is the basis on which this production rate is computed? What is the tendency of the material to grow on the walls of the crystallizer? What materials of construction can be used in contact with the solution at various temperatures? What utilities will be available at the crystallizer location, and what are the costs associated with the use of these utilities? Is the final product to be blended or mixed with other crystalline materials or solids? What size of product and what shape of product are required to meet these requirements? How can the crystalline material be separated from the mother liquor and dried? Are there temperature requirements or wash requirements which must be met? How can these solids or mixtures of solids be handled and stored without undue breakage and caking? Is polymorphism an issue?

Another basic consideration is whether crystallization is best carried out on a batch basis or on a continuous basis. The present tendency in most processing plants is to use continuous equipment whenever possible. Continuous equipment permits adjusting of the operating variables to a relatively fine degree in order to achieve the best results in terms of energy usage and product characteristics. It allows the use of a smaller labor force and results in a continuous utility demand, which minimizes the size of boilers, cooling towers, and power-generation facilities. It also minimizes the capital investment required in the crystallizer and in the feed-storage and product-liquor-storage facilities.

Materials that have a tendency to grow readily on the walls of the crytallizer require periodic washout, and therefore an otherwise continuous operation would be interrupted once or even twice a week for the removal of these deposits. The impact that this contingency may have on the processing-equipment train ahead of the crystallizer must be considered.

A batch operation usually has economic application only on small scale, or when multiple products are produced in common facilities.

CRYSTALLIZER OPERATION

Crystal growth is a layer-by-layer process, and the retention time required in most commercial equipment to produce crystals of the size normally desired is often on the order of 2 to 6 h. Growth rates are usually limited to less than 1 to 2 μm/min. On the other hand, nucleation in a supersaturated solution can be generated in a fraction of a second. The influence of any upsets in operating conditions, in terms of the excess nuclei produced, is very short-term in comparison with the total growth period of the product removed from the crystallizer. A worst-case scenario for batch or continuous operation occurs when the explosion of nuclei is so severe that it is impossible to grow an acceptable crystal size distribution, requiring redesolution or washout of the system. In a practical sense, this means that steadiness of operation is much more important in crystallization equipment than it is in many other types of process equipment.

It is to be expected that six to nine retention periods will pass before the effects of an upset will be damped out. Thus, the recovery period may last from 12 to 54 h.

The **rate of nuclei formation** required to sustain a given product size decreases exponentially with increasing size of the product. Although when crystals in the range of 100 to 50 mesh are produced, the system may react quickly, the system response when generating large crystals in the 14-mesh size range is quite slow. This is because a single pound of 150-mesh seed crystals is sufficient to provide the total number of particles in a ton of 14-mesh product crystals. In any system producing relatively large crystals, nucleation must be carefully controlled with respect to all internal and external sources. Particular attention must be paid to preventing seed crystals from entering with the incoming feed stream or being returned to the crystallizer with recycle streams of mother liquor coming back from the filter or centrifuge.

Experience has shown that in any given body operating at a given production rate, control of the magma (slurry) density is important to the control of crystal size. Although in some systems a change in slurry density does not result in a change in the rate nucleation, the more general case is that an increase in the magma density increases the product size through reduction in nucleation and increased retention time of the crystals in the growing bed. The reduction in supersaturation at longer retention times together with the increased surface area at higher percent solids appears to be responsible for the larger product.

A reduction in the magma density will generally increase nucleation and decrease the particle size. This technique has the disadvantage that crystal formation on the equipment surfaces increases because lower slurry densities create higher levels of supersaturation within the equipment, particularly at the critical boiling surface in a vaporization-type crystallizer.

High levels of supersaturation at the liquid surface or at the tube walls in a surface-cooled crystallizer are the dominant cause of wall salting. Although some types of crystallizers can operate for several months continuously when crystallizing KCl or $(NH_4)_2SO_4$, most machines have much shorter operating cycles. Second only to control of particle size, the extension of operating cycles is the most difficult operating problem to be solved in most installations.

In the forced-circulation-type crystallizer (Fig. 19-71) primary control over particle size is exercised by the designer in selecting the circulating system and volume of the body. From the operating standpoint there is little that can be done to an existing unit other than supply external seed, classify the discharge crystals, or control the slurry density. Nevertheless, machines of this type are frequently carefully controlled by these techniques and produce a predictable and desirable product-size distribution.

When crystals cannot be grown sufficiently large in forced-circulation equipment to meet product-size requirements, it is common to employ one of the designs that allow some influence to be exercised over the population density of the finer crystals. In the DTB design (Fig. 18-76) this is done by regulating the flow in the circulating pipe so as to withdraw a portion of the fines in the body in the amount of about 0.05 to 0.5 percent by settled volume. The exact quantity of solids depends on the size of the product crystals and on the capacity of the fines-dissolving system. If the machine is not operating stably, this quantity of solids will appear and then disappear, indicating changes in the nucleation rate within the circuit. At steady-state operation, the quantity of solids overflowing will remain relatively constant, with some solids appearing at all times. Should the slurry density of product crystals circulated within the machine rise to a value higher than about 50 percent settled volume, large quantities of product crystals will appear in the overflow system, disabling the fines-destruction equipment. Too high a circulating rate through the fines trap will produce this same result. Too low a flow through the fines circuit will remove insufficient particles and result in a smaller product-size crystal. To operate effectively, a crystallizer of the type employing fines-destruction techniques requires more sophisticated control than does operation of the simpler forced-circulation equipment.

The classifying crystallizer (Fig. 18-77) requires approximately the same control of the fines-removal stream and, in addition, requires control of the fluidizing flow circulated by the main pump. This flow must be adjusted to achieve the proper degree of fluidization in the suspension chamber, and this quantity of flow varies as the crystal size varies between start-up operation and normal operation. As with the draft-tube-baffle machine, a considerably higher degree of skill is required for operation of this equipment than of the forced-circulation type.

While most of the industrial designs in use today are built to reduce the problems due to excess nucleation, it is true that in some crystallizing systems a deficiency of seed crystals is produced and the product crystals are larger than are wanted or required. In such systems nucleation can be increased by increasing the mechanical stimulus created by the circulating devices or by seeding through the addition of fine crystals from some external source.

CRYSTALLIZER COSTS

Because crystallizers can come with such a wide variety of attachments, capacities, materials of construction, and designs, it is very difficult to present an accurate picture of the costs for any except certain specific types of equipment, crystallizing specific compounds. This is illustrated in Fig. 18-82, which shows the prices of equipment for crystallizing two different compounds at various production rates, one of the compounds being produced in two alternative crystallizer modes. Installed cost (including cost of equipment and accessories, foundations and supporting steel, utility piping, process piping and pumps, electrical switchgear, instrumentation, and labor, but excluding cost of a building) will be approximately twice these price figures.

Most crystallization equipment is custom-designed, and costs for a particular application may vary greatly from those illustrated in Fig. 18-82. Realistic estimation of installation costs also requires reference to local labor rates, site-specific factors, and other case specifics.

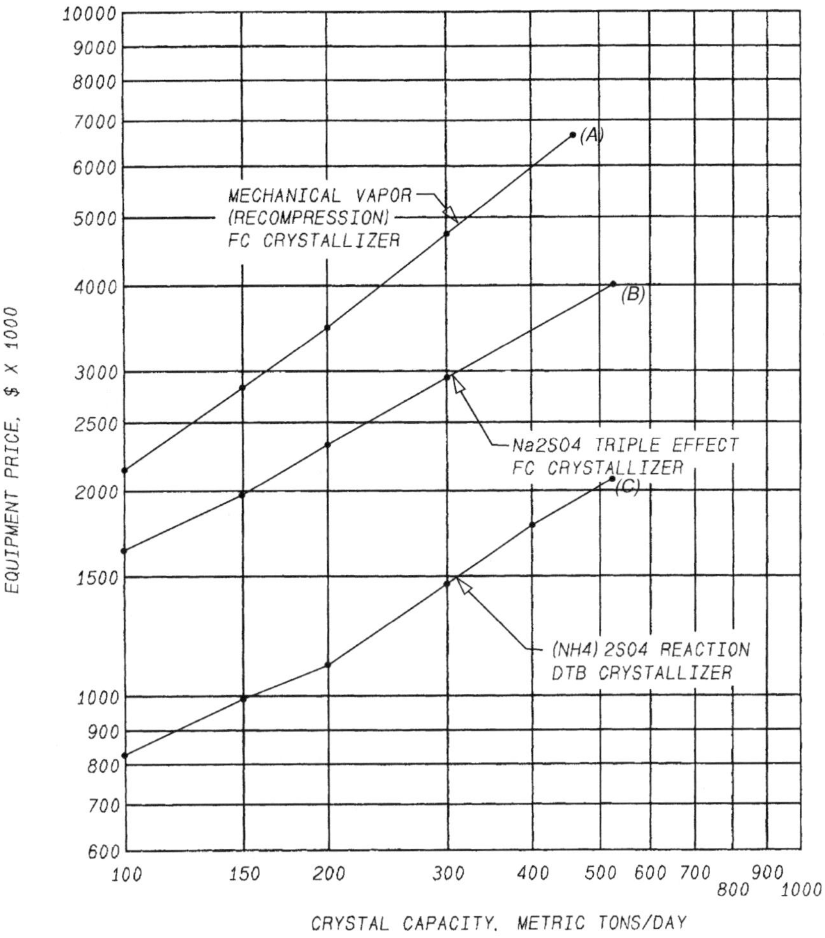

FIG. 18-82 Equipment prices, FOB point of fabrication, for typical crystallizer systems. Prices are for crystallizer plus accessories including vacuum equipment (2005). (A), (B) Na_2SO_4 production from Glauber's salt. Melting tank included. (C) Reaction of $NH_3 + H_2SO_4$ to make $(NH_4\ 2SO_4)$.

LEACHING

GENERAL REFERENCES: Coulson and Richardson, *Chemical Engineering,* 5th ed., vol. 2, Butterworth-Heinemann Publisher, 2002, chap. 10, "Leaching," pp. 502–541. Prabhudesai in Schweitzer, *Handbook of Separation Techniques for Chemical Engineers,* 3d ed., McGraw-Hill, New York, 1996, sec. 5.1. Wilkes, *Fluid Mechanics for Chemical Engineers,* Prentice-Hall, 1999. Wakeman, "Extraction (Liquid-Solid)" in *Kirk-Othmer Encyclopedia of Chemical Technology,* 4th ed., vol. 10, Wiley, New York, 1993, p. 186. McCabe, Smith, and Harriott, *Unit Operations of Chemical Engineering,* 7th ed., McGraw-Hill, New York, 2005. Harriott, *Chemical Reactor Design,* Marcel Dekker, 2003, pp. 89–99. Mular, Halbe, and Barratt, *Mineral Processing Plant Design, Practice, and Control,* vols. 1 and 2, Society for Mining, Metallurgy, and Exploration, Inc., 2002. Section on reactors in annual issues of *Chemical Engineering Buyers' Guide.*

DEFINITION

Leaching is the removal of a soluble fraction, in the form of a solution, from an insoluble, usually permeable, solid phase with which it is associated. Leaching generally involves selective dissolution with or without diffusion; in the extreme case of simple washing, it requires only displacement (with some mixing) of one interstitial liquid by another with which it is miscible. The soluble constituent may be solid or liquid, and it may be incorporated within, chemically combined with, adsorbed upon, or bound mechanically in the pore structure of the insoluble material. Sometimes, the insoluble phase may be massive and porous, but usually it is particulate; the particles may be openly porous, cellular with selectively permeable cell walls, or surface-activated.

By convention, elution of a surface-adsorbed solute is treated as a special case of adsorption, rather than leaching. The washing of filter cakes is also excluded.

Due to its great breadth of application and its importance to some ancient processes, leaching is known by many names including *extraction, solid-liquid extraction, lixiviation, percolation, infusion, washing,* and *decantation-settling.* If the stream of solids being leached is densified by settling, it is often called *underflow* and hydrometallurgists may refer to it as *pulp.* Oil seed processors may refer to the solids as *marc.* The liquid stream containing the leached solute is called *overflow, extract, solution, lixiviate, leachate,* or *miscella.*

Mechanism Leaching may simply result from the solubility of a substance in a liquid, or it may be enabled by a chemical reaction. The rate of transport of solvent into the mass to be leached, or of the soluble fraction into the solvent, or of extract solution out of the insoluble material, or of some combination of these rates may influence overall leaching kinetics, as may an interfacial resistance or a chemical reaction rate.

Inasmuch as the overflow and underflow streams are not immiscible phases but streams based on the same solvent, the concept of equilibrium for leaching is not the one applied in other mass-transfer separations. If the solute is not adsorbed on the inert solid, true equilibrium is reached only when all the solute is dissolved and distributed uniformly throughout the solvent in both underflow and overflow (or when the solvent is uniformly saturated with the solute, a condition never encountered in a properly designed extractor). The practical interpretation of leaching equilibrium is the state in which the overflow and underflow liquids are of the same composition; on a y-x diagram, the equilibrium line will be a straight line through the origin with a slope of unity. It is customary to calculate the number of ideal (equilibrium) stages required for a given leaching task and to adjust the number by applying a stage efficiency factor, although local efficiencies, if known, can be applied stage by stage.

Usually, however, it is not feasible to establish a stage or overall efficiency or a leaching rate index (e.g., overall coefficient) without testing small-scale models of likely apparatus. In fact, the results of such tests may have to be scaled up empirically, without explicit evaluation of rate or quasi-equilibrium indices.

Methods of Operation Leaching systems are distinguished by operating cycle (batch, continuous, or multibatch intermittent); by direction of streams (cocurrent, countercurrent, or hybrid flow); by staging (single-stage, multistage, or differential-stage); and by method of contacting (sprayed percolation, immersed percolation, or solids dispersion). In general, descriptors from all four categories must be assigned to stipulate a leaching system completely (e.g., the Bollman-type extractor is a continuous hybrid-flow multistage sprayed percolator).

Whatever the mechanism and the method of operation, it is clear that the leaching process will be favored by increased surface per unit volume of solids to be leached and by decreased radial distances that must be traversed within the solids, both of which are favored by decreased particle size. Fine solids, on the other hand, cause slow percolation rate, difficult solids separation, and possible poor quality of solid product. The basis for an optimum particle size is established by these characteristics.

LEACHING EQUIPMENT

There are two primary categories of contacting method according to which leaching equipment is classified: (1) leaching may be accomplished by percolation and (2) the particulate solids may be dispersed into a liquid phase and then separated from it. Each may be operated in a batch or continuous manner. Materials that disintegrate during leaching are treated in the second class of equipment. An important exception to this classification is in-situ leaching, as discussed below.

Percolation *Heap leaching,* as shown in Fig. 18-83 (see Mular et al. pp. 1571–1630, loc. cit.) is very widely applied to the ores of copper and precious metals, but percolation is also conducted on a smaller scale in batch tanks or vats and in continuous or dump extractors. In the heap leaching of low-grade oxidized gold ores, for instance, a dilute alkaline solution of sodium cyanide is distributed over a heap of ore that typically has been crushed finer than 1 in and the fines agglomerated with the addition of Portland cement at conveyor transfer points. Heap leaching of very low-grade gold ores and many oxide copper ores is conducted on run-of-mine material. Heap leaching is the least expensive form of leaching. In virtually all cases, an impervious polymeric membrane is installed before the heap is constructed.

In situ leaching, depicted in Fig. 18-84, depends on the existing permeability of a subsurface deposit containing minerals or compounds that are to be dissolved and extracted. Holes ("wells") are drilled into the rock or soil surrounding the deposit and are lined with tubing that is perforated at appropriate depth intervals. The leaching solution is pumped down the injection wells and flows through the deposit or "formation," and the "pregnant" solution is extracted from production wells, treated for solute recovery, reconstituted, and reinjected. In situ leaching is used for extraction of halite (NaCl) and uranium, as well as for the removal of toxic or hazardous constituents from contaminated soil or groundwater.

Batch Percolators The **batch tank** is not unlike a big nutsche filter; it is a large circular or rectangular tank with a false bottom. The solids to be leached are dumped into the tank to a uniform depth. They are sprayed with solvent until their solute content is reduced to an economic minimum and are then excavated. Countercurrent flow of the solvent through a series of tanks is common, with fresh solvent entering the tank containing most nearly exhausted material. So-called *vat leaching* was practiced in oxide copper ore processing prior to 1980, and the vats were typically 53 by 20 by 5.5 m (175 by 67 by 18 ft) and extracted about 8200 Mg (9000 U.S. tons) of ore on a 13-day cycle. Some tanks operate under pressure, to contain volatile solvents or increase the percolation rate. A series of pressure tanks operating with countercurrent solvent flow is called a **diffusion battery.**

Continuous Percolators Coarse solids are also leached by percolation in moving-bed equipment, including single-deck and multideck

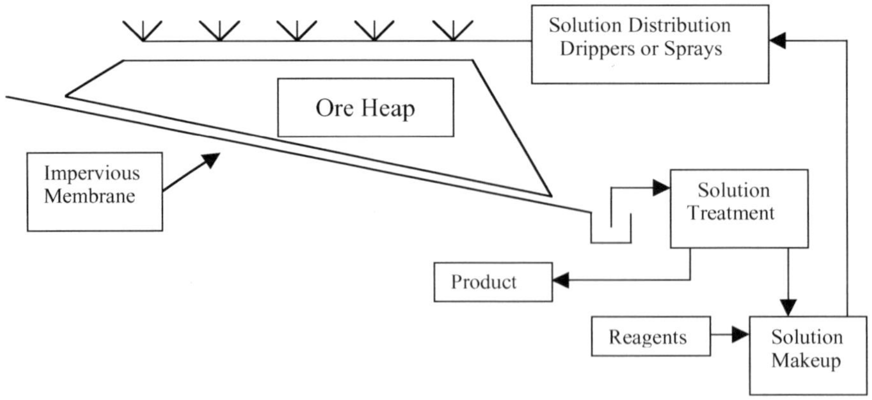

FIG. 18-83 Heap leaching for copper or precious metals.

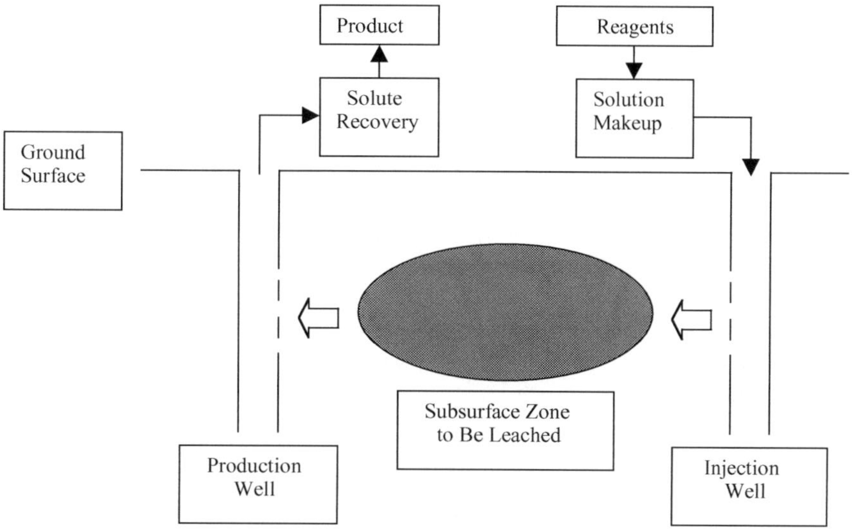

FIG. 18-84 In situ leaching.

rake classifiers, bucket-elevator contactors, and horizontal-belt conveyors.

The Bollman-type extractor shown in Fig. 18-85 is a bucket-elevator unit designed to handle about 2000 to 20,000 kg/h (50 to 500 U.S. tons/day) of flaky solids (e.g., soybeans). Buckets with perforated bottoms are held on an endless moving belt. Dry flakes, fed into the descending buckets, are sprayed with partially enriched solvent ("half miscella") pumped from the bottom of the column of ascending buckets. As the buckets rise on the other side of the unit, the solids are sprayed with a countercurrent stream of pure solvent. Exhausted flakes are dumped from the buckets at the top of the unit into a paddle conveyor; enriched solvent, the "full miscella," is pumped from the bottom of the casing. Because the solids are unagitated and because the final miscella moves cocurrently, the Bollman extractor permits the use of thin flakes while producing extract of good clarity. It is only partially a countercurrent device, however, and it sometimes permits channeling and consequent low stage efficiency. Perhaps for this reason, it is being displaced in the oil extraction industry by horizontal basket, pan, or belt percolators (Schwartzberg, loc. cit.).

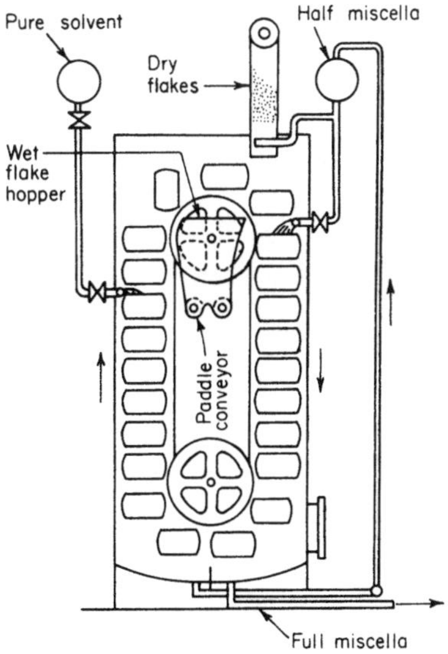

FIG. 18-85 Bollman-type extractor. (*McCabe, Smith, and Harriott,* Unit Operations of Chemical Engineering, *5th ed., p. 616. Copyright 1993 by McGraw-Hill, Inc. and used with permission.*)

In the **horizontal-basket design,** illustrated by the **Rotocel extractor** (Fig. 18-86), walled compartments in the form of annular sectors with liquid-permeable floors revolve about a central axis. The compartments successively pass a feed point, a number of solvent sprays, a drainage section, and a discharge station (where the floor opens to discharge the extracted solids). The discharge station is circumferentially contiguous to the feed point. Countercurrent extraction is achieved by feeding fresh solvent only to the last compartment before dumping occurs and by washing the solids in each preceding compartment with the effluent from the succeeding one. The Rotocel is simple and inexpensive, and it requires little headroom. This type of equipment is made by a number of manufacturers. Horizontal table and tilting-pan vacuum filters, of which it is the gravity counterpart, are used as extractors for leaching processes involving difficult solution-residue separation. Detailed descriptions of the Bollman-type and Rotocel extractors are presented on pp. 765 and 766 of McCabe. et al., 7th ed., loc. cit.

The **endless-belt percolator** (Wakeman, loc. cit.) is similar in principle, but the successive feed, solvent spray, drainage, and dumping stations are linearly rather than circularly disposed. Examples are the **de Smet belt extractor** (uncompartmented) and the **Lurgi frame belt** (compartmented), the latter being a kind of linear equivalent of the Rotocel. Horizontal-belt vacuum filters, which resemble endless-belt extractors, are sometimes used for leaching.

The **Kennedy extractor** (Fig. 18-87), also requiring little headroom, operates substantially as a percolator that moves the bed of solids through the solvent rather than the conventional opposite. It comprises a nearly horizontal line of chambers through each of which in succession the solids being leached are moved by a slow impeller enclosed in that section. There is an opportunity for drainage between stages when the impeller lifts solids above the liquid level before dumping them into the next chamber. Solvent flows countercurrently from chamber to chamber. Because the solids are subjected to mechanical action somewhat more intense than in other types of

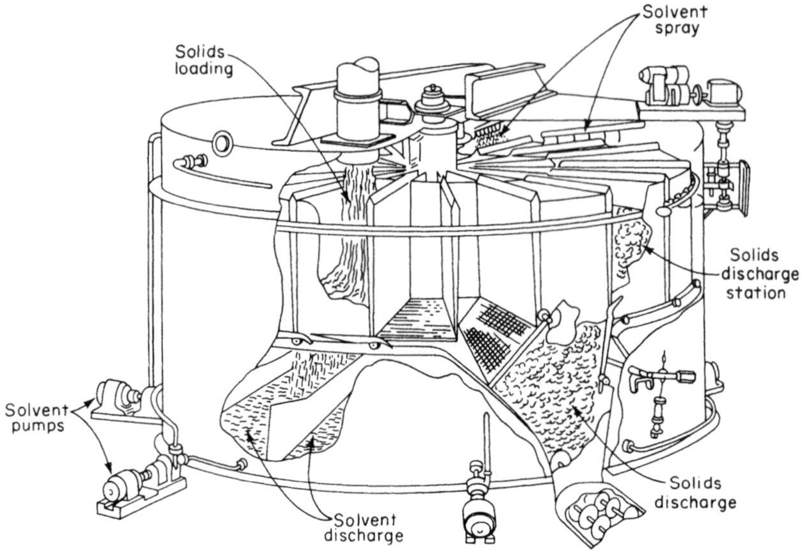

FIG. 18-86 Rotocel extractor. [*Rickles,* Chem. Eng. *72(6): 164 (1965). Used with permission of McGraw-Hill, Inc.*]

continuous percolator, the Kennedy extractor is now little used for fragile materials such as flaked oil seeds.

Dispersed-Solids Leaching Equipment for batch leaching of fine solids in a liquid suspension is now confined mainly to batch tanks with rotating impellers. For a detailed discussion of all aspects of the suspension of solid particles in fluids, refer to agitation of particle suspensions at the beginning of this subsection.

Batch Stirred Tanks Tanks agitated by coaxial impellers (turbines, paddles, or propellers) are commonly used for batch dissolution of solids in liquids and may be used for leaching fine solids. Insofar as the controlling rate in the mass transfer is the rate of transfer of material into or from the interior of the solid particles rather than the rate of transfer to or from the surface of particles, the main function of the agitator is to supply unexhausted solvent to the particles while they reside in the tank long enough for the diffusive process to be completed. The agitator does this most efficiently if it just gently circulates the solids across the tank bottom or barely suspends them above the bottom. However, if the slurry contains particles having significantly different settling velocities, it is usually necessary to introduce sufficient mixing power to ensure full suspension of all particles. Failure to do so will result in an accumulation of the larger or denser particles unless provision is made to drain the settled material continuously.

The leached solids must be separated from the extract by settling and decantation or by external filters, centrifuges, or thickeners, all of which are treated elsewhere in Sec. 18. The difficulty of solids-extract separation and the fact that a batch stirred tank provides only a single equilibrium stage are its major disadvantages.

Pachuca Tanks Air-agitated Pachuca tanks were widely used in mineral processing until the 1960s when the industry concluded that mechanical agitation was more economical and more effective for solids suspension. A description of Pachuca tanks can be found in previous editions of *Perry's Handbook.*

Impeller-agitated ("Stirred") tanks Often called continuous stirred-tank reactors (CSTRs), they can be operated singly or in series. Figure 18-89 illustrates three tanks in series, each with a mechanical agitator. Nearly all stirred tanks are equipped with vertical baffles to prevent swirling and ineffective energy utilization by the agitator. Advancing of slurry from one stage to the next may be by overflow if successive stages are lower, or interstage pumps may be used.

Autoclaves Autoclaves, as shown in Fig. 18-90, are closed, usually multi-compartmented, vessels often designed for operation at pressures in excess of 600 psig (40 bar) and temperatures of 600°F or higher. The purpose of some autoclaves is simply to effect aqueous oxidation, e.g., of organic wastes or sulfide minerals. In the latter case, an example is oxidation of pyrite, followed by cyanide leaching of precious metals under ambient conditions. Other autoclaves are designed to effect leaching, as in the case of sulfuric acid leaching of nickel and cobalt from lateritic ores. The feed stream is preheated by steam from the flash cooling tower(s) and delivered to the autoclave by one or

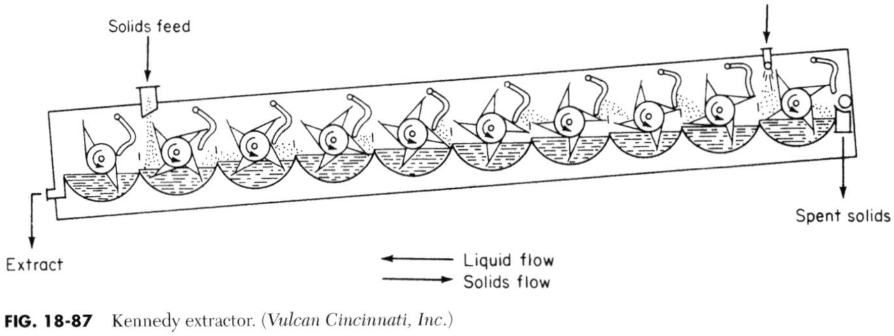

FIG. 18-87 Kennedy extractor. (*Vulcan Cincinnati, Inc.*)

more positive displacement pumps, usually of the piston diaphragm type. If oxidation is required, oxygen may be used instead of air to reduce operating pressure or to improve kinetics. Flash cooling of the autoclave product is usually accomplished in one or more pressure reduction vessels with abrasion-resistant nozzles and targets.

Continuous Dispersed-Solids Leaching

Vertical-plate extractor. Exemplified by the **Bonotto extractor** (Fig. 18-88), this consists of a column divided into cylindrical compartments by equispaced horizontal plates. Each plate has a radial opening staggered 180° from the openings of the plates immediately above and below it, and each is wiped by a rotating radial blade. Alternatively, the plates may be mounted on a coaxial shaft and rotated past stationary blades. The solids, fed to the top plate, thus are caused to fall to each lower plate in succession. The solids fall as a curtain into solvent which flows upward through the tower. They are discharged by a screw conveyor and compactor. Like the Bollman extractor, the Bonotto has been virtually displaced by horizontal belt or tray percolators for the extraction of oil seeds.

Gravity sedimentation tanks. Operated as thickeners, these tanks can serve as continuous contacting and separating devices in which fine solids may be leached continuously. A series of such units properly connected permit true continuous countercurrent washing of fine solids. If appropriate, a mixing tank may be associated with each thickener to improve the contact between the solids and liquid being fed to that stage. Gravity sedimentation thickeners are described under "Gravity Sedimentation Operations." Of all continuous leaching equipment, gravity thickeners require the most area, and they are limited to relatively fine solids.

Impeller-agitated ("stirred") tanks. Often called continuous stirred-tank reactors (CSTR), they can be operated singly or in series. Figure 18-89 illustrates three tanks in series, each with a mechanical agitator. Nearly all stirred tanks are equipped with vertical baffles to prevent swirling and ineffective energy utilization by the agitator. Advancing of slurry from one stage to the next may be by overflow if successive stages are lower, or interstage pumps may be used.

Autoclaves. Autoclaves, as shown in Fig. 18-90, are closed, usually multicompartmented, vessels often designed for operation at pressures in excess of 600 psig (40 bars) and temperatures of 600°F or higher. The purpose of some autoclaves is simply to effect aqueous oxidation, e.g., of organic wastes or sulfide minerals. In the latter case, an example is oxidation of pyrite, followed by cyanide leaching of precious metals under ambient conditions. Other autoclaves are designed to effect leaching, as in the case of sulfuric acid leaching of nickel and cobalt from lateritic ores. The feed stream is preheated by steam from the flash cooling tower(s) and delivered to the autoclave by one or more

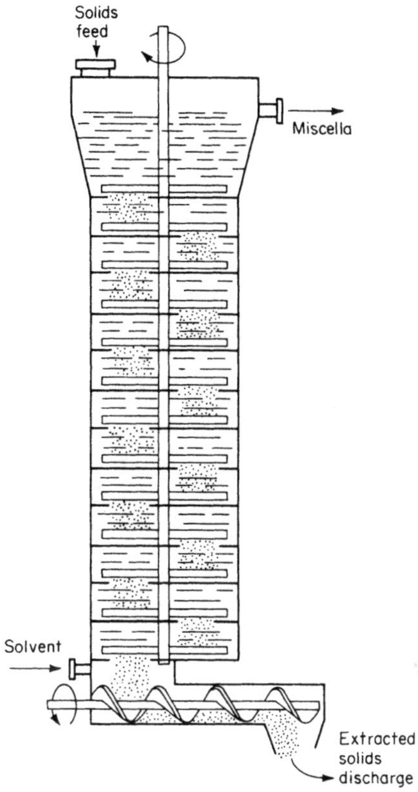

FIG. 18-88 Bonotto extractor. [*Rickles,* Chem. Eng. **72***(6): 163 (1965); copyright 1965 by McGraw-Hill, Inc., New York. Excerpted with special permission of McGraw-Hill.*]

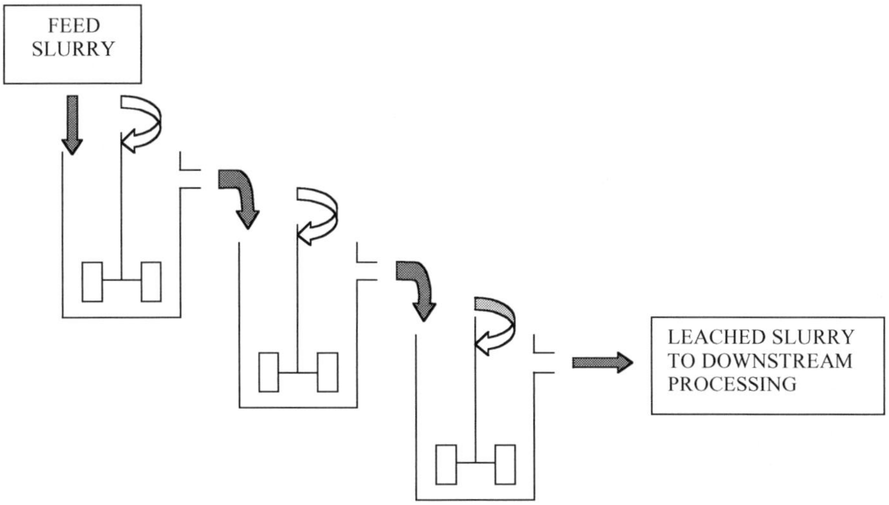

FIG. 18-89 Stirred tanks, three in series, with gravity overflow.

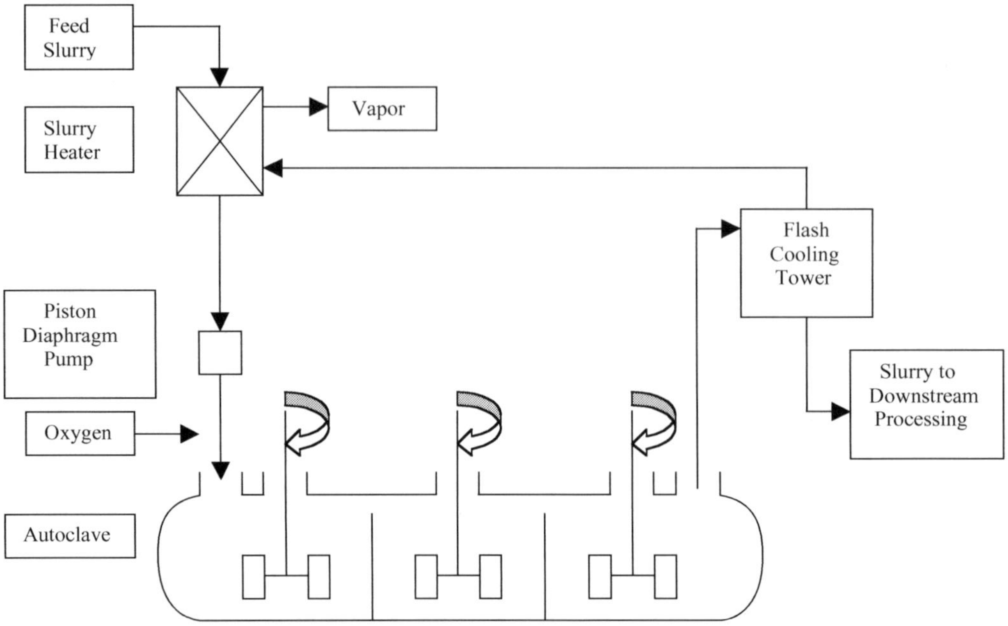

FIG. 18-90 Three-compartment autoclave.

positive displacement pumps, usually of the piston diaphragm type. If oxidation is required, oxygen may be used instead of air to reduce operating pressure or to improve kinetics. Flash cooling of the autoclave product is usually accomplished in one or more pressure-reduction vessels with abrasion-resistant nozzles and targets.

Screw-Conveyor Extractors One type of continuous leaching equipment, employing the screw-conveyor principle, is strictly speaking neither a percolator nor a dispersed-solids extractor. Although it is often classed with percolators, there can be sufficient agitation of the solids during their conveyance by the screw that the action differs from an orthodox percolation.

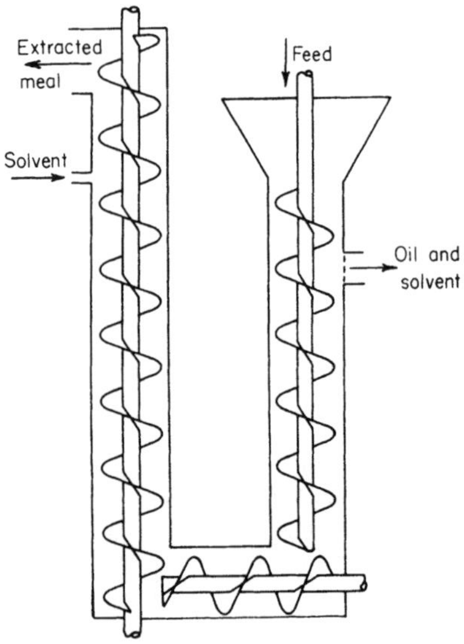

FIG. 18-91 Hildebrandt extractor. (*McCabe, Smith, and Harriott,* Unit Operations of Chemical Engineering, *5th ed., p. 616. Copyright 1993 by McGraw-Hill, Inc. and used with permission.*)

The **Hildebrandt total-immersion extractor** is shown schematically in Fig. 18-91. The helix surface is perforated so that solvent can pass through countercurrently. The screws are so designed to compact the solids during their passage through the unit. The design offers the obvious advantages of countercurrent action and continuous solids compaction, but there are possibilities of some solvent loss and feed overflow, and successful operation is limited to light, permeable solids.

A somewhat similar but simpler design uses a horizontal screw section for leaching and a second screw in an inclined section for washing, draining, and discharging the extracted solids.

In the **De Danske Sukkerfabriker,** the axis of the extractor is tilted to about 10° from the horizontal, eliminating the necessity of two screws at different angles of inclination.

Sugar-beet cossettes are successfully extracted while being transported upward in a vertical tower by an arrangement of inclined plates or wings attached to an axial shaft. The action is assisted by staggered guide plates on the tower wall. The shell is filled with water that passes downward as the beets travel upward. This configuration is employed in the **BMA diffusion tower** (Wakeman, loc. cit.).

Schwartzberg (loc. cit.) reports that screw-conveyor extractors, once widely employed to extract flaked oil seeds, have fallen into disuse for this application because of their destructive action on the fragile seed flakes.

Tray Classifier A hybrid like the screw-conveyor classifier, the tray classifier rakes pulp up the sloping bottom of a tank while solvent flows in the opposite direction. The solvent is forced by a baffle to the bottom of the tank at the lower end before it overflows. The solids must be rugged enough to stand the stress of raking.

SELECTION OR DESIGN OF A LEACHING PROCESS*

At the heart of a leaching plant design at any level—conceptual, preliminary, firm engineering, or whatever—is unit-operations and process design of the extraction unit or line. The major aspects that are particular for the leaching operation are the selection of process

*Portions of this subsection are adaptations from the still-pertinent article by Rickles (loc. cit.).

and operating conditions and the sizing of the extraction equipment.

Process and Operating Conditions The major parameters that must be fixed or identified are the solvent to be used, the temperature, the terminal stream compositions and quantities, leaching cycle (batch or continuous), contact method, and specific extractor choice.

Choice of Solvent The solvent selected will offer the best balance of a number of desirable characteristics: high saturation limit and selectivity for the solute to be extracted, capability to produce extracted material of quality unimpaired by the solvent, chemical stability under process conditions, low viscosity, low vapor pressure, low toxicity and flammability, low density, low surface tension, ease and economy of recovery from the extract stream, and price. These factors are listed in an approximate order of decreasing importance, but the specifics of each application determine their interaction and relative significance, and any one can control the decision under the right combination of process conditions.

Temperature The temperature of the extraction should be chosen for the best balance of solubility, solvent-vapor pressure, solute diffusivity, solvent selectivity, and sensitivity of product. In some cases, temperature sensitivity of materials of construction to corrosion or erosion attack may be significant.

Terminal Stream Compositions and Quantities These are basically linked to an arbitrary given: the production capacity of the leaching plant (rate of extract production or rate of raw-material purification by extraction). When options are permitted, the degree of solute removal and the concentration of the extract stream chosen are those that maximize process economy while sustaining conformance to regulatory standards.

Leaching Cycle and Contact Method As is true generally, the choice between continuous and intermittent operation is largely a matter of the size and nature of the process of which the extraction is a part. The choice of a percolation or solids-dispersion technique depends principally on the amenability of the extraction to effective, sufficiently rapid percolation.

Type of Reactor The specific type of reactor that is most compatible (or least incompatible) with the chosen combination of the preceding parameters seldom is clearly and unequivocally perceived without difficulty, if at all. In the end, however, that remains the objective. As is always true, the ultimate criteria are reliability and profitability.

Extractor-Sizing Calculations For any given throughput rate (which fixes the cross-sectional area and/or the number of extractors), the size of the units boils down to the number of stages required, actual or equivalent. In calculation, this resolves into determination of the number of ideal stages required and application of appropriate stage efficiencies. The methods of calculation resemble those for other mass-transfer operations (see Secs. 13, 14, and 15), involving equilibrium data and contact conditions, and based on material balances. They are discussed briefly here with reference to countercurrent contacting.

Software Packages Since the late 1990s, increasing use has been made of software developed for modeling and simulation of all types of unit operations, including leaching. Packages currently available for mineral processing applications can be found, for instance, in Mular et al., loc. cit., pp. 479 and 495. Monthly issues of *Chemical Engineering Progress* (*CEP*) usually contain a page entitled *Software* that announces new packages for various applications, and the same publication usually contains a summary each year of all packages of use to chemical engineers and mineral processors.

Composition Diagrams In its elemental form, a leaching system consists of three components: inert, insoluble solids; a single nonadsorbed solute, which may be liquid or solid; and a single solvent.* Thus, it is a ternary system, albeit an unusual one, as already mentioned, by virtue of the total mutual "insolubility" of two of the phases and the simple nature of equilibrium.

The composition of a typical system is satisfactorily presented in the form of a diagram. Those diagrams most frequently employed are a right-triangular plot of mass fraction of solvent against mass fraction of solute (Fig. 18-92*a*) and a plot suggestive of a Ponchon-Savarit diagram, with inerts taking the place of enthalpy (Fig. 18-92*b*). A third diagram, less frequently used, is a modified McCabe-Thiele plot in which the overflow solution (inerts-free) and the underflow solution (traveling out of a stage with the inerts) are treated as pseudo phases, the mass fraction of solute in overflow, y, being plotted against the mass fraction of solute in underflow, x. (An additional representation, the equilateral-triangular diagram frequently employed for liquid-liquid ternary systems, is seldom used because the field of leaching data is confined to a small portion of the triangle.)

*The solubility of the inert, adsorption of solute on the inert, and complexity of solvent and extracted material can be taken into account if necessary. Their consideration is beyond the scope of this treatment.

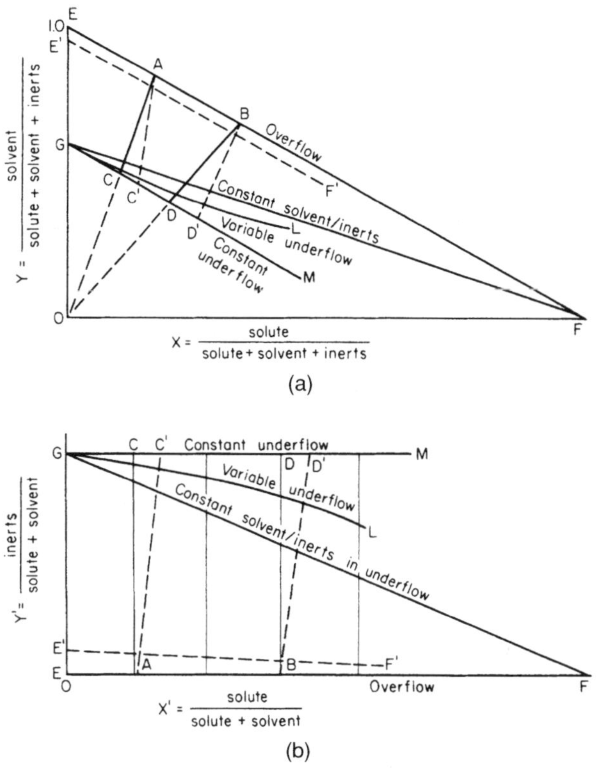

FIG. 18-92 Composition diagrams for leaching calculations: (*a*) right-triangular diagram; (*b*) modified Ponchon-Savarit diagram.

With reference to Fig. 18-92 (both graphs), *EF* represents the locus of overflow compositions for the case in which the overflow stream contains no inert solids. *E′F′* represents the overflow streams containing some inert solids, either by entrainment or by partial solubility in the overflow solution. Lines *GF, GL,* and *GM* represent the loci of underflow compositions for the three different conditions indicated on the diagram. In Fig. 18-92*a*, the constant underflow line *GM* is parallel to *EF*, the hypotenuse of the triangle, whereas *GF* passes through the right-hand vertex representing 100 percent solute. In Fig. 18-92*b*, underflow line *GM* is parallel to the abscissa, and *GF* passes through the point on the abscissa representing the composition of the clear solution adhering to the inert solids.

Compositions of overflow and underflow streams leaving the same stage are represented by the intersection of the composition lines for those streams with a tie line (*AC, AC′, BD, BD′*). Equilibrium tie lines (*AC, BD*) pass through the origin (representing 100 percent inerts) in Fig. 18-92*a*, and are vertical (representing the same inert-free solution composition in both streams) in Fig. 18-92*b*. For nonequilibrium conditions with or without adsorption or for equilibrium conditions with selective adsorption, the tie lines are displaced, such as *AC′* and *BD′*. Point *C′* is to the right of *C* if the solute concentration in the overflow solution is less than that in the underflow solution adhering to the solids. Unequal concentrations in the two solutions indicate insufficient contact time and/or preferential adsorption of one of the components on the inert solids. Tie lines such as *AC′* may be considered as "practical tie lines" (i.e., they represent actual rather than ideal stages) if data on underflow and overflow composition have been obtained experimentally under conditions simulating actual operation, particularly with respect to contact time, agitation, and particle size of solids.

The illustrative construction lines of Fig. 18-92 have been made with the assumption of constant underflow. In the more realistic case of variable underflow, the points C, C', D, D' would lie along line GL. Like the practical tie lines, GL is a representation of experimental data.

Algebraic Computation This method starts with calculation of the quantities and compositions of all the terminal streams, using a convenient quantity of one of the streams as the basis of calculation. Material balance and stream compositions are then computed for a terminal ideal stage at either end of an extraction battery (i.e., at point A or point B in Fig. 18-92), using equilibrium and solution-retention data. Calculations are repeated for each successive ideal stage from one end of the system to the other until an ideal stage which corresponds to the desired conditions is obtained. Any solid-liquid extraction problem can be solved by this method.

For certain simplified cases it is possible to calculate directly the number of stages required to attain a desired product composition for a given set of feed conditions. For example, if equilibrium is attained in all stages and if the underflow mass rate is constant, both the equilibrium and operating lines on a modified McCabe-Thiele diagram are straight, and it is possible to calculate directly the number of ideal stages required to accommodate any rational set of terminal flows and compositions (McCabe, Smith, and Harriott, op. cit.):

$$N = \frac{\log [(y_b - x_b)/(y_a - x_a)]}{\log [(y_b - y_a)/(x_b - x_a)]} \tag{18-45}$$

Even when the conditions of equilibrium in each stage and constant underflow obtain, Eq. (18-45) normally is not valid for the first stage because the unextracted solids entering that stage usually are not premixed with solution to produce the underflow mass that will leave. This is easily rectified by calculating the exit streams for the first stage and using those values in Eq. (18-45) to calculate the number of stages required after stage 1.

Graphical Method This method of calculation is simply a diagrammatic representation of all the possible compositions in a leaching system, including equilibrium values, on which material balances across ideal (or, in some cases, nonideal) stages can be evaluated in the graphical equivalent of the stage-by-stage algebraic computation. It normally is simpler than the hand calculation of the algebraic solution, and it is viewed by many as helpful because it permits visualization of the process variables and their effect on the operation. Any of the four types of composition diagrams described above can be used, but modified Ponchon-Savarit or right-triangular plots (Fig. 18-92) are most convenient for leaching calculations.

The techniques of graphical solution, in fact, are not unlike those for distillation and absorption (binary) problems using McCabe-Thiele, Ponchon-Savarit, and right-triangular diagrams and are similar to those described in Sec. 15 for solvent-extraction (ternary) systems. More detailed explanations of the application of the several graphical conventions to leaching are presented by: Coulson and Richardson, right triangle; Rickles, modified Ponchon-Savarit; McCabe, Smith, and Harriott, modified McCabe-Thiele; and Schwartzberg, equilateral ternary diagram; all in the publications cited as general references. (See also Treybal, *Mass Transfer Operations,* 3d ed., McGraw-Hill, New York, 1980.)

GRAVITY SEDIMENTATION OPERATIONS

GENERAL REFERENCES: Albertson, *Fluid/Particle Sep. J.,* **7,** IS (1994). Jewell, Fourie, and Lord, *Paste and Thickened Tailings—A Guide,* pp. 49–79, Australian Centre for Geomechanics, 2002. Mular, Halbe, and Barratt, *Mineral Processing Plant Design, Practice, and Control,* vol. 2, pp. 1295–1312 and 2164–2173, SME, 2002. Sankey and Payne, *Chemical Reagents in the Mineral Processing Industry,* p. 245, SME, 1985. Schweitzer, *Handbook of Separation Techniques for Chemical Engineers,* 2d ed., pp. 4-121 to 4-147, McGraw-Hill, 1988. Wilhelm and Naide, *Min. Eng.* (Littleton, Colo.), 1710 (1981).

Sedimentation is the partial separation or concentration of suspended solid particles from a liquid by gravity settling. This field may be divided into the functional operations of thickening and clarification. The primary purpose of thickening is to increase the concentration of suspended solids in a feed stream, while that of clarification is to remove a relatively small quantity of suspended particles and produce a clear effluent. These two functions are similar and occur simultaneously, and the terminology merely makes a distinction between the primary process results desired. Generally, thickener mechanisms are designed for the heavier-duty requirements imposed by a large quantity of relatively concentrated pulp, while clarifiers usually will include features that ensure essentially complete suspended-solids removal, such as greater depth, special provision for coagulation or flocculation of the feed suspension, and greater overflow-weir length.

CLASSIFICATION OF SETTLEABLE SOLIDS AND THE NATURE OF SEDIMENTATION

The types of sedimentation encountered in process technology will be greatly affected not only by the obvious factors—particle size, liquid viscosity, solid and solution densities—but also by the characteristics of the particles within the slurry. These properties, as well as the process requirements, will help determine both the type of equipment which will achieve the desired ends most effectively and the testing methods to be used to select the equipment.

Figure 18-93 illustrates the relationship between solids concentration, interparticle cohesiveness, and the type of sedimentation that may exist. "Totally discrete" particles include many mineral particles (usually greater in diameter than 20 μm), salt crystals, and similar substances that have little tendency to cohere. "Flocculent" particles generally will include those smaller than 20 μm (unless present in a dispersed state owing to surface charges), metal hydroxides, many chemical precipitates, and most organic substances other than true colloids.

At low concentrations, the type of sedimentation encountered is called

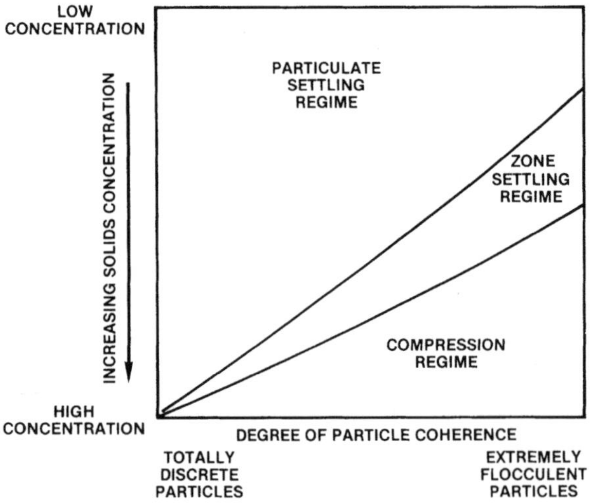

FIG. 18-93 Combined effect of particle coherence and solids concentration on the settling characteristics of a suspension.

particulate settling. Regardless of their nature, particles are sufficiently far apart to settle freely. Faster-settling particles may collide with slower-settling ones and, if they do not cohere, continue downward at their own specific rate. Those that do cohere will form floccules of a larger diameter that will settle at a rate greater than that of the individual particles.

There is a gradual transition from particulate settling into the zone-settling regime, where the particles are constrained to settle as a mass. The principal characteristic of this zone is that the settling rate of the mass, as observed in batch tests, will be a function of its solids concentration (for any particular condition of flocculation, particle density, etc.).

The solids concentration ultimately will reach a level at which particle descent is restrained not only by hydrodynamic forces but also partially by mechanical support from the particles below; therefore, the weight of particles in mutual contact can influence the rate of sedimentation of those at lower levels. This compression, as it is termed, will result in further solids concentration because of compaction of the individual floccules and partial filling of the interfloc voids by the deformed floccules. Accordingly, the rate of sedimentation in the compression regime is a function of both the solids concentration and the depth of pulp in this particular zone. As indicated in Fig. 18-93, granular, nonflocculent particles may reach their ultimate solids concentration without passing through this regime.

As an illustration, coarse-size (45 μm) the aluminum oxide trihydrate particles produced in the Bayer process would be located near the extreme left of Fig. 18-93. These solids settle in a particulate manner, passing through a zone-settling regime only briefly, and reach a terminal density or ultimate solids concentration without any significant compressive effects. At this point, the solids concentration may be as much as 80 percent by weight. The same compound, but of the gelatinous nature it has when precipitated in water treatment as aluminum hydroxide, would be on the extreme right-hand side of the figure. This flocculent material enters into a zone-settling regime at a low concentration (relative to the ultimate concentration it can reach) and gradually thickens. With sufficient pulp depth present, preferably aided by gentle stirring or vibration, the compression-zone effect will occur; this is essential for the sludge to attain its maximum solids concentration, around 10 percent. Certain fine-size (1- to 2-μm) precipitates of this compound will possess characteristics intermediate between the two extremes.

A feed stream to be clarified or thickened can exist at any state represented within this diagram. As it becomes concentrated owing to sedimentation, it may pass through all the regimes, and the settling rate in any one may be the size-determining factor for the required equipment.

Sedimentation Testing To design and size sedimentation equipment, reference information from similar applications is preferred. Data from full-scale sedimentation equipment, operating in the application under consideration, are always a first choice for sizing new equipment. However, quite often the application under question deviates sufficiently from reference installations. The characteristics of the feed stream for the new application (i.e., solids characteristics, particle size, viscosities, pH, use of flocculants, etc.) must be identical to the existing application. It is also necessary to know how close to "capacity" the existing equipment is operating. If the feed characteristics and operating conditions are different for the application under question, bench- or pilot-scale testing is recommended to size and design a new sedimentation unit.

To properly design and size sedimentation equipment, several pieces of information are required. Some information is unique to the job site (application, feed rate, etc.), while other data are supplied from similar references or from test work. Site-specific information from the plant site includes

- Application – objectives (underflow, TSS, hardness, etc.)
- Feed rate—design and maximum
- Feed characteristics—solids concentration, chemistry
- Site-specific requirements: seismic zone, weather-related specifications, local mechanical design codes, and the user's preferred design specifications
- Local operating practices

In the event testing is required to design either a thickener or a clarifier, the testing must be structured to produce all or some of the following information:

- Feed stream characteristics
- Chemical treatment (type, solution concentration, dose, etc.): coagulants and flocculants (organic or inorganic); acid/base for treatment and pH correction
- Coagulation and flocculation (mixing time, energy requirements, solids concentration)
- Expected sedimentation objectives: underflow slurry density or concentration; overflow solids concentration (suspended solids and/or turbidity); chemical treatment for soluble components (i.e., hardness, metals, anions, pH, etc.)
- Vessel area and depth
- Settled solids rheology (for raking mechanism design and drive torque specification)

There are three basic approaches to testing for sedimentation equipment:

- **Batch bench-scale settling tests** The most common procedure requires a relatively small amount of sample tested in a controlled environment using laboratory equipment under static conditions.
- **Semicontinuous bench-scale tests** Laboratory pumps are used which pump feed slurry and chemicals into settling cylinders from which overflow liquor and underflow slurry are continuously collected.
- **Continuous piloting** A small-diameter thickener or clarifier of the same design as the full-scale equipment being considered is used.

TESTING COMMON TO CLARIFIERS AND THICKENERS

Feed Characterization Sample characterization is necessary for both thickening and clarification testing. Without these data included in the basis of design, the sizing and predicted performance cannot be validated for the specified feed stream. Characterization requires the following measurements as a minimum:

- General chemical makeup of the solids and liquor phases
- Feed solids concentration
- Particle size distribution—include coarse (+100 μm) and fine (−20 μm) particle diameters
- Particle specific gravity
- Liquid specific gravity
- Liquid-phase dissolved materials concentration
- Temperature
- pH

Coagulant and/or Flocculant Selection Coagulants and flocculants are widely used to enhance the settling rate which reduces thickener and clarifier size and improves overflow clarity and/or underflow slurry density. The terms *coagulation* and *flocculation* are sometimes used interchangeably; however, each term describes separate functions in the particle agglomeration process.

Coagulation is a preconditioning step that may be required to destabilize the solids suspension to allow complete flocculation to occur in clarification applications. Flocculation is the bridging and binding of destabilized solids into larger particles. As particle size increases, settling rate generally increases. The science of flocculation is not discussed here but can be found in numerous texts and literature which are readily available from flocculant vendors.

Both coagulation and flocculation are typically considered in designing clarifiers, whereas flocculation is normally the only step in designing thickeners.

Coagulants may be either organic such as polyelectrolytes or inorganic such as alum. Coagulants can be used alone or in conjunction with flocculants to improve the performance of the flocculant or reduce the quantity of the flocculant required. In some systems, where a flocculant has been used in an upstream process, a coagulant may be needed to allow additional flocculant to be effective.

There are two primary types of flocculants:

- *Natural flocculants* Starch, guar, and other natural materials have historically been used for sedimentation flocculation, but have been replaced by more effective synthetic polymers.
- *Synthetic polymeric flocculants* There are hundreds of synthetic polymers available developed for specific applications.

Because of the many available flocculants, a screening program is necessary to choose an effective flocculant. The choice of flocculant can be narrowed by considering the following:

- Prior experience with flocculants on the feed stream under evaluation is always a good source of data.
- Test one each of the major types of flocculant charge: anionic, nonionic, and cationic.
- Test one each of the synthetic polymer length: long chain, short chain.

The purpose of the screening test is to select a coagulant or flocculant whose generic type will most likely be effective in plant operation, and therefore, suitable for clarifier or thickener testing. Although a thickener or clarifier may be started up on the flocculant selected in the testing, it is very common to conduct further tests on the full-scale machine to further optimize dosage or flocculant type. The flocculant manufacturer can be a source of great assistance in both the testing and the full-scale optimization of flocculant use.

Coagulant or flocculant solutions should be made up according to the manufacturer's instructions and used within the shelf life recommended. The solution concentration recommended for testing is typically more dilute than the "neat" concentration so that the viscosity is lower to make dispersion more rapid during testing.

In the screen tests, each coagulant or flocculant is added to the beaker samples of representative slurry or liquor in a dropwise fashion, while the sample is mixed with a spatula, stirrer, or 3-6 jar stirrer mechanism. The amount of coagulant or flocculant required to initiate floc particle formation is noted along with relevant notes as to the size of the floc, capture of fines, resultant liquor clarity, and stability of the floc structure. The dosage is typically noted in g/t solids if the sample is primarily solids (thickener design), or in mg/L liquor if the sample is primarily for clarification and the solids concentration is low.

TESTING SPECIFIC TO CLARIFICATION

Detention Test This test utilizes a 1- to 4-L beaker or similar vessel. The sample is placed in the container, flocculated by suitable means if required, and allowed to settle. Small samples for suspended-solids analysis are withdrawn from a point approximately midway between liquid surface and settled solids interface, taken with sufficient care that settled solids are not resuspended. Sampling times may be at consecutively longer intervals, such as 5, 10, 20, 40, and 80 min.

The suspended-solids concentration can be plotted on log-log paper as a function of the sampling (detention) time. A straight line usually will result, and the required static detention time t to achieve a certain suspended-solids concentration C in the overflow of an ideal basin can be taken directly from the graph. If the plot is a straight line, the data are described by the equation

$$C = Kt^m \tag{18-46}$$

where the coefficient K and exponent m are characteristic of the particular suspension.

Should the suspension contain a fraction of solids which can be considered "unsettleable," the data are more easily represented by using the so-called second-order procedure. This depends on the data being reasonably represented by the equation

$$Kt = \frac{1}{C - C_\infty} - \frac{1}{C_0 - C_\infty} \tag{18-47}$$

where C_∞ is the unsettleable-solids concentration and $C0$ is the concentration of suspended solids in the unsettled (feed) sample. The residual-solids concentration remaining in suspension after a sufficiently long detention time (C_∞) must be determined first, and the data then plotted on linear paper as the reciprocal concentration function $1/(C - C_\infty)$ versus time.

Bulk Settling Test After the detention test is completed, a bulk settling test is done to determine the maximum overflow rate. This is done by carrying out a settling test in which the solids are first concentrated to a level at which zone settling just begins. This is usually marked by a very diffuse interface during initial settling. Its rate of descent is measured with a graduated cylinder of suitable size, preferably at least 1 L, and the initial straight-line portion of the settling curve is used for specifying a bulk-settling rate. The design overflow rate generally should not exceed half of the bulk settling rate. From the two clarifier tests, detention time and bulk settling rate, the more conservative results will govern the size of the clarifier.

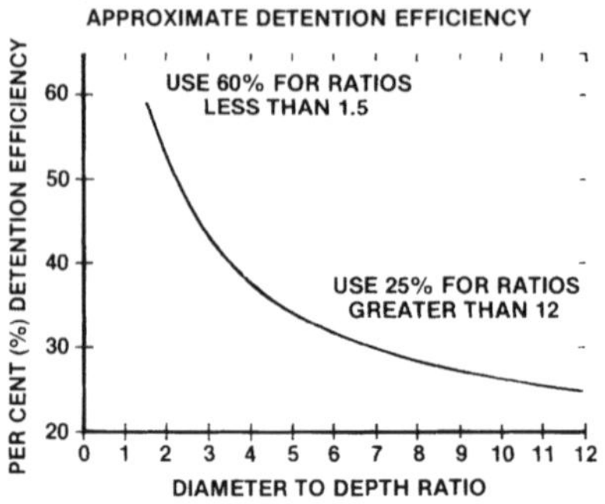

FIG. 18-94 Efficiency curve for scale-up of batch clarification data to determine nominal detention time in a continuous clarifier.

Clarification with Solids Recycle In many instances, the rate of clarification is enhanced by increasing the solids concentration in the flocculation zone of the clarifier. This is done in a full-scale operation by internally or externally recycling previously settled solids into the flocculation zone where they are mixed with fresh, coagulated feed. The higher population of solids improves the flocculation efficiency and clarification rate.

To conduct these tests, a sample of feed is first treated at the chemical dosages and mixing intensity determined in the screening tests and flocculated according to the screening test. The solids are allowed to settle, and the supernatant is carefully decanted. The settled solids are then transferred to a new fresh sample, and tests are conducted again, using the same chemical dosages and mixing intensity. Recycle can continue with subsequent tests until the suspended solids in the sample can have concentrations of 1, 2, 3, and 5 g/L. Bulk settling rate, suspended solids, and other effluent parameters are measured with each test until an optimal treatment scenario is found.

In some suspensions, very fine colloidal solids are present and are very difficult to coagulate. In these cases, it is typically necessary to adjust for coagulation mixing intensity and time to obtain coagulated solids that are more amenable to flocculation.

Detention Efficiency Conversion from the ideal basin sized by detention-time procedures to an actual clarifier requires the inclusion of an efficiency factor to account for the effects of turbulence and nonuniform flow. Efficiencies vary greatly, being dependent not only on the relative dimensions of the clarifier and the means of feeding but also on the characteristics of the particles. The curve shown in Fig. 18-94 can be used to scale up laboratory data in sizing circular clarifiers. The static detention time determined from a test to produce a specific effluent solids concentration is divided by the efficiency (expressed as a fraction) to determine the nominal detention time, which represents the volume of the clarifier above the settled pulp interface divided by the overflow rate. Different diameter-depth combinations are considered by using the corresponding efficiency factor. In most cases, area may be determined by factors other than the bulk-settling rate, such as practical tank-depth limitations.

TESTING SPECIFIC TO THICKENING

Optimization of Flocculation Conditions After a flocculant type is selected, the next step is to conduct a range of tests using the selected flocculant, to gather data on the effects of feed slurry solids concentrations on flocculant dosage and settling rate. There are a range of solids concentrations for which flocculation effectiveness is maximized, resulting in improved settling characteristics. Operating within this feed solids range results in smaller equipment sizes, higher underflow slurry densities, better overflow liquor clarity, and lower flocculant dosages.

The tests are conducted using a series of samples prepared at solids concentrations decreasing incrementally in concentration from the expected thickener feed concentration. Typically, the samples are prepared in 250- to 500-mL graduated cylinders which give some distance

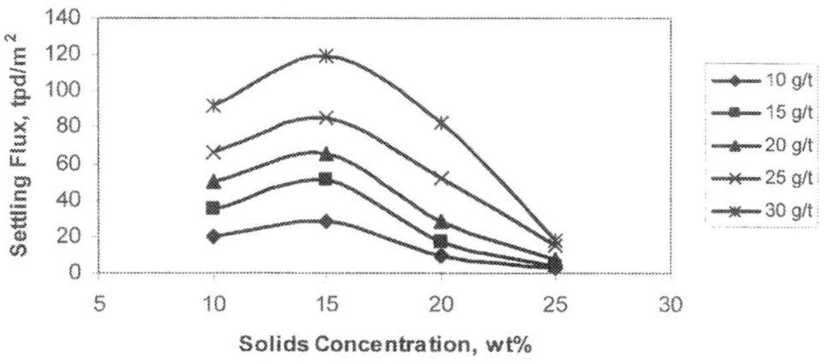

FIG. 18-95 Data showing that slurry solids concentration affects flocculation efficiency, thus improving solids settling flux.

to measure the settling rate more accurately. For some very fine solids samples (e.g., alumina red mud, clays, leached nickel laterites, etc), it is recommended to also check a sample diluted to 2 to 3 wt % solids. Begin adding the flocculant solution dropwise; make notes on the dosage at which flocculation begins and the settling velocity. Continue adding flocculant incrementally and noting the floc structure, fines capture, liquor clarity, and settling velocity. Once the settling velocity remains constant for a few tests, sample testing can be stopped. From the tests, the plot shown in Fig. 18-95 can be drawn and the results used to set conditions for the larger and final tests for sizing the thickening equipment. The test procedure for the design tests should be structured to span the optimum solids concentration and two points slightly higher and lower. The flocculant dosage should be checked at the optimum and at dosages slightly higher and slightly lower than that determined in the above tests.

Determination of Thickener Basin Area The area requirements for thickeners frequently are based on the solids flux rates measured in the zone-settling regime. Theory holds that, for any specific sedimentation condition, a critical concentration will exist in the thickener which will limit the solids throughput rate. As the concentration in this critical zone represents a steady-state condition, its depth in the settling bed of solids may vary, responding to changes in the feed rate, underflow withdrawal rate, or flocculant dosage. In thickeners operating at relatively high solids retention times and/or low throughput rates, this zone generally does not exist.

Many batch-test methods which are based on determining the solids flux rate at this critical concentration have been developed. Most methods recognize that as the solids enter compression, thickening behavior is no longer a function only of solids concentration. Hence, these methods attempt to utilize the "critical" point dividing these two zones and size the area on the basis of the settling rate of a layer of pulp at this concentration. The difficulty lies in discerning where this point is located on the settling curve.

Many procedures have been developed, but two have been more widely used: the Coe and Clevenger approach and the Kynch method as defined by Talmage and Fitch (op. cit.).

The former requires measurement of the initial settling rate of a pulp at different solids concentrations varying from feed to final underflow value. The area requirement for each solids concentration tested is calculated by equating the net overflow rate to the corresponding interfacial settling rate, as represented by the following equation for the unit area:

$$\text{Unit area} = \frac{1/C_i - 1/C_u}{v_i} \tag{18-48}$$

where C_i is the solids concentration at the interfacial settling velocity v_i and C_u is the underflow concentration, both concentrations being expressed in terms of mass of solids per unit volume of slurry. Using kg/L for the concentrations and m/day for the settling velocity yields a unit area value in m^2/(ton/day).

These unit area values, plotted as a function of the feed concentration, will describe a maximum value that can be used to specify the thickener design unit area for the particular underflow concentration C_u employed in Eq. (18-48).

The method is applicable for *unflocculated* pulps or those in which the ionic characteristics of the solution produce a flocculent structure. If polymeric flocculants are used, an approach based on the Kynch theory is preferred. In this method, the test is carried out at the optimum feed solids and flocculant dose (as determined in tests described earlier) and continued until underflow concentration is achieved in the cylinder. The flocculant solution should be added to the slurry under conditions which promote rapid dispersion and uniform, complete mixing with a minimum of shear. In cylinder tests, this is accomplished by simultaneously injecting and mixing flocculant with the slurry, using an apparatus consisting of a syringe, a tube, and an inverted rubber stopper. The rubber stopper, having a diameter approximately 75 percent that of the cylinder diameter, provides sufficient turbulence as it is moved gently up and down through the slurry to cause good blending of the flocculant and pulp. To determine the unit area, Talmage and Fitch (op. cit.) proposed an equation derived from a relationship equivalent to that shown in Eq. (18-49):

$$\text{Unit area} = \frac{t_u}{C_0 H_0} \tag{18-49}$$

where t_u is the time, days; C_0 is the initial solids concentration in the feed, t/m^3; and H_0 is the initial height of the slurry in the test cylinder, m. The term t_u is taken from the intersection of a tangent to the curve at the critical point and a horizontal line representing the depth of pulp at underflow concentration. There are various means for selecting this critical point, all of them empirical, and the unit area value determined cannot be considered precise. The review by Pearse (op. cit.) presents many of the different procedures used in applying this approach to laboratory settling test data.

Two other approaches avoid using the critical point by computing the area requirements from the settling conditions existing at the underflow concentration. The Wilhelm and Naide procedure (op. cit.) applies zone-settling theory (Kynch) to the entire thickening regime. Tangents drawn to the settling curve are used to calculate the settling velocity at all concentrations obtained in the test. This permits construction of a plot (Fig. 18-96) showing unit area as a function of underflow concentration.

A second, "direct" approach which yields a similar result, since it also takes compression into account, utilizes the value of settling time t_x taken from the settling curve at a particular underflow concentration. This value is used to solve the Talmage and Fitch equation (18-49) for unit area.

Compression bed depth will have a significant effect on the overall settling rate (increasing compression zone depth reduces unit area). Therefore, in applying either of these two procedures it is necessary to run the test in a vessel having an average bed depth close to that expected in a full-scale thickener. This requires a very large sample, and it is more convenient to carry out the test in a cylinder having a volume of 1 to 4 liters. The calculated unit area value from this test can be extrapolated to full-scale depth by carrying out similar tests at

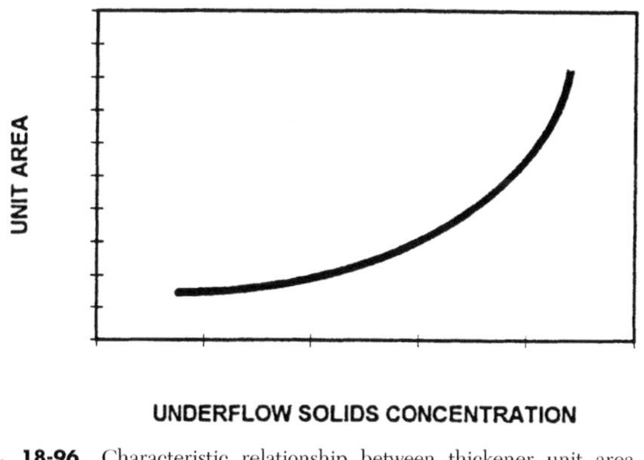

FIG. 18-96 Characteristic relationship between thickener unit area and underflow solids concentration (fixed flocculant dosage and pulp depth).

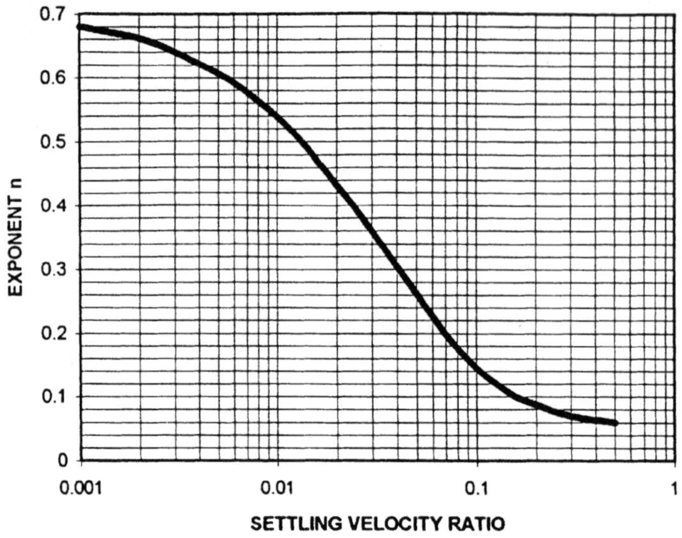

FIG. 18-97 Depth correction factor to be applied to unit areas determined by Wilhelm-Naide and "direct" methods. Velocity ratio calculated using tangents to settling curve at a particular settled solids concentration and at start of test.

different depths to determine the effect on unit area. Alternatively, an empirical relationship can be used which is effective in applying a depth correction to laboratory cylinder data over normal operating ranges. The unit area calculated by either the Wilhelm and Naide approach or the direct method is multiplied by a factor equal to $(h/H)^n$, where h is the average depth of the pulp in the cylinder, H is the expected full-scale compression zone depth, usually taken as 1 m, and n is the exponent calculated from Fig. 18-97. For conservative design purposes, the minimum value of this factor that should be used is 0.25.

It is essential to use a slow-speed (approximately 0.1 r/min) picket rake in all cylinder tests to prevent particle bridging and allow the sample to attain the underflow density which is obtainable in a full-scale thickener.

Continuously operated, small-scale or pilot-plant thickeners, ranging from 75 mm diameter by 400 mm depth to several meters in diameter, are also effectively used for sizing full-scale equipment. This approach requires a significantly greater volume of sample, such as would be available in an operating installation or a pilot plant. Continuous units and batch cylinders will produce equivalent results if proper procedures are followed with either system.

Thickener-Basin Depth The pulp depth required in the thickener will be greatly affected by the role that compression plays in determining the rate of sedimentation. If the zone-settling conditions define the area needed, then depth of pulp will be unimportant and can be largely ignored, as the "normal" depth found in the thickener will be sufficient. On the other hand, with the compression zone controlling, depth of pulp will be significant, and it is essential to measure the sedimentation rate under these conditions. This is true for the new deep-bed, high-density thickeners.

To determine the compression-zone requirement in a thickener, a test should be run in a deep cylinder in which the average settling pulp depth approximates the depth anticipated in the full-scale basin. The average density of the pulp in compression is calculated and used in Eq. (18-50) to determine the required compression-zone volume:

$$V = \frac{\theta_c(\rho_s - \rho_l)}{\rho_s(\rho_{sl} - \rho_l)} \tag{18-50}$$

where V is the volume, m^3, required per ton of solids per day; θ_c is the compression time, days, required in the test to reach underflow concentration; and ρ_s, ρ_l, ρ_{sl}, are the densities of the solids, liquid, and slurry (average), respectively, ton/m^3. This value divided by the average depth of the pulp during the period represents the unit area defined by compression requirements. If it exceeds the value determined from the zone-settling tests, it is the quantity to be used.

The side depth of the thickener is determined as the sum of the depths needed for the compression zone and for the clear zone. Normally, 1.5 to 2 m of clear liquid depth above the expected pulp level in a thickener will be sufficient for stable, effective operation. When the location of the pulp level cannot be predicted in advance or it is expected to be relatively low, a thickener sidewall depth of 2 to 3 m is usually safe. Greater depth may be used in order to provide better clarity, although in most thickener applications the improvement obtained by this means will be marginal.

Scale-up Factors Factors used in thickening will vary, but, typically, a 1.2 to 1.3 multiplier applied to the unit area calculated from laboratory data is sufficient if proper testing procedures have been followed and the samples are representative.

Torque Requirements Sufficient torque must be available in the raking mechanism of a full-scale thickener to allow it to move through the slurry and assist solids movement to the underflow outlet. Granular, particulate solids that settle rapidly and reach a terminal solids concentration without going through any apparent compression or zone-settling region require a maximum raking capability, as they must be moved to the outlet solely by the mechanism. At the other end of the spectrum, extremely fine materials, such as clays and precipitates, require a minimum of raking, for most of the solids may reach the underflow outlet hydrodynamically. The rakes prevent a gradual buildup of some solids on the bottom, however, and the gentle stirring action from the rake arm often aids the thickening process. As the underflow concentration approaches its ultimate limit, the consistency will increase greatly, resulting in a higher raking requirement and an increase in torque.

For most materials, the particle size lies somewhere between these two extremes, and the torques required in two properly designed thickeners of the same size but in distinct applications can differ greatly. Unfortunately, test methods to specify torque from small-scale tests are of questionable value, since it is difficult to duplicate actual conditions. Manufacturers of sedimentation equipment select torque ratings from experience with similar substances and will recommend a torque capability on this basis. Definitions of *operating torque* vary with the manufacturer, and the user should ask the supplier to specify the *B-10 life* for bearings and to reference appropriate mechanical standards for continuous operation of the selected gear set at specific torque levels. This will provide guidelines for plant operators and help avoid premature failure of the mechanism. Abnormal conditions above the normal operating torque are inevitable, and a thickener should be provided with sufficient torque capability for short-term operation at higher levels in order to ensure continuous performance.

Underflow Pump Requirements Many suspensions will thicken to a concentration higher than that which can be handled by conventional slurry pumps. Thickening tests should be performed with this in mind, for, in general, the unit area to produce the maximum concentration that can be pumped is the usual design basis. Determination of this ultimate pumpable concentration is largely a judgmental decision requiring some experience with slurry pumping; however, the behavior of the thickened suspension can be used as an approximate guide to pumpability. The supernatant should be decanted following a test and the settled solids repulped in the cylinder to a uniform consistency. Repulping is done easily with a rubber stopper fastened to the end of a rigid rod. If the bulk of the repulped slurry can be poured from the cylinder when it is tilted 10 to 30° above the horizontal, the corresponding thickener underflow can be handled by most types of slurry pumps. But if the slurry requires cylinder shaking or other mechanical means for its removal, it should be diluted to a more fluid condition, if conventional pump systems are to be employed. The best suggestion is to consult with pump experts and suppliers who have amassed large databases for a wide range of materials.

THICKENERS

The primary function of a continuous thickener is to concentrate suspended solids by gravity settling so that a steady-state material balance

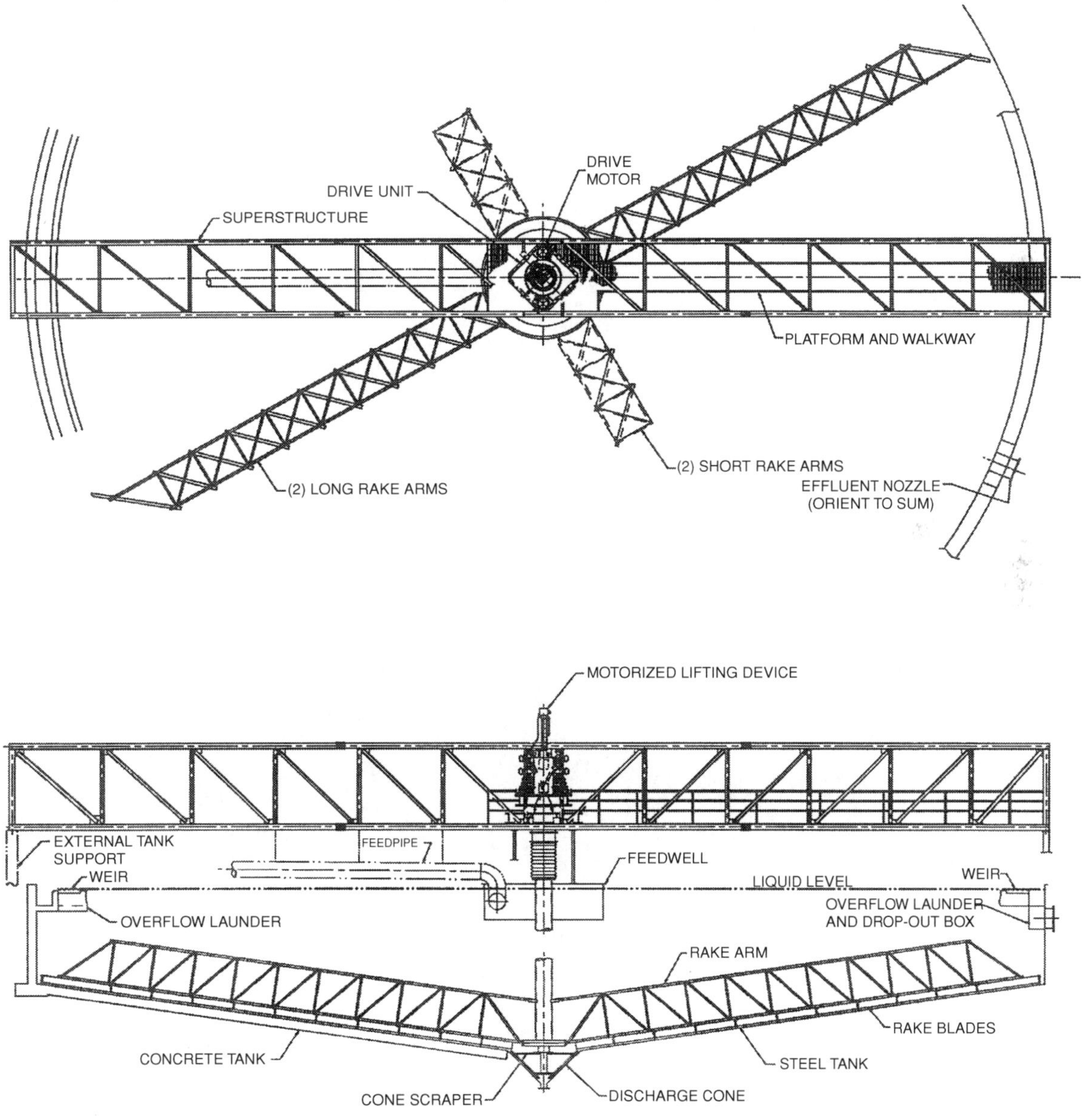

FIG. 18-98 Unit thickener bridge-mounted mechanism. *(Dorr-Oliver EIMCO.)*

is achieved, solids being withdrawn continuously in the underflow at the rate they are supplied in the feed. Normally, an inventory of pulp is maintained in order to achieve the desired concentration. This volume will vary somewhat as operating conditions change; on occasion, this inventory can be used for storage of solids when feed and underflow rates are reduced or temporarily suspended.

A thickener has several basic components: a tank to contain the slurry, feed piping and a feedwell to allow the feed stream to enter the tank, a rake mechanism to assist in moving the concentrated solids to the withdrawal points, an underflow solids-withdrawal system, and an overflow launder. The basic design of a bridge-supported thickener mechanism is illustrated in Fig. 18-98.

Thickener Types Flocculants are commonly used in thickeners, and this practice has resulted in thickener classifications as *conventional, high-rate, ultrahigh-rate,* or *high-density.* These designations can be confusing in that they imply sharp distinction between each type, which is *not* the case.

High-Rate Thickeners The greater capacity expected from a high-rate thickener is due solely to the effective use of flocculant to maximize throughput. In most applications there is a threshold dosage and feed solids concentration at which a noticeable increase in capacity begins to occur, as shown in Fig. 18-99. This effect will continue up to a limit, at which point the capacity will be a maximum unless a lower underflow solids concentration is accepted, as illustrated in Fig. 18-95. Since flocculant is usually added to a thickener in either the feed line or the feedwell, there are a number of proprietary feedwell designs which are used in high-rate thickeners to help optimize flocculation. Deaeration systems may be included in some cases to avoid air entrainment in the flocculated slurry. The other components of these units are not materially different from those of a conventional thickener.

Ultrahigh-Rate Thickeners This type of thickener uses a tall, deep tank with a steep bottom cone and may be used with or without a raking mechanism. This combines the functions of a thickener (to provide a dense underflow) and a clarifier (to provide a clear overflow or supernatant) but is considerably taller. It is generally one-half to one-third the diameter of a conventional or high-rate thickener. Figure 18-100 illustrates the internals of these units, showing the use of dewatering cones, whose function is similar to that of the lamella inclined plates of the titled-plate thickeners.

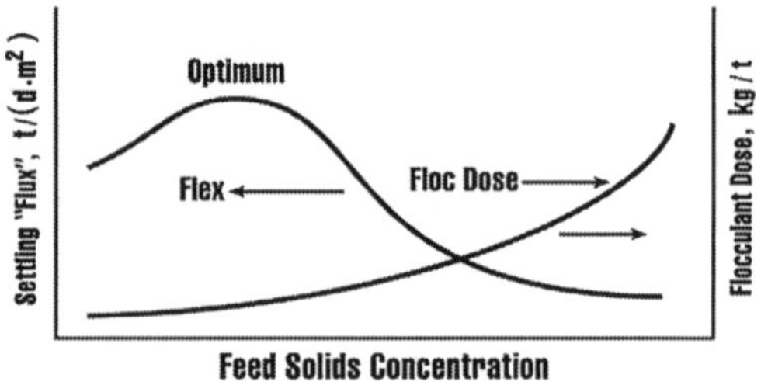

FIG. 18-99 Settling flux curve.

High-Density Thickeners Thickeners can be designed to produce underflows having very high apparent viscosity, permitting disposal of waste slurries at a concentration that avoids segregation of fines and coarse particles or formation of a free-liquid pond on the surface of the deposit. This practice is applied in *dry-stacking* systems and underground *paste-fill* operations for disposal of mine tailings and similar materials. The thickener mechanism generally will require a special rake design and provide a torque capability much higher than normal for that particular diameter thickener (Fig. 18-101). Underflow slurries will be at a higher concentration than for conventional or high-rate thickeners, being 5 to 10 percent lower than vacuum filter cake from the same material. Special pumping requirements are necessary if the slurry is to be

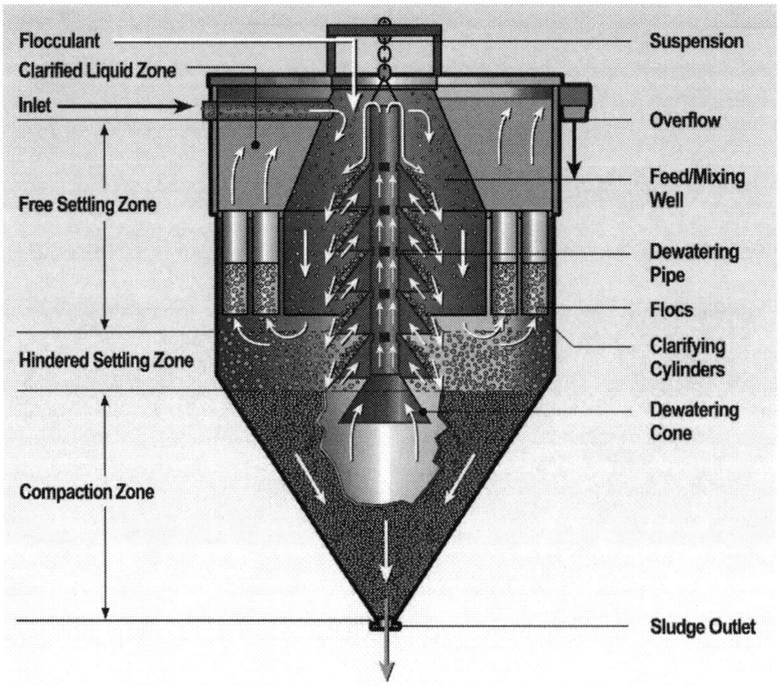

FIG. 18-100 Ultrahigh-rate thickener. (*Dorr-Oliver EIMCO.*)

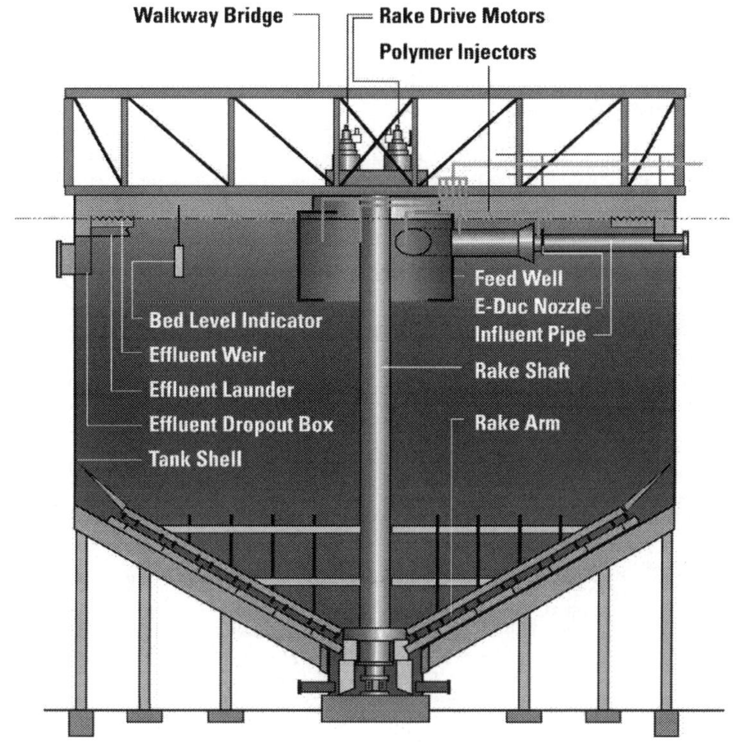

FIG. 18-101 Deep Cone™ paste thickener. *(Courtesy of Dorr-Oliver EIMCO.)*

transported a significant distance, with line pressure drop typically in the range of 3 to 4 kPa/m of pipeline.

Design Features There are four classes of thickeners, each differentiated by its drive mechanism: (1) bridge-supported, (2) center-column supported, (3) traction drives, and (4) without drives. The diameter of the tank will range from 2 to 150 m (6.5 to 492 ft), and the support structure often is related to the size required. These classes are described in detail in the subsection "Components and Accessories for Sedimentation Units."

Operation When operated correctly, thickeners require a minimum of attention and, if the feed characteristics do not change radically, can be expected to maintain design performance consistently. In this regard, it is usually desirable to monitor feed and underflow rates and solids concentrations, flocculant dosage rate, and pulp interface level, preferably with dependable instrumentation systems. Process variations are then easily handled by changing the principal operating controls—underflow rate and flocculant dose—to maintain stability.

Starting up a thickener is usually the most difficult part of the operation, and there is more potential for mechanical damage to the mechanism at this stage than at any other time. In general, two conditions require special attention at this point: underflow pumping and mechanism torque. If possible, the underflow pump should be in operation as soon as feed enters the system, recirculating underflow slurry at a reduced rate if the material is relatively fine or advancing it to the next process step (or disposal) if the feed contains a considerable quantity of coarse solids, e.g., more than 20 percent + 75 μm particles. At this stage of the operation, coarse solids separate from the pulp and produce a difficult raking and pumping situation. Torque can rise rapidly if this material accumulates faster than it is removed. If the torque reaches a point where the automatic control system raises the rakes, it is usually preferable to reduce or cut off the feed completely until the torque drops and the rakes are returned to the lowest position. As the fine fraction of the feed slurry begins to thicken and accumulate in the basin, providing both buoyancy and fluidity, torque will drop and normal feeding can be continued. This applies whether the thickener tank is empty at start-up or filled with liquid. The latter approach contributes to coarse-solids raking problems but at the same time provides conditions more suited to good flocculation, with the result that the thickener will reach stable operation much sooner.

As the solids inventory in the thickener reaches a normal level—usually about 0.5 to 1.0 m below the feedwell outlet—with underflow slurry at the desired concentration, the torque will reach a normal operating range. Special note should be made of the torque reading at this time. Subsequent higher torque levels while operating conditions remain unchanged can almost always be attributed to island formation, and corrective action can be taken early, before serious problems develop. *Island* is the name given to a mass of semisolidified solids that have accumulated on or in front of the rakes, often as a result of excessive flocculant use. This mass usually will continue to grow in size, eventually producing a torque spike that can shut down the thickener and often resulting in lower underflow densities than would otherwise be achievable.

An island is easily detected, usually by the higher-than-normal, gradually increasing torque reading. Probing the rake arms near the thickener center with a rigid rod will confirm this condition—the mass is easily distinguished by its cohesive, claylike consistency. At an early stage, the island is readily removed by raising the rakes until the torque drops to a minimum value. The rakes are then lowered gradually, a few centimeters at a time, so as to shave off the mass of solids and discharge this gelled material through the underflow. This operation can take several hours, and if island formation is a frequent occurrence, the procedure should be carried out on a regular basis, typically once a day, preferably with an automatic system to control the entire operation.

Stable thickener performance can be maintained by carefully monitoring operating conditions, particularly the pulp interface level and the underflow rate and concentration. As process changes occur, the pulp level can vary; regulation of the underflow pumping rate will keep the level within the desired range. If the underflow varies in concentration, this can be corrected by adjusting the flocculant dosage. Response will not be immediate, of course, and care should be taken to make only small step changes at any one time. Procedures for use of automatic control are described in the section on instrumentation.

CLARIFIERS

Continuous clarifiers generally are employed with dilute suspensions, principally industrial process streams and domestic municipal wastes, and their primary purpose is to produce a relatively clear overflow. They are basically identical to thickeners in design and layout except that they employ a mechanism of lighter construction and a drive head with a lower torque capability. These differences are permitted in clarification applications because the thickened pulp produced is smaller in volume and appreciably lower in suspended solids concentration, owing in part to the large percentage of relatively fine (smaller than 10 μm) solids. The installed cost of a clarifier, therefore, is approximately 5 to 10 percent less than that of a thickener of equal tank size, as given in Fig. 18-106.

Rectangular Clarifiers Rectangular clarifiers are employed primarily in municipal water and waste treatment plants, as well as in certain industrial plants, also for waste streams. The raking mechanism employed in many designs consists of a chaintype drag, although suction systems are used for light-duty applications. The drag moves the deposited pulp to a sludge hopper located on one end by means of scrapers fixed to endless chains. During their return to the sludge raking position, the flights may travel near the water level and thus act as skimming devices for removal of surface scum. Rectangular clarifiers are available in widths of 2 to 10 m (6 to 33 ft). The length is generally 3 to 5 times the width. The larger widths have multiple raking mechanisms, each with a separate drive.

This type of clarifier is used in applications such as preliminary oil-water separations in refineries and clarification of waste streams in steel mills. When multiple units are employed, common walls are possible, reducing construction costs and saving on floor space. Overflow clarities, however, generally are not as good as with circular clarifiers, due primarily to reduced overflow weir length for equivalent areas.

Circular Clarifiers Circular units are available in the same three basic types as single-compartment thickeners: bridge, center-column, and peripheral-traction. Because of economic considerations, the bridge-supported type is limited generally to tanks less than 40 m in diameter.

A circular clarifier often is equipped with a surface-skimming device, which includes a rotating skimmer, scum baffle, and scum-box assembly. In sewage and organic-waste applications, squeegees normally are provided for the rake-arm blades, as it is desirable that the bottom be scraped clean to preclude accumulation of organic solids, with resultant septicity and flotation of decomposing material.

Center-drive mechanisms are also installed in square tanks. This mechanism differs from the standard circular mechanism in that a hinged corner blade is provided to sweep the corners which lie outside the path of the main mechanism.

Clarifier-Thickener Clarifiers can serve as thickeners, achieving additional densification in a deep sludge sump adjacent to the center that extends a short distance radially and provides adequate retention time and pulp depth to compact the solids to a high density. Drive mechanisms on this type of clarifier usually must have higher torque capability than would be supplied on a standard clarifier.

Industrial Waste Secondary Clarifiers Many plants which formerly discharged organic wastes to the sewer have turned to using their own treatment facilities in order to reduce municipal treatment plant charges. For organic wastes, the waste-activated sludge process is a preferred approach, using an aeration basin for the bio-oxidation step and a secondary clarifier to produce a clear effluent and to concentrate the biomass for recycling to the basin. To produce an acceptable effluent and achieve sufficient concentration of the low-density solids that make up the biomass, certain design criteria must be followed. Typical design parameters include the following:

Feed pipe velocity: ≤ 1.2 m/s.
Energy-dissipating feed entry velocity (tangential): ≤ 0.5 m/s.
Downward velocity from feedwell: ≤ 0.5–0.75 (peak) m/min.
Feedwell depth: Entry port depth +1–3 m.
Tank depth: typically 3–5 m.
Radial velocity below feedwell: ≤90% of downward velocity.

Overflow rate can range between 0.68 and 2.0 m/h depending on the application. Consult an equipment supplier and manual of practices for recommended overflow rates for specific applications.

Inclined-Plate Clarifiers Lamella or inclined-plate separators have achieved increased use for clarification. They contain a multiplicity of plates inclined at 45 to 60° from the horizontal. Various feed methods are employed so that the influent passes into each inclined channel. The geometry of the plates results in the solids having to settle only a short distance in each channel before sliding down the base to the collection zone beneath the plates. The clarified liquid passes in the opposite direction beneath the ceiling of each channel to the overflow connection.

The area that is theoretically available for separation is equal to the sum of the projected areas or all channels on the horizontal plane. Figure 18-102 shows the horizontally projected area A_S, of a single channel in a clarifier of unit width. For a settling length L and width W, inclined at angle α, the horizontally projected area A_S can be calculated as

$$A_S = LW \cos \alpha \tag{18-51}$$

Multiply this area by the total number of plates in the clarifier to calculate the total clarification area available. However, α must be larger than the angle of repose of the sludge so that it will slide down the plate, and the most common range is 55 to 60°. Plate spacing must be large enough to accommodate the opposite flows of liquid and sludge while reducing interference and preventing plugging and to provide enough residence time for the solids to settle to the bottom plate. Usual X values are 50 to 75 mm (2 to 3 in).

Many different designs are available, the major difference among them being in feed-distribution methods and plate configurations. Operating capacities range from 0.2 to 1.2 gpm/ft² projected horizontal area.

The principal advantage of the inclined-plate clarifier is the increased solids recirculation capacity per unit of plane area. Major disadvantages are an underflow solids concentration that generally is lower than in other gravity clarifiers and difficulty of cleaning when scaling or deposition occurs. The lower underflow composition is due primarily to the reduced compression-zone volume relative to the large settling area. When flocculants are employed, flocculating equipment and tankage preceding the separator are required, as the design does not permit internal flocculation.

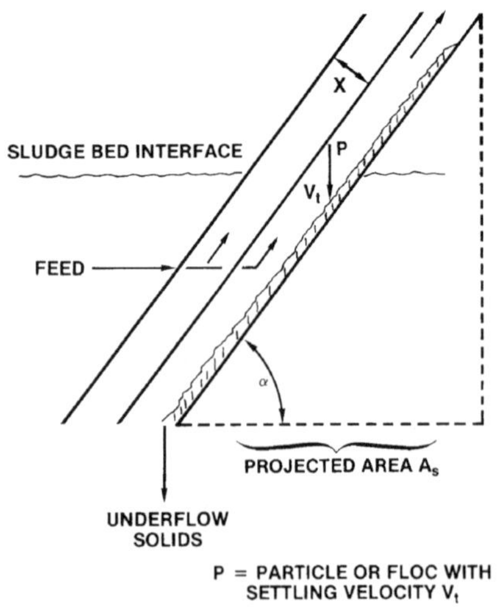

FIG. 18-102 Basic concept of the inclined-plate type of clarifier.

The **ultrahigh-rate** (rakeless) thickener uses internal cones to achieve the tilted-plate effect. The design allows internal flocculation. The tank is tall, with a 60° bottom cone, providing sludge compression height and volume, resulting in a high-density underflow.

Solids-Contact Clarifiers When desirable, mixing, flocculation, and sedimentation all may be accomplished in a single tank. Of the various designs available, those employing mechanically assisted mixing in the reaction zone are the most efficient. They generally permit the highest overflow rate at a minimum chemical dosage while producing the best effluent quality. The unit illustrated in Fig. 18-103*a* consists of a combination dual drive which has a low-speed rake mechanism and a high-rate low-shear turbine located in the top portion of the center well for internal solids recirculation. The influent, dosed with chemicals, is contacted with previously settled solids in a recirculation draft tube within the reaction well by means of the pumping action of the turbine. Owing to the higher concentration of solids being recirculated, chemical reactions are more rapid, and flocculation is improved. Outside of the feed well the flocculated particles settle to the bottom and are raked to the center to be used again in the recirculation process. When particles are too heavy to be circulated up through the draft tube (as in the case of metallurgical pulps), a modified design (see Fig. 18-103) using external recirculation of a portion of the thickened underflow is chosen. These units employ a special mixing impeller in a feed well with a controlled outlet.

Solids-contact clarifiers are advantageous for clarifying turbid waters or slurries that require coagulation and flocculation for the removal of bacteria, suspended solids, or color. Applications include softening water by lime addition; clarifying industrial-process streams, sewage, and industrial wastewaters; tertiary treatment for removal of phosphates, BOD_5, and turbidity; and silica removal from produce water, cooling tower makeup, and geothermal brines.

COMPONENTS AND ACCESSORIES FOR SEDIMENTATION UNITS

Sedimentation systems consist of a collection of components, each of which can be supplied in a number of variations. The basic components are the same, whether the system is for thickening or clarifying: tank, drive-support structure, drive unit and lifting device, rake structure, feedwell, overflow arrangement, underflow arrangement, instrumentation, and flocculation facilities.

Tanks Tanks or basins are constructed of such materials as steel, concrete, wood, compacted earth, plastic sheeting, and soil cement. The selection of the materials of construction is based on environmental cost, availability, topography, water table, ground conditions, climate, operating temperature, and chemical-corrosion resistance. Typically, industrial tanks up to 30 m (100 ft) in diameter are made of steel. Concrete generally is used in municipal applications and in larger industrial applications. Extremely large units employing earthen basins with impermeable liners have proved to be economical.

Drive-Support Structures There are three basic drive mechanisms. These are (1) the bridge-supported mechanism, (2) the center-column-supported mechanism, and (3) the traction-drive thickener containing a center-column-supported mechanism with the driving arm attached to a motorized carriage at the tank periphery.

Bridge-Supported Thickeners These thickeners (Fig. 18-98) are common in diameters up to 30 m, the maximum being about 45 m (150 ft). They offer the following advantages over a center-column-supported design: (1) ability to transfer loads to the tank periphery; (2) ability to give a denser and more consistent underflow concentration with the single draw-off point; (3) a less complicated lifting device; (4) fewer structural members subject to mud accumulation; (5) access to the drive from both ends of the bridge; and (6) lower cost for units smaller than 30 m in diameter.

Center-Column-Supported Thickeners These thickeners are usually 20 m (65 ft) or more in diameter. The mechanism is supported by a stationary steel or concrete center column, and the raking arms are attached to a driving cage which rotates around the center column.

Traction Thickeners These thickeners are most adaptable to tanks larger than 60 m (200 ft) in diameter. Maintenance generally is less difficult than with other types of thickeners, which is an advantage in remote locations. The drive may be supported on the concrete wall (the wall would be a structural member) or supported outside the wall on the ground (a standard tank wall could be used). Disadvantages of the traction thickener are that (1) no practical lifting device can be used, (2) operation may be difficult in climates where snow and ice are common, and (3) the driving-torque effort must be transmitted from the tank periphery to the center, where the heaviest raking conditions occur.

The rakeless ultrahigh-rate thickeners use elevated tanks up to 20-m diameter. Advantages are no drive, high throughput rate, and the small footprint. Disadvantages are the height of the elevated tank.

Drive Assemblies The drive assembly is the key component of a sedimentation unit. The drive assembly provides (1) the force to move the rakes through the thickened pulp and to move settled solids to the point of discharge, (2) the support for the mechanism which permits it to rotate, (3) adequate reserve capacity to withstand upsets and temporary overloads, and (4) a reliable control which protects the mechanism from damage when a major overload occurs.

Drives usually have steel or iron main spur gears mounted on bearings, alloy-steel pinions, or a planetary gear. Direct-drive hydraulic systems are also employed. The gearing components preferably are enclosed for maximum service life. The drive typically includes a torque-measuring system with torque indicated on the mechanism and often transmitted to a remote indicator. If the torque becomes excessive, it can automatically activate such safeguards against structural damage as sounding an alarm, raising the rakes, and stopping the drive.

Rake-Lifting Mechanisms These should be provided when abnormal thickener operation is probable. Abnormal thickener operation or excessive torque may result from insufficient underflow pumping, surges in the solids feed rate, excessive amounts of large particles, sloughing of solids accumulated between the rakes and the bottom of the tank or on structural members of the rake mechanism, or miscellaneous obstructions falling into the thickener. The lifting mechanism may be set to raise the rakes automatically when a specific torque level (e.g., 40 percent of design) is encountered, continuing to lift until the torque returns to normal or until the maximum lift height is reached. Generally, corrective action must be taken to eliminate the cause of the upset. Once the torque returns to normal, the rake mechanism is lowered slowly to "plow" gradually through the excess accumulated solids until these are removed from the tank.

Motorized rake-lifting devices typically are designed to allow for a vertical lift of the rake mechanism of up to 90 cm (3 ft).

The cable arm design uses cables attached to a truss above or near the liquid surface to move the rake arms, which are hinged to the drive structure, allowing the rakes to raise when excessive torque is encountered. A major advantage of this design is the relatively small surface area of the raking mechanism, which reduces the solids accumulation and downtime in applications in which scaling or island formation can occur.

One disadvantage of this or any hinged-arm or other self-lifting design is that there is very little lift at the center, where the overload usually occurs. A further disadvantage is the difficulty of returning the rakes to the lowered position in settlers containing solids that compact firmly.

Rake Mechanism The rake mechanism assists in moving the settled solids to the point of discharge. It also aids in thickening the pulp by disrupting bridged floccules, permitting trapped fluid to escape and allowing the floccules to become more consolidated. Rake mechanisms are designed for specific applications, usually having two long rake arms with an option for two short rake arms for bridge-supported and center-column-supported units. Traction units usually have one long arm, two short arms, and one intermediate arm.

Figure 18-104 illustrates types of rake-arm designs. The conventional design typically is used in bridge-supported units, while the dual-slope design is used for units of larger diameter.

Rake blades can have attached spikes or serrated bottoms to cut into solids that have a tendency to compact. Lifting devices typically are used with these applications.

Rake-speed requirements depend on the type of solids entering the thickener. Peripheral speed ranges used are, for slow-settling solids, 3 to 8 m/min (10 to 25 ft/min); for fast-settling solids, 8 to 12 m/min

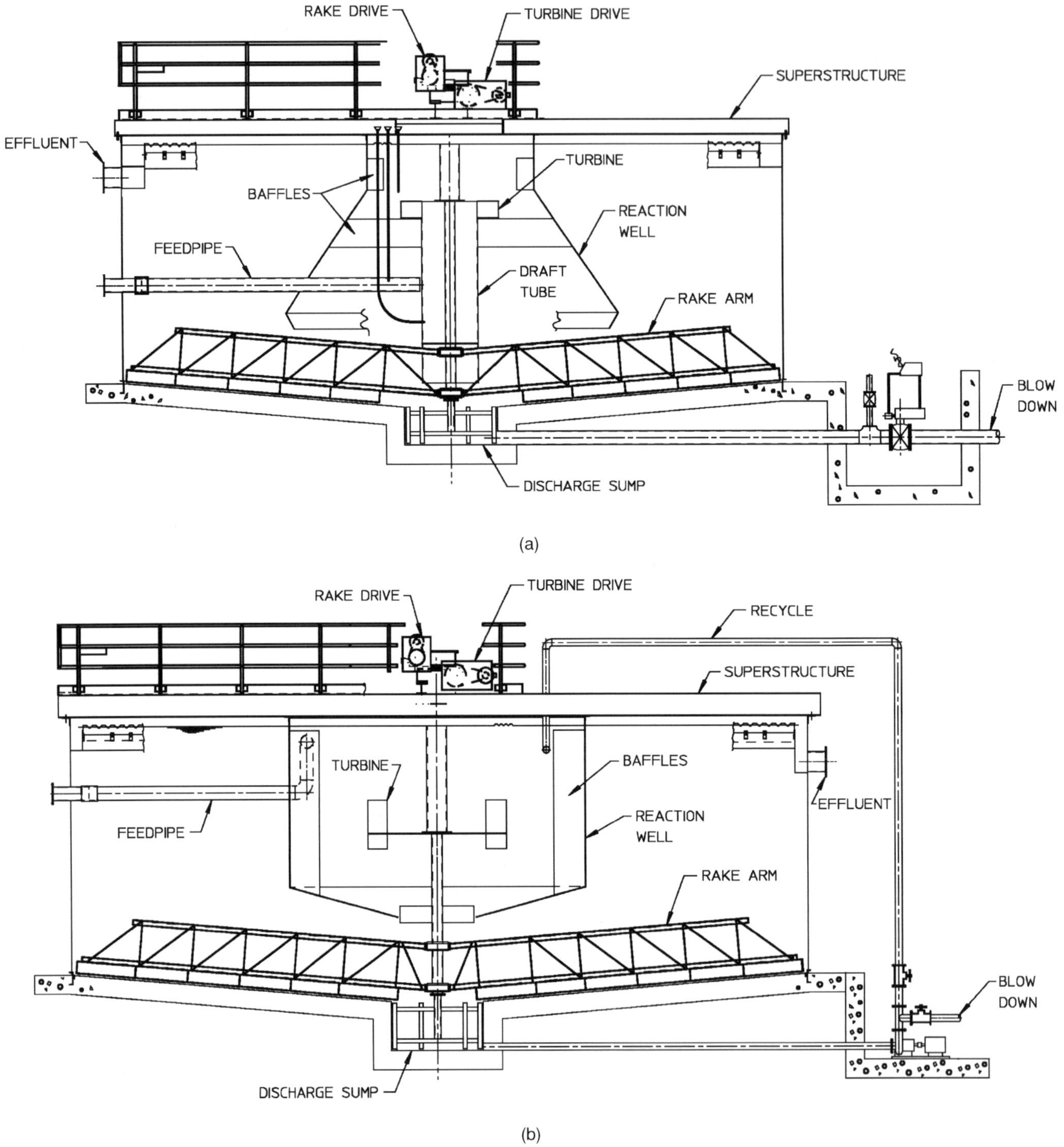

FIG. 18-103 Solids-contact reactor clarifiers. *(Dorr-Oliver EIMCO.)*

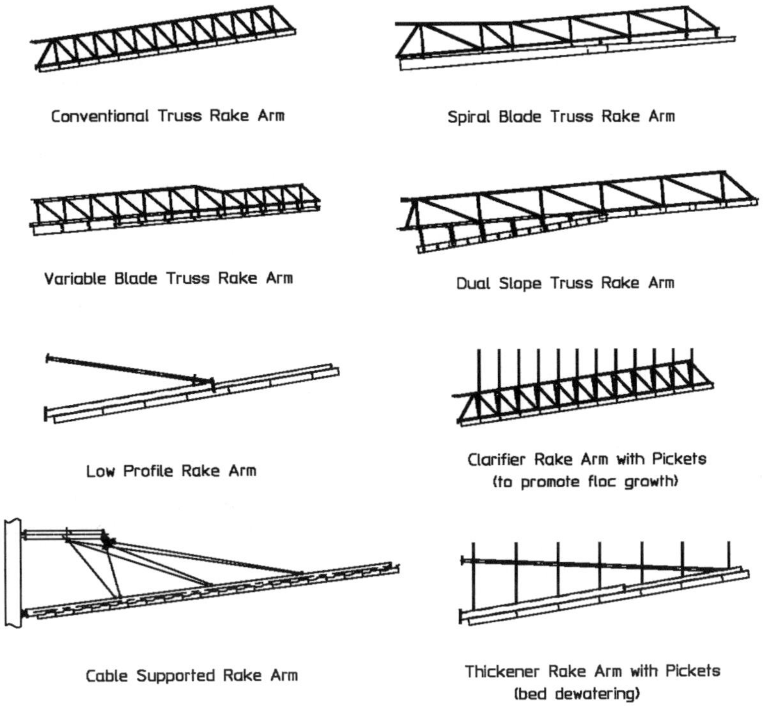

FIG. 18-104 Rake-mechanism designs for specific applications and duties. *(Dorr-Oliver EIMCO.)*

(25 to 40 ft/min); and for coarse solids or crystalline materials, 12 to 30 m/min (40 to 100 ft/min).

Feedwell The feedwell is designed to allow the feed to enter the thickener with minimum turbulence and uniform distribution while dissipating most of its kinetic energy. Feed slurry enters the feedwell, which is usually located in the center of the thickener, through a pipe or launder suspended from the bridge. To avoid excess velocity, an open launder normally has a slope no greater than 1 to 2 percent. Pulp should enter the launder at a velocity that prevents sanding at the inlet. With nonsanding pulps, the feed may also enter upward through the center column from a pipeline installed beneath the tank.

The standard feedwell for a thickener is designed for a maximum vertical outlet velocity of about 1.5 m/min (5 ft/min). High turbidity caused by short-circuiting the feed to the overflow can be reduced by increasing the depth of the feedwell. When overflow clarity is important or the solids specific gravity is close to the liquid specific gravity, deep feedwells of large diameter are used, and measures are taken to reduce the velocity of the entering feed slurry.

Shallow feedwells may be used when overflow clarity is not important, the overflow rate is low, and/or solids density is appreciably greater than that of water. Some special feedwell designs used to dissipate entrance velocity and create quiescent settling conditions split the feed stream and allow it to enter the feedwell tangentially on opposite sides. The two streams shear or collide with one another to dissipate kinetic energy.

When flocculants are used, often it will be found that the optimum solids concentration for flocculation is considerably less than the normal concentration, and significant savings in reagent cost will be made possible by dilution of the feed prior to flocculation. This can be achieved by recycling overflow or more efficiently by feedwell modifications, including openings in the feedwell rim. These will allow supernatant to enter the feedwell, and flocculant can be added at this point or injected below the surface of the pulp in the feedwell. Another effective means of achieving this dilution prior to flocculant addition is illustrated in Fig. 18-105. This approach utilizes the energy available in the incoming feed stream to achieve the dilution by momentum transfer and requires no additional energy expenditure to dilute this slurry by as much as three to four times.

Overflow Arrangements Clarified effluent typically is removed in a peripheral launder located inside or outside the tank. The effluent enters the launder by overflowing a V-notch or level flat weir, or through submerged orifices in the bottom of the launder. Uneven overflow rates caused by wind blowing across the liquid surface in large thickeners can be better controlled when submerged orifices or V-notch weirs are used. Radial launders are used when uniform upward liquid flow is desired in order to improve clarifier detention efficiency. This arrangement provides an additional benefit in reducing the effect of wind, which can seriously impair clarity in applications that employ basins of large diameter.

The hydraulic capacity of a launder must be sufficient to prevent flooding, which can cause short-circuiting of the feed and deterioration of overflow clarity. Standards are occasionally imposed on weir overflow rates for clarifiers used in municipal applications; typical rates are 3.5 to 15 $m^3/(h \cdot m)$ [7000 to 30,000 gal/(day·ft)], and they are highly dependent on clarifier side-water depth. Industrial clarifiers may have higher overflow rates, depending on the application and the desired overflow clarity. Launders can be arranged in a variety of configurations to achieve the desired overflow rate. Several alternatives to improve clarity include an annular launder inside the tank (the liquid overflows both sides), radial launders connected to the peripheral launder (providing the very long weir that may be needed when abnormally high overflow rates are encountered and overflow clarity is important), and Stamford baffles, which are located below the launder to direct flow currents back toward the center of the clarifier.

In many thickener applications, on the other hand, complete peripheral launders are not required, and no difference in either overflow clarity or underflow concentration will result through the use of launders extending over only a fraction (e.g., one-fifth) of the perimeter.

FIG. 18-105 E-Duc® Feed dilution system installed on a 122-m diameter thickener. (*Dorr-Oliver EIMCO.*)

For design purposes, a weir-loading rate in the range of 7.5 to 30.0 $m^3/(h \cdot m)$ [10 to 40 gpm/ft] can be used, the higher values being employed with well-flocculated, rapidly settling slurries. The overflow launder required may occupy only a single section of the perimeter rather than consisting of multiple, shorter segments spaced uniformly around the tank.

Underflow Arrangements Concentrated solids are removed from the thickener by use of centrifugal or positive displacement pumps or, particularly with large-volume flows, by gravity discharge through a flow control valve or orifice suitable for slurry applications. Due to the risk to the thickener operation of a plugged underflow pipe, it is recommended that duplicate underflow pipes and pumps be installed in all thickening applications. Provision to recycle underflow slurry back to the feedwell is also useful, particularly if solids are to be stored in the thickener.

There are three basic underflow arrangements: (1) the underflow pump adjacent to the thickener sidewall with buried piping from the discharge cone, (2) the underflow pump under the thickeners or adjacent to the sidewall with the piping from the discharge cone in a tunnel, and (3) the underflow pump located in the center of the thickener on the bridge, or using piping up through the center column.

Pump Adjacent to Thickener with Buried Piping This arrangement of buried piping from the discharge cone is the least expensive system but the most susceptible to plugging. It is used only when the solids do not compact to an unpumpable slurry and can be easily backflushed if plugging occurs. Typically, two or more underflow pipes are installed from the discharge cone to the underflow pump so that solids removal can continue if one of the lines plugs. Valves should be installed to permit flushing with water and compressed air in both directions to remove blockages.

Tunnel A tunnel may be constructed under the thickener to provide access to the discharge cone when underflow slurries are difficult to pump and have characteristics that cause plugging. The underflow pump may be installed underneath the thickener or at the perimeter. Occasionally thickeners are installed on legs or piers, making tunneling for access to the center unnecessary. A tunnel or an elevated thickener is more expensive than the other underflow arrangements, but there are certain operational and maintenance advantages. Of course, the hazards of working in a tunnel (flooding and interrupted ventilation, for example) and related safety regulations must be considered.

Center-Column Pumping This arrangement may be used instead of a tunnel. Several designs are available. One is a bridge-mounted pump with a suction line through a wet or dry center column. The pump selection may be limiting, requiring special attention to priming, net positive suction head, and the maximum density that the pump can handle. Another design has the underflow pump located in a room under the thickener mechanism and connected to openings in the column. Access is through the drive gear at the top of the column.

INSTRUMENTATION

Thickener Thickener control philosophies are usually based on the idea that the underflow density obtained is the most important performance criterion. The overflow clarity is also a consideration, but this is generally not as critical. Additional factors which must be considered are optimization of flocculant usage and protection of the raking mechanism.

Automated control schemes employ one or more sets of controls, which will fit into three categories: (1) control loops which are used to regulate the addition of flocculant, (2) control loops to regulate the withdrawal of underflow, and (3) rake drive controls. Frequently, the feed to a thickener is not controlled and most control systems have been designed with some flexibility to deal with changes in feed characteristics.

Flocculant addition rate can be regulated in proportion to the thickener volumetric feed rate or solids mass flow in a feed-forward mode, or in a feed-back mode on either rake torque, underflow density, settling solids (sludge) bed level, or solids settling rate.

Underflow is usually withdrawn continuously on the bases of bed level, rake torque, or underflow solids concentration in a feedback mode. Some installations incorporate two or more of these parameters in their underflow withdrawal control philosophy. For example, the continuous withdrawal may be based on underflow solids density with an override to increase the withdrawal rate if either the rake torque or the bed level reaches a preset value. In some cases, underflow withdrawal has been regulated in a feed-forward mode on the basis of thickener feed solids mass flow rate. Any automated underflow pumping scheme should incorporate a lower limit on volumetric flow rate as a safeguard against line pluggage.

It is also important to consider the level of the sludge bed in the thickener. Although this can be allowed to increase or decrease within moderate limits, it must be controlled enough to prevent solids from overflowing the thickener or from falling so low that the underflow density becomes too dilute. The settling slurry within the sludge bed is normally free flowing and will disperse to a consistent level across the thickener diameter.

Rake drive controls protect the drive mechanism from damage and usually incorporate an alarm to indicate high torque with an interlock to shut down the drive at a higher torque level. They can have an automated rake raising and lowering feature with a device to indicate the elevation of the rakes.

A complete automated control scheme incorporates controls from each of the three categories. It is important to consider the interaction of the various controls, especially of the flocculant addition and underflow withdrawal control loops, when designing a system. The lag and dead times of any feedback loops as well as the actual response of the system to changes in manipulated variables must be considered. For example, in some applications it is possible that excessive flocculant addition may produce an increase in the rake torque (due to island formation or viscosity increase) without a corresponding increase in underflow density. Additionally, sludge bed level sensors generally require periodic cleaning to produce a reliable signal. In many cases, it has not been possible to effectively maintain the sludge bed level sensors, requiring a change in the thickener control logic after start-up. Some manufacturers offer complete thickener control packages.

Clarifiers Control philosophies for clarifiers are based on the idea that the overflow is the most important performance criterion. Underflow density or suspended solids content is a consideration, as is optimal use of flocculation and pH control reagents. Automated controls are of three basic types: (1) control loops that optimize coagulant, flocculant, and pH control reagent additions; (2) those that regulate underflow removal; and (3) rake drive controls. Equalization of the feed is provided in some installations, but the clarifier feed is usually not a controlled variable with respect to the clarifier operation.

Automated controls for flocculating reagents can use a feed-forward mode based on feed turbidity and feed volumetric rate, or a feed-back mode incorporating a streaming current detector on the flocculated feed. Attempts to control coagulant addition on the basis of overflow turbidity generally have been less successful. Control for pH has been accomplished by feed-forward modes on the feed pH and by feed-back modes on the basis of clarifier feedwell or external reaction tank pH. Control loops based on measurement of feedwell pH are useful for control in applications in which flocculated solids are internally recirculated within the clarifier feedwell.

Automated sludge withdrawal controls are usually based on the sludge bed level or pressure. These can operate in on-off or continuous modes and can use either single-point or continuous sludge level indication sensors. In many applications, automated control of underflow withdrawal does not provide an advantage, since so few settled solids are produced that it is only necessary to remove sludge for a short interval once a day, or even less frequently. In applications in which the underflow is recirculated internally within the feedwell, it is necessary to maintain sufficient sludge inventory for the recirculation turbine to pull from. This can be handled in an automated system with a single-point *low sludge* bed level sensor in conjunction with a low-level alarm or pump shutoff solenoid. Some applications require continuous external recirculation of the underflow direct to the feedwell or external reaction tanks, and an automated control loop can be used to maintain recirculation based on flow measurement, with a manually adjusted setpoint.

Control philosophies applied to continuous countercurrent decantation (CCD) thickeners are similar to those used for thickeners in other applications, but have emphasis on maintaining the CCD circuit in balance. It is important to prevent any one of the thickeners from pumping out too fast, otherwise an upstream unit could be starved of wash liquor while at the same time too much underflow could be placed in a downstream unit, disrupting the operation of both units as well as reducing the circuit washing efficiency. Several control configurations have been attempted, and the more successful schemes have linked the solids mass flow rate of underflow pumping to that of the upstream unit or to the CCD circuit solids mass feed rate. Wide variations in the solids feed rate to a CCD circuit will require some means of dampening these fluctuations if design wash efficiency is to be maintained.

The following types of devices are commonly applied to measure the various operational parameters of thickeners and clarifiers. They have been used in conjunction with automatic valves and variable-speed pumps to achieve automatic operation as well as to simply provide local or remote indications.

INSTRUMENTATION AND CONTROLS

Torque Rake torque is an indication of the force necessary to rotate the rakes. Higher rake torque is an indication of higher underflow density or viscosity, deeper mud bed, higher fraction of coarse material, island formation, or heavy scale buildup on the rake arms.

Rake torque measurement is usually provided by the thickener manufacturer as part of the rake drive mechanism. Typical methods involve load cells, motor power measurement, hydraulic pressure, or mechanical displacement against a spring. Torque-measuring devices are designed to produce a signal that may be utilized for alarming or control.

Rake Height Rake lifting devices are used to minimize the torque on the arms by lifting them out of heavy bed solids and enable the rake to continue running during upset conditions. It is desirable to prevent the rake drives from running for extended periods at torques above 50 to 60 percent, to prevent accelerated wear on the drive. Lifting the rakes a small distance is usually effective in relieving the pressure on the rakes, thus reducing the torque. Because of this, in using "torque indication" in a control strategy one must also consider the rake height to effectively control the thickener. The two most common rake height indicators are the ultrasonic or potentiometer type with a reeling cable. Lifting of the rakes allows a short period of time to make corrections before one is forced to shut down the thickener.

Bed Level There are several general types of bed level detection instruments: ultrasonic, nuclear, float and rod, and reeling (with various sensors). Each has advantages and disadvantages, which are discussed below. There is not a standard bed level sensor that is recommended for all applications.

- **Ultrasonic** bed level sensors work by sending a pulse down from just under the surface, which bounces off the bed and back to the receiver. Elapsed time is used to calculate the distance. Advantages are noninterfering location, measurement over a large span, and relatively low cost. The downside is that they do not work on all applications. If the overflow is cloudy, it can interfere with the transmission or cause too much reflection to give a reliable signal. Scaling affects accuracy and can cause drifting or loss of signal. Using them on concentrate thickeners has proved to be particularly troublesome.
- **Nuclear** bed level sensors work by sensing either the background radiation level or attenuation between a source and detector, depending on whether the solids have a natural background radiation level. The sensor is comprised of a long rod that extends down into the bed with radiation detectors spaced along the length. If the ore changes from not having radiation to having it, there will be problems. The advantages are that it is relatively reliable when properly applied. The downside is that it measures over a limited range, may interfere with rakes (a hinged version that will swing out of the way when the rakes passby is available), and is relatively expensive.
- **Float and rod** types work with a ball with a hollow sleeve that slides up and down on a rod that extends down into the bed. The ball weight can be adjusted to float on top of the bed of solids.

TABLE 18-7 Typical Thickener and Clarifier Design Criteria and Operating Conditions

	Percent solids		Unit area, $m^2/(t/d)$*†	Overflow rate, $m^3/(m^2·h)$*
	Feed	Underflow		
Alumina Bayer process				
Red mud, primary	3–4	10–25	2–5	
Red mud, washers	6–8	15–35	1–4	
Hydrate, fine or seed	1–10	20–50	1.2–3	0.07–0.12
Brine purification	0.2–2.0	8–15		0.5–1.2
Coal, refuse	0.5–6	20–40	0.5–1	0.7–1.7
Coal, heavy-media (magnetic)	20–30	60–70	0.05–0.1	
Cyanide, leached-ore	16–33	40–60	0.3–1.3	
Flue dust, blast-furnace	0.2–2.0	40–60		1.5–3.7
Flue dust, BOF	0.2–2.0	30–70		1–3.7
Flue-gas desulfurization sludge	3–12	20–45	0.3–3†	
High-density paste thickeners				
Red mud, washers‡	3–4	45–50	0.05–0.08	
Coal, refuse†	6–8	50–54	0.08–0.1	
Cyanide, leached ore†	10–15	65–70	0.05–0.08	
Copper tailings†	10–20	50–75	0.07–0.15	
Tailings (magnetic)†	10–20	60–75	0.07–0.1	
Tailings (nonmagnetic)†	10–20	60–70	0.07–0.1	
Magnesium hydroxide from brine	8–10	25–40	5–10	
Magnesium hydroxide from seawater	1–4	15–20	3–10	0.5–0.8
Metallurgical				
Copper concentrates	14–50	40–75	0.2–2	
Copper tailings	10–30	45–65	0.4–1	
Iron ore				
Concentrate (magnetic)	20–35	50–70	0.01–0.08	
Concentrate (nonmagnetic), coarse: 40–65% −325	25–40	60–75	0.02–0.1	
Concentrate (nonmagnetic), fine: 65–100% −325	15–30	60–70	0.15–0.4	
Tailings (magnetic)	2–5	45–60	0.6–1.5	1.2–2.4
Tailings (nonmagnetic)	2–10	45–50	0.8–3	0.7–1.2
Lead concentrates	20–25	60–80	0.5–1	
Molybdenum concentrates	10–35	50–60	0.2–0.4	
Nickel, $(NH_4)_2CO_3$ leach residue	15–25	45–60	0.3–0.5	
Nickel, acid leach residue	20	60	0.8	
Zinc concentrates	10–20	50–60	0.3–0.7	
Zinc leach residue	5–10	25–40	0.8–1.5	
Municipal waste				
Primary clarifier	0.02–0.05	0.5–1.5		1–1.7
Thickening				
Primary sludge	1–3	5–10	8	
Waste-activated sludge	0.2–1.5	2–3	33	
Anaerobically digested sludge	4–8	6–12	10	
Phosphate slimes	1–3	5–15	1.2–18	
Pickle liquor and rinse water	1–8	9–18	3.5–5	
Plating waste	2–5	5–30		1.2
Potash slimes	1–5	6–25	4–12	
Potato-processing waste	0.3–0.5	5–6		1
Pulp and paper				
Green-liquor clarifier	0.2	5		0.8
White-liquor clarifier	8	35–45	0.8–1.6	
Kraft waste	0.01–0.05	2–5		0.8–1.2
Deinking waste	0.01–0.05	4–7		1–1.2
Paper-mill waste	0.01–0.05	2–8		1.2–2.2
Sugarcane defecation			0.5§	
Sugar-beet carbonation	2–5	15–20	0.03–0.07‡	
Uranium				
Acid-leached ore	10–30	25–65	0.02–1	
Alkaline-leached ore	20	60	1	
Uranium precipitate	1–2	10–25	5–12.5	
Water treatment				
Clarification (after 30-min flocculation)				1–1.3
Softening lime-soda (high-rate, solids-contact clarifiers)				3.7
Softening lime-sludge	5–10	20–45	0.6–2.5	

*$m^2/(t/d) \times 9.76 = ft^2/(short\ ton/day)$; $m^3/(m^2·h) \times 0.41 = gal/(ft^2·min)$; 1 t = 1 metric ton.
†High-rate thickeners using required flocculant dosages operate at 10 to 50 percent of these unit areas.
‡Typical design using high Density/Paste thickeners. Feed per cent solids are that diluted for flocculation.
§Basis: 1 t of cane or beets.

These are subject to fouling and sticking, and can be installed and measured only in the area above the rakes; however, they are relatively inexpensive.

- **Reeling devices** work by dropping a sensor down on a cable and sensing the bed level by optical or conductivity sensors. In theory they are nonfouling and get out of the way of the rakes, but in practice, they frequently become entangled with the rakes. The price is mid-range. Freezing wind and cold temperatures can lead to icing problems.
- **Vibrating or tuning fork sensors** are designed to sense a difference in the vibrating frequency in different masses of solids. These are used in Europe and Africa.

- **Bubble tube or differential pressure** is an old, but tried and true, method of bed level detection. There may be some plugging or fouling of the tube over time.
- **External density** through sample ports. Slurry samples are taken from nozzles on the side of the tank and pass through a density meter to determine the presence of solids. This system can be set up with automated valves to measure several different sample points. This system requires external piping and disposal of the sample stream. Line pluggage is often a problem.

Bed Pressure Because thickeners maintain a constant liquid level, the pressure at the bottom of the thickener is an indication of the overall specific gravity in the tank. If the liquor specific gravity is constant, the overall specific gravity is an indication of the amount of solids in the tank and can be converted to a rough solids inventory. This can be a very effective tool for thickener control. Because of relative height-to-diameter ratios, it is considered less useful for large-diameter thickeners.

Differential pressure sensors are used to measure the bed pressure, leaving one leg open to the atmosphere to compensate for barometric pressure variations. Care must be taken in the installation to minimize the plugging with solids. This is normally done by tilting the tank nozzle on which the DP cell is mounted downward from the sensor, so the solids tend to settle away from the sensor surface. A shutoff valve and washer flush tap are recommended to allow easy maintenance.

Flow Rate Flow rates for feed and underflow lines are useful, particularly when combined with density measurements to generate solids mass flow rates. Since flocculant is usually dosed on a solids mass basis, knowing the mass flow rate is very useful for flocculant control, providing a fast response system. Flow rate measurement is an absolute necessity for the newer generation of ultrahigh-rate and ultrahigh-density thickeners. The streams being measured are usually slurries, and the flow rate is usually measured by either magnetic flowmeters or Doppler-type flowmeters. As long as these instruments are properly installed in suitable full straight pipe sections, avoiding air if possible, they are accurate and reliable. If the feed stream is in an open launder, flow measurement is more difficult but can be accomplished using ultrasonic devices.

Density Nuclear gauges are the norm for density measurement. Nuclear density instruments require nuclear handling permits in most countries. Note that there are now some types that use very low level sources that do not require nuclear licensing. Density gauges should be recalibrated regularly as they are subject to drifting. Small flow applications may be able to use a coriolis meter to measure both mass and percent solids with one instrument.

Settling Rate The settling rate in the feed well is a good indication of the degree of flocculation, and it can be used to maintain consistent flocculation over widely varying feed conditions. A settleometer is a device that automatically pulls a sample from the feed well and measures the settling rate. The flocculant can then be adjusted to maintain a constant settling rate.

Overflow Turbidity Overflow turbidity can be used to control flocculant or coagulant. There may be some significant lag time between the actual flocculation process and when the clarified liquor reaches the overflow discharge point where the sensor is typically positioned. These sensors and meters are generally used as alarms or for trim only.

Note: It is critical that all instruments be well maintained and serviced on a regular basis in order to get the best results.

CONTINUOUS COUNTERCURRENT DECANTATION

The system of separation of solid-phase material from an associated solution by repeated stages of dilution and gravity sedimentation is adapted for many industrial-processing applications through an operation known as continuous countercurrent decantation (CCD). The flow of solids proceeds in a direction countercurrent to the flow of solution diluent (water, usually), with each stage composed of a mixing step followed by settling of the solids from the suspension. The number of stages ranges from 2 to as many as 10, depending on the degree of separation required, the amount of wash fluid added (which influences the final solute concentration in the first-stage overflow), and the underflow solids concentration attainable. Applications include processes in which the solution is the valuable component (as in alumina extraction), or in which purified solids are sought (magnesium hydroxide from seawater), or both (as frequently encountered in the chemical-processing industry and in base-metal hydrometallurgy).

The factors which may make CCD a preferred choice over other separation systems include the following: rapidly settling solids, assisted by flocculation: relatively high ratio of solids concentration between underflow and feed; moderately high wash ratios allowable (2 to 4 times the volume of liquor in the thickened underflows); large quantity of solids to be processed; and the presence of fine-size solids that are difficult to concentrate by other means. A technical feasibility and economic study is desirable in order to make the optimum choice.

Flow-Sheet Design Thickener-sizing tests, as described earlier, will determine unit areas, flocculant dosages, and underflow densities for the various stages. For most cases, unit areas will not vary significantly throughout the circuit; similarly, underflow concentrations should be relatively constant. In practice, the same unit area is generally used for all thickeners in the circuit to simplify construction. Serious consideration should be given to the design underflow density since operating at the higher, manageable densities will offer the benefits of improved wash efficiency. Many CCD installations, alumina in particular, have installed paste thickeners and reduced the number of stages or lowered the required volume of wash water.

Equipment The equipment selected for CCD circuits may consist of multiple-compartment washing-tray thickeners or a train of individual unit thickeners. The washing-tray thickener consists of a vertical array of coaxial trays connected in series, contained in a single tank. The advantages of this design are smaller floor-area requirements, less pumping equipment and piping, and reduced heat losses in circuits operating at elevated temperatures. However, operation is generally more difficult, and user preference has shifted toward ultrahigh-rate and ultrahigh-density thickeners.

Underflow Pumping Diaphragm pumps with open discharge are employed in some low-volume cases, primarily because underflow densities are readily controlled with these units. Disadvantages include the generally higher maintenance and initial costs than for other types and their inability to transfer the slurry any great distance. Large flows often are best handled with variable-speed, rubber-lined centrifugal pumps, utilizing automatic control to maintain the underflow rate and density.

Overflow Pumps These can be omitted if the thickeners are located at increasing elevations from first to last so that overflows are transferred by gravity or if the mixture of underflow and overflow is to be pumped. Overflow pumps are necessary, however, when maximum flexibility and control are sought.

Interstage Mixing Efficiencies Mixing or stage efficiencies rarely achieve the ideal 100 percent, in which solute concentrations in overflow and underflow liquor from each thickener are identical. Part of the deficiency is due to insufficient blending of the two streams, and attaining equilibrium will be hampered further by heavily flocculated solids. In systems in which flocculants are used, interstage efficiencies often will drop gradually from first to last thickener, and typical values will range from 98 percent to as low as 70 percent. In some cases, operators will add the flocculant to an overflow solution which is to be blended with the corresponding underflow. While this is very effective for good flocculation, it can result in reflocculation of the solids before the entrained liquor has had a chance to blend completely with the overflow liquor. The preferable procedure is to recycle a portion of the overflow back to the feed line of the same thickener, adding the reagent to this liquor.

The usual method of interstage mixing consists of a relatively simple arrangement in which the flows from preceding and succeeding stages are added to a feed box at the thickener periphery. A nominal detention time in this mixing tank of 30 to 60 s and sufficient energy input to avoid solids settling will ensure interstage efficiencies greater than 95 percent.

The performance of a CCD circuit can be estimated through use of the following equations, which assume 100 percent stage efficiency:

$$R = \frac{O/U[(O/U)^N - 1]}{[(O/U)^{N+1} - 1]} \tag{18-52}$$

$$R = 1 - \left(\frac{U}{O'}\right)^N \tag{18-53}$$

for O/U and $U/O' \neq 1$. R is the fraction of dissolved value in the feed which is recovered in the overflow liquor from the first thickener, O and U are the overflow and underflow liquor volumes per unit weight of underflow solids, and N is the number of stages. Equation (18-49) applies to a system in which the circuit receives dry solids with which the second-stage thickener overflow is mixed to extract the soluble component. In this instance, O' refers to the overflow volume from the thickeners following the first stage.

For more precise values, computer programs can be used to calculate soluble recovery as well as solution compositions for conditions that are typical of a CCD circuit, with varying underflow concentrations, stage efficiencies, and solution densities in each of the stages. The calculation sequence is easily performed by utilizing material-balance equations around each thickener.

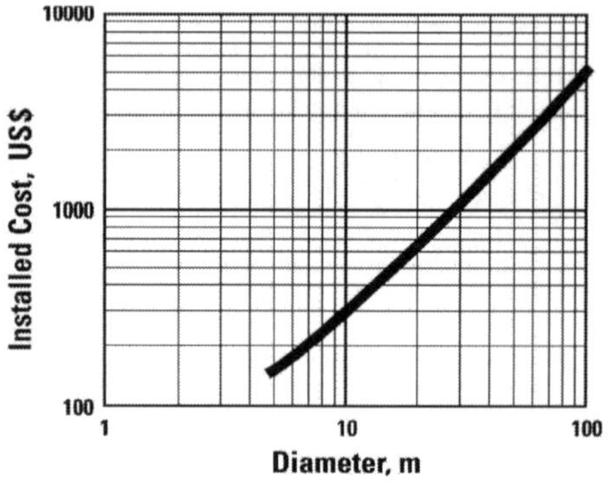

FIG. 18-106 Approximate installed cost of single-compartment thickeners (2005 US $).

DESIGN SIZING CRITERIA

Table 18-7 has the typical design sizing criteria and operating conditions for a number of applications. It is presented for purposes of illustration or preliminary estimate. Actual thickening and classification performance is dependent on particle-size destribution, specific gravity, sludge bed compaction characteristics, and other factors. Final design should be based on bench scale tests.

THICKENER COSTS

Equipment Costs vary widely for a given diameter because of the many types of construction. As a general rule, the total installed cost will be about 3 to 4 times the cost of the raking mechanism (including drivehead and lift), plus walkways and bridge or centerpier cage, railings, and overflow launders. Figure 18-106 shows the approximate installed costs of thickeners up to 107 m (350 ft) in diameter. These costs are to be used only as a guide. They include the erection of mechanism and tank plus normal uncomplicated site preparation, excavation, reinforcing bar placement, backfill, and surveying. The price does not include any electrical work, pumps, piping, instrumentation, walkways, or lifting mechanisms. Special design modifications, which are not in the price, could include elevated tanks (for underflow handling); special feedwell designs to control dilution, entrance velocity, and turbulence; electrical and drive enclosures required because of climatic conditions; and mechanism designs required because of scale buildup tendencies.

Operating Costs Power cost for a continuous thickener is an almost insignificant item. For example, a unit thickener 60 m (200 ft) in diameter with a torque rating of 1.0 MN·m (8.8 Mlbf·in) will normally require 12 kW (16 hp). The low power consumption is due to the very slow rotative speeds. Normally, a mechanism will be designed for a peripheral speed of about 9 m/min (0.5 ft/s), which corresponds to only 3 r/h for a 60-m (200-ft) unit. This low speed also means very low maintenance costs. Operating labor is low because little attention is normally required after initial operation has balanced the feed and underflow. If chemicals are required for flocculation, the chemical cost frequently dwarfs all other operating costs.

FILTRATION

GENERAL REFERENCES: Moir, *Chem. Eng.,* **89**(15), 46 (1982). Brown, ibid., 58; also published as McGraw-Hill Repr. A078. Cheremisinoff and Azbel, *Liquid Filtration,* Ann Arbor Science, Woburn, Mass., 1983. Orr (ed.), *Filtration: Principles and Practice,* part I, Marcel Dekker, New York, 1977; part II, 1979. Purchas (ed.), *Solid/Liquid Separation Equipment Scale-Up,* Uplands Press, Croydon, England, 1977. Schweitzer (ed.), *Handbook of Separation Techniques for Chemical Engineers,* part 4, McGraw-Hill, New York, 1979. Shoemaker (ed.), "What the Filter Man Needs to Know about Filtration," *Am. Inst. Chem. Eng. Symp. Ser.,* **73**(171), (1977). Talcott et al., in *Kirk-Othmer Encyclopedia of Chemical Technology,* 3d ed., vol. 10, Wiley, New York, 1980, p. 284. Tiller et al., *Chem. Eng.,* **81**(9), 116–136 (1974); also published as McGraw-Hill Repr. R203.

DEFINITIONS AND CLASSIFICATION

Filtration is the separation of a fluid-solids mixture involving passage of most of the fluid through a porous barrier which retains most of the solid particulates contained in the mixture. This subsection deals only with the filtration of solids from liquids; gas filtration is treated in Sec. 17. **Filtration** is the term for the unit operation. A filter is a piece of unit-operations equipment by which filtration is performed. The **filter medium** or **septum** is the barrier that lets the liquid pass while retaining most of the solids; it may be a screen, cloth, paper, or bed of solids. The liquid that passes through the filter medium is called the **filtrate.**

Filtration and filters can be classified several ways:

1. *By driving force.* The filtrate is induced to flow through the filter medium by hydrostatic head (gravity), pressure applied upstream of the filter medium, vacuum or reduced pressure applied downstream of the filter medium, or centrifugal force across the medium. Centrifugal filtration is closely related to centrifugal sedimentation, and both are discussed later under "Centrifuges."

2. *By filtration mechanism.* Although the mechanism for separation and accumulation of solids is not clearly understood, two models are generally considered and are the basis for the application of theory to the filtration process. When solids are stopped at the surface of a filter medium and pile upon one another to form a cake of increasing thickness, the separation is called **cake filtration.** When solids are trapped within the pores or body of the medium, it is termed **depth, filter-medium,** or **clarifying filtration.**

3. *By objective.* The process goal of filtration may be dry solids (the cake is the product of value), clarified liquid (the filtrate is the product of value), or both. Good solids recovery is best obtained by cake filtration, while clarification of the liquid is accomplished by either depth or cake filtration.

4. *By operating cycle.* Filtration may be intermittent (batch) or continuous. Batch filters may be operated with constant-pressure driving force, at constant rate, or in cycles that are variable with respect to both pressure and rate. Batch cycle can vary greatly, depending on filter area and solids loading.

5. *By nature of the solids.* Cake filtration may involve an accumulation of solids that is compressible or substantially incompressible, corresponding roughly in filter-medium filtration to particles that are deformable and to those that are rigid. The particle or particle-aggregate size may be of the same order of magnitude as the minimum pore size of most filter media (1 to 10 μm and greater), or may be smaller (1 μm down to the dimension of bacteria and even large molecules). Most filtrations involve solids of the former size range; those of the latter range can be filtered, if at all, only by filter-medium-type filtration or by ultrafiltration unless they are converted to the former range by aggregation prior to filtration.

These methods of classification are not mutually exclusive. Thus filters usually are divided first into the two groups of cake and clarifying equipment, then into groups of machines using the same kind of driving force, then further into batch and continuous classes. This is the scheme of classification underlying the discussion of filters of this subsection. Within it, the other aspects of operating cycle, the nature of the solids, and additional factors (e.g., types and classification of filter media) will be treated explicitly or implicitly.

FILTRATION THEORY

While research has developed a significant and detailed filtration theory, it is still so difficult to define a given liquid-solid system that it is both faster and more accurate to determine filter requirements by performing small-scale tests. Filtration theory does, however, show how the test data can best be correlated, and extrapolated when necessary, for use in scale-up calculations.

In cake or surface filtration, there are two primary areas of consideration: continuous filtration, in which the resistance of the filter cake (deposited process solids) is very large with respect to that of the filter media and filtrate drainage, and batch pressure filtration, in which the resistance of the filter cake is not very large with respect to that of the filter media and filtrate drainage. Batch pressure filters are generally fitted with heavy, tight filter cloths plus a layer of precoat and these represent a significant resistance that must be taken into account. Continuous filters, except for precoats, use relatively open cloths that offer little resistance compared to that of the filter cake.

Simplified theory for both batch and continuous filtration is based on the time-honored Hagen-Poiseuille equation:

$$\frac{1}{A}\frac{dV}{d\Theta} = \frac{P}{\mu(\alpha w V/A + r)} \tag{18-54}$$

where V is the volume of filtrate collected, Θ is the filtration time, A is the filter area, P is the total pressure across the system, w is the weight of cake solids/unit volume of filtrate, μ is the filtrate viscosity, α is the cake-specific resistance, and r is the resistance of the filter cloth plus the drainage system.

CONTINUOUS FILTRATION

Since testing and scale-up are different for batch and continuous filtration, discussion in this section will be limited to continuous filtration.

It is both convenient and reasonable in continuous filtration, except for precoat filters, to assume that the resistance of the filter cloth plus filtrate drainage is negligible compared to the resistance of the filter cake and to assume that both pressure drop and specific cake resistance remain constant throughout the filter cycle. Equation (18-54), integrated under these conditions, may then be manipulated to give the following relationships:

$$W = \sqrt{\frac{2wP\Theta_f}{\mu(\alpha w V/A + r)}} \tag{18-55}$$

$$V_f = \sqrt{\frac{2P\Theta_f}{\mu\alpha w}} \tag{18-56}$$

$$\Theta_w = \frac{WV_W\mu\alpha}{P_w} \tag{18-57}$$

$$\Theta_W \propto NW^2 \tag{18-58}$$

$$\frac{\Theta_W}{\Theta_f} = 2\frac{V_W}{V_f} \tag{18-59}$$

where W is the weight of dry filter cake solids/unit area, V_f is the volume of cake formation filtrate/unit area, V_w is the volume of cake wash filtrate/unit area, Θ_f is the cake formation time, Θ_w is the cake wash time, and N is the wash ratio, the volume of cake wash/volume of liquid in the discharged cake.

As long as the suspended solids concentration in the feed remains constant, these equations lead to the following convenient correlations:

$$\log W \text{ vs. } \log \Theta_f \tag{18-60}$$

$$\log V_f \text{ vs. } \log \Theta_f \tag{18-61}$$

$$\Theta_w \text{ vs. } WV_w \tag{18-62}$$

$$\Theta_w \text{ vs. } NW^2 \tag{18-63}$$

$$\Theta_{w/f} \text{ vs. } V_w/V_f \tag{18-64}$$

There are two other useful empirical correlations as follows:

$$W \text{ vs. cake thickness} \tag{18-65}$$

$$\log R \text{ vs. } N \tag{18-66}$$

where R is percent remaining—the percent of solute in the unwashed cake that remains after washing.

FACTORS INFLUENCING SMALL-SCALE TESTING

[Purchas (ed.), *Solid/Liquid Separation Equipment Scale-Up,* Uplands Press, Croydon, England, 1977.]

Vacuum or Pressure The vast majority of all continuous filters use vacuum to provide the driving force for filtration. However, if the feed slurry contains a highly volatile liquid phase, or if it is hot, saturated, and/or near the atmospheric pressure boiling point, the use of pressure for the driving force may be required. Pressure filtration might also be used where the required cake moisture content is lower than that obtainable with vacuum.

The objective of most continuous filters is to produce a dry or handleable cake. Most vacuum filters easily discharge a "dry" consolidated cake as they are usually operated in an open or semiopen environment. However, whenever the filter must operate under pressure or within a vapor-tight enclosure, either because of the need for a greater driving force or because of the vapor pressure of the liquid phase, a dry cake discharge becomes difficult. The problem of removing a dry cake from a pressurized enclosure has precluded the use of continuous-pressure filters in many cases where this is a requirement. Applications which do discharge dry cake from a "sealed" enclosure are restricted to relatively dry, friable cakes that will "flow" through double valves which form a "vapor lock."

Cake Discharge For any filter application to be practical, it must be possible to produce a cake thick enough to discharge. Table 18-8 tabulates the minimum acceptable cake thickness required for discharge for various types of filters and discharge mechanisms. The experimenter, when running small-scale tests, should decide early in the test program which type of discharge is applicable and then tailor the data collected to fit the physical requirements of that type of unit. Note, however, that the data correlations recommended later are sufficiently general in nature to apply to most equipment types.

TABLE 18-8 Minimum Cake Thickness for Discharge

Filter type	Minimum design thickness	
	mm	in
Drum		
Belt	3–5	$\frac{1}{8}$–$\frac{3}{16}$
Roll discharge	1	$\frac{1}{32}$
Std. scraper	6	$\frac{1}{4}$
Coil	3–5	$\frac{1}{8}$–$\frac{3}{16}$
String discharge	6	$\frac{1}{4}$
Precoat	0–3 max.	0–$\frac{1}{8}$ max.
Horizontal belt	3–5	$\frac{1}{8}$–$\frac{3}{16}$
Horizontal table	20	$\frac{3}{4}$
Tilting pan	20–25	$\frac{3}{4}$–1
Disc	10–13	$\frac{3}{8}$–$\frac{1}{2}$

Feed Slurry Temperature Temperature can be both an aid and a limitation. As temperature of the feed slurry is increased, the viscosity of the liquid phase is decreased, causing an increase in filtration rate and a decrease in cake moisture content. The limit to the benefits of increased temperature occurs when the vapor pressure of the liquid phase starts to materially reduce the allowable vacuum. If the liquid phase is permitted to flash within the filter internals, various undesired results may ensue: disruption in cake formation adjacent to the medium, scale deposit on the filter internals, a sharp rise in pressure drop within the filter drainage passages due to increased vapor flow, or decreased vacuum pump capacity. In most cases, the vacuum system should be designed so that the liquid phase does not boil.

In some special cases, steam filtration can be used to gain the advantages of temperature without having to heat the feed slurry. Where applicable, dry steam is passed through the deliquored cake to raise the temperature of the residual moisture, reduce its viscosity, and lower its content. The final drying or cooling period which follows steam filtration uses the residual heat left in the cake to evaporate some additional moisture.

Cake Thickness Control Sometimes the rate of cake formation with bottom feed–type filters is rapid enough to create a cake too thick for subsequent operations. Cake thickness may be controlled by adjusting the bridge-blocks in the filter valve to decrease the effective submergence, by reducing the slurry level in the vat, and by reducing the vacuum level in the cake formation portion of the filter valve. If these measures are inadequate, it may be necessary to use a top-loading filter.

Cake thickness must frequently be restricted when cake washing is required or the final cake moisture content is critical. Where the time required for cake washing is the rate-controlling step in the filter cycle, maximum filtration rate will be obtained when using the minimum cake thickness that gives good cake discharge. Where minimum cake moisture content is the controlling factor, there is usually some leeway with respect to cake thickness, although the minimum required for cake discharge is controlling in some cases. Since a relatively constant quantity of moisture is transferred from the medium to the filter cake when the vacuum is released prior to cake discharge, very thin cakes will sometimes be wetter than thicker cakes.

The effect of an increase in cake thickness on the time required for washing is easy to see if one considers what happens when the cake thickness is doubled. Assume that two cakes have the same permeability and that the quantity of washing fluid to cake solids is to remain constant. Doubling of the cake thickness doubles the resistance to flow of the washing fluid through the cake. At the same time, the quantity of washing fluid per unit area is also doubled. Thus, the time required for the washing fluid to pass through the cake is increased by the square of the ratio of the cake thicknesses. In this particular example, the washing time would be increased by a factor of four, while cake production would only be doubled.

Filter Cycle Each filter cycle is composed of cake formation plus one or more of the following operations: deliquoring (dewatering or drying), washing, thermal drying, steam drying, and cake discharge. The number of these operations required by a given filtration operation depends upon the process flowsheet. It is neither possible nor necessary to consider all of these operations at once. The basic testing program is designed to look at each operation individually. The requirements for each of the steps are then fit into a single filter cycle.

All filters utilizing a rotary filter valve have their areas divided into a number of sections, sectors, or segments (see Fig. 18-134). When a drainage port passes from one portion of the filter valve to another, the change at the filter medium does not occur instantaneously nor does it occur at some precise location on the filter surface. The change is relatively gradual and occurs over an area, as the drainage port at the filter valve first closes by passing onto a stationary bridge-block and then opens as it passes off that bridge-block on the other side.

On a horizontal belt filter, the equivalent sections extend across the filter in narrow strips. Therefore, changes in vacuum do occur rapidly and may be considered as happening at a particular point along the length of the filter.

Representative Samples The results which are obtained in any bench-scale testing program can be only as good as the sample which is tested. It is absolutely essential that the sample used be representative of the slurry in the full-scale plant and that it be tested under the conditions that prevail in the process. If there is to be some significant time between taking or producing the sample and commencing the test program, due consideration must be given to what effect this time lapse may have on the characteristics of the slurry. If the slurry is at a temperature different from ambient, the subsequent heating and/or cooling could change the particle size distribution. Even sample age itself may exert a significant influence on particle size. If there is likely to be an effect, the bench-scale testing program should be carried out at the plant or laboratory site on fresh material.

Whenever a sample is to be held for some time or shipped to a distant laboratory for testing, some type of characterizing filtration test should be run on the fresh sample and then duplicated at the time of the test program. A comparison of the results of the two tests will indicate how much of a change there has been in the sample. If the change is too great, there would be no point in proceeding with the tests, and it would be necessary to make arrangements to work on a fresh sample. Any shipped sample, especially during the winter months, must be protected from freezing, as freezing can substantially change the filtration characteristics of a slurry, *particularly hydrated* materials.

The slurry should always be defined as completely as possible by noting suspended solids concentration, particle size distribution, viscosity, density of solids and liquid, temperature, chemical composition, and so on.

Feed Solids Concentration Feed slurries that are so dilute that they settle rapidly usually yield reduced solids filtration rates and produce stratified cakes with higher moisture contents than would normally be obtained with a homogeneous cake. It is well known that an increase in feed solids concentration is generally an effective means of increasing solids filtration rate, assisting in forming a homogeneous suspension and thereby minimizing cake moisture content, and so on. Equipment required to concentrate a slurry sample and the tests needed to predict how far a slurry will thicken are discussed elsewhere in this section.

Pretreatment Chemicals Even though the suspended solids concentration of the slurry to be tested may be correct, it is frequently necessary to modify the slurry in order to provide an acceptable filtration rate, washing rate, or final cake moisture content. The most common treatment, and one which may provide improvement in all three of these categories, is the addition of flocculating agents, either inorganic chemicals or natural or synthetic polymers. The main task at this point is to determine which is the most effective chemical and the quantity of chemical which should be used.

It is usually difficult to observe visually a change in floc structure in a concentrated slurry. The two best indications that an effective quantity of chemical has been added is a sudden thickening or increase in viscosity of the slurry and the formation of *riverlets* on the surface of a spatula when treated slurry is shaken from it. It is generally necessary to exceed a threshold quantity of chemical before there is a measurable improvement. The proper dosage becomes an economic balance between the cost of additional chemicals and the savings resulting from a reduction in filter area.

Screening tests are used to determine the best chemical and its approximate dosage. It is usually convenient to use small, graduated beakers and sample quantities in the range of 50 to 150 mL. The chemical may be added with a syringe or medicine dropper and a note made of the quantity used, together with the results. The experimenter should filter and wash the flocculated sample on any convenient, small, top-loading filter with good filtrate drainage. The cake formation and wash times obtained from these micro tests are not intended to provide sizing data, but they do provide an excellent indication of the relative effectiveness of various chemicals and treatment levels.

With any chemical treatment system, the main task is one of getting the chemical thoroughly mixed with the solids without degrading the flocs which are formed. For those slurries that are relatively fluid, the chemical can frequently be added and mixed satisfactorily using a relatively wide spatula. However, for those thick, relatively viscous slurries, a power mixer will be required. In this case, the mixer should be stopped about one second after the last of the flocculant is added. Should this approach be required, it means that a suitably designed addition system must be supplied with the full-scale installation in order to do an effective job of flocculation.

While the volume of chemical used should be minimized, the experimenter must use good judgment based on the viscosities of both the slurry to be treated and the chemical used. If both are relatively viscous, use of a power mixer is indicated.

There are a number of commercially available surfactants that can be employed as an aid in filter cake moisture reduction. These reagents can be added to the filter feed slurry or to the filter cake wash water, if washing is used. Since these reagents have a dispersing effect, flocculation may be required subsequently. Typical moisture reductions of 2 to 4 percentage points are obtained at reagent dosages of 200 to 500 g/mt of solids.

Cloth Blinding Continuous filters, except for precoats, generally use some type of medium to effect the separation of the solid and filtrate phases. Since the medium is in contact with the process solids, there is always the danger, and almost invariably the actual occurrence, of medium blinding. The term **blinding** refers to blockage of the fabric itself, either by the wedging of process solids or by solids precipitated in and around the yarn.

The filter medium chosen should be as open as possible yet still able to maintain the required filtrate clarity. Those fabrics which will produce a clear filtrate and yet do not have rapid blinding tendencies are frequently light in weight (woven from thin filaments or yarn) and will not wear as long as some of the heavier, more open fabrics (woven from heavy filaments or yarn). Whenever the filter follows a gravity thickening or clarification step, it is advisable to return the filtrate to the thickener or clarifier so that the filtrate clarity requirements may be relaxed in favor of using a heavier, more open cloth with reduced blinding tendencies. Excessively dirty filtrates should be avoided as the solids may be abrasive and detrimental to the internals of the filter or perhaps may cut the fabric yarn.

It should be noted at this point that an **absolutely** clear filtrate can rarely be obtained on a cloth-covered continuous filter. The passages through the medium are invariably larger than some of the solids in the slurry, and there will be some amount of solids passing through the medium. Once the pores of the fabric have been bridged, the solids themselves form the septum for the remaining particles, and the filtrate becomes clear. It is this bridging action of the solids that permits the use of a relatively open filter medium, while at the same time maintaining a reasonably clear filtrate.

Filters with media in the form of an endless belt have greatly reduced the concern about blinding. Most synthetic fabrics can be successfully cleaned of process solids by washing the medium after cake discharge, and the rate of blinding due to chemical precipitation also can be drastically reduced. Current practice suggests that the belt-type filter with continuous-medium washing be the first choice unless experience has shown that medium blinding is not a factor or if the belt-type system cannot be successfully applied.

Sealing of the belt along the edges of the filter drum is never perfect, and some leakage should be expected. If good clarity is essential, it may be preferable to use a drum filter with the cloth caulked in place and design the system to contend with the effects of blinding.

The one exception to the points noted above is the continuous precoat filter. Here the purpose of the filter medium is to act as a support for the sacrificial bed of precoat material. Thus, the medium should be tight enough to retain the precoat solids and prevent bleeding of the precoat solids through the filter medium during operation, yet open enough to permit easy cleaning at the end of each cycle. Lightweight felt media work well in these respects.

Homogeneous Cake Accurate test results and optimum filter performance require the formation of a homogeneous cake and thus the maintenance of a similarly homogeneous suspension. Settling in the sample container during a bottom-feed test program can usually be detected by comparing the back-calculated feed solids concentration (based on filtrate, wet cake, and dry cake weights) with the slurry solids concentration as prepared. It is normal to find that the back-calculated concentration is slightly lower than the prepared concentration. This difference is normally within 2 percentage points and should never be greater than 5 percentage points. Since this difference does exist, it means that the slurry sample will concentrate to some extent as the tests continue. Adding fresh slurry to the sample container after each test can counteract this condition, as the system will reach an equilibrium similar to that found in a full-scale machine.

If a more positive check is required on the quality of the filter cake, particle size analyses may be run and compared with the sample as prepared.

In a top-feed filter test, the filter cake will contain all of the solids, provided they are all emptied from the sample container. The danger in this type of test is that the solids will stratify, particularly if the cake formation time is prolonged. Close examination of the filter cake will indicate whether or not this has happened. If there has been significant stratification, the feed slurry should be modified by thickening and/or flocculation in order to increase the dry solids filtration rate and permit formation of a homogeneous cake. Another possibility, but not necessarily the best, is to use a thinner, but still dischargeable, cake to avoid stratification.

Agitation of Sample All slurries used in bottom-feed tests must be agitated by hand (if slurry characteristics permit) to check whether or not the solids are settling out around the edges of the container and to determine the degree of agitation required to maintain the solids in suspension. Generally speaking, if the solids can be maintained in suspension by hand agitation, the slurry can be processed by a bottom-feed-type filter.

Agitation by a wide spatula may be substituted for hand agitation, but only after it has been determined by feel that the spatula will provide the needed agitation. If this cannot be done, then confirmation of proper agitation must be based on back-calculated feed solids concentrations and/or particle size analyses of the filter cakes.

If it is not possible to maintain a uniform suspension, the sample should be thickened and the flowsheet modified to provide the required thickening.

Mechanical agitation of a sample is very difficult to use effectively. Generally speaking, if enough room is left in the sample container for the leaf and the agitator, the agitation is not sufficient to prevent settling out in the corners of the container. If sufficient agitation is used to maintain suspension in all parts of the container, then it is highly probable that the velocity of the slurry across the face of the test leaf will wash the solids from the leaf and give very erroneous results. Furthermore, the tendency is to leave the agitator running continuously, or at least for so long a period of time that there is attrition of the solids and, therefore, inaccurate results. Mechanical agitation during testing can generally be justified only in a most unusual and exceptional circumstance.

Use of Steam or Hot Air It was indicated earlier that the cycle might include steam filtration or thermal drying using hot air. While effective use is made of both steam and hot air, the applications are rather limited, and testing procedures are difficult and specialized. As a general rule, steam application will reduce cake moisture 2 to 4 percentage points. Hot-air drying can produce a bone-dry cake, but generally it is practical only if the air rate is high, greater than about 1800 $m^3/m^2 \cdot h$ (98 cfm/ft^2). Both systems require a suitable hood which must contact the dam on the leaf during the drying cycle, allowing the steam or hot air to pass through the cake without dilution by cold air. The end of the operation can be determined by a noticeable increase in the temperature of the gas leaving the leaf.

SMALL-SCALE TEST PROCEDURES

[Purchas (ed.), *Solid/Liquid Separation Equipment Scale-Up*, Uplands Press, Croydon, England, 1977.]

Apparatus There are several variations of the bench-scale test leaf that may be used, but they all have features similar to the one discussed below.

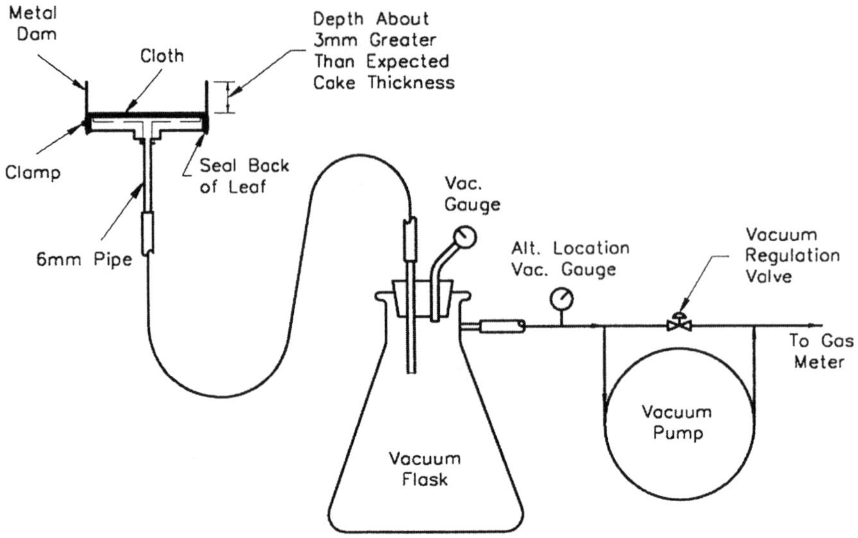

FIG. 18-107 Typical bottom-feed leaf test setup.

One typical test leaf is a circular disc with a plane area of 92.9 cm^2 (0.1 ft^2). One face of the leaf is grooved to provide large filtrate drainage passages and a support for the filter medium. A threaded drainage connection is provided on the center of the other face of the leaf. The test leaf is fitted with a filter medium and a dam, and the assembly clamped together as shown in Fig. 18-107. The depth of the dam for bottom-feed tests should be no greater than the depth of the maximum cake thickness, except where cake washing tests are to be performed. In this case, the dam depth should be about 3 mm (⅛ in) greater than the maximum expected cake thickness. Excessive dam depth will interfere with slurry agitation and can result in the formation of a nonhomogeneous cake.

It is absolutely necessary that a dam be used in all cases, except for roll discharge applications which do not involve cake washing or where the maximum cake thickness is on the order of 2 mm or less. If a dam is not used, filter cake will form past the edge of the leaf in the general shape of a mushroom. When this happens, the total filter area is some unknown value, greater than the area of the leaf, that constantly increases with time during cake formation.

The back of the leaf assembly and the joint where the dam overlaps must be sealed with some suitable material so that the filtrate volume collected accurately represents the liquid associated with the deposited cake solids.

Figure 18-108 also contains a schematic layout of the equipment which is required for all bottom-feed leaf tests. Note that there are no valves in the drainage line between the test leaf and the filtrate receiver, nor between the filtrate receiver and the vacuum pump.

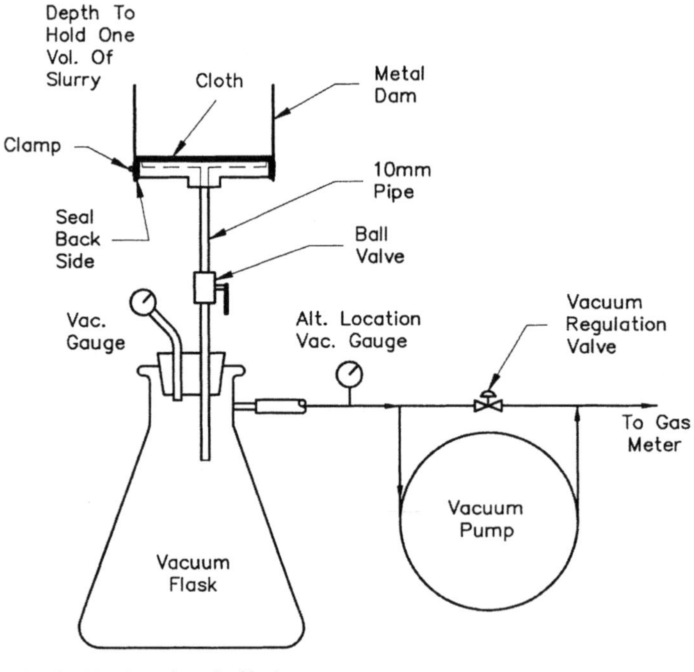

FIG. 18-108 Typical top-feed leaf test setup.

At the start of the leaf test run, the hose between the test leaf and filtrate receiver should be crimped by hand to bring the filtrate receiver to the operating vacuum level. The use of a valve at this point is not only less convenient but very frequently results in a hydraulic restriction. The net result, then, is a measurement of flow through the valve rather than the rate at which the filter cake is capable of forming. Hydraulic restriction is something which should always be kept in mind. If the filtrate runs at a high and full pipe flow rate into the filtrate receiver, it is quite likely that there is some degree of hydraulic restriction, and larger tubing and piping should be considered. When very high air flow rates are obtained, the experimenter must be satisfied that the rates being measured are limited by cake resistance and not by pressure drop through the equipment.

There will be many times when the quantity of sample is limited. While it is best to use the 92.9 cm^2 (0.1 ft^2) area leaf in order to minimize edge effects and improve accuracy, when the sample volume is limited it is much better to have several data points with a smaller leaf than only one or two using the larger leaf. Data from leaves as small as 23.2 cm^2 (0.025 ft^2) are reasonably accurate and can be used to scale up to commercially sized units. However, it is usually prudent to employ a more conservative scale-up factor.

For top-feed applications, the most convenient assembly is that shown in Fig. 18-108. The depth of the dam must, of course, be sufficient to contain the total quantity of feed slurry required for the test. Since the test leaf is mounted on top of the vacuum receiver, it is necessary to provide a valve between the test leaf and the receiver so that the desired operating vacuum may be obtained in the receiver before the start of a test run. It is imperative, however, that there be no restriction in this valve. The preferred choice is a ball valve with the full bore of the drainage piping.

Test Program Figure 18-109 is a suggested data sheet which contains spaces for most of the information which should be taken during a leaf test program, together with space for certain calculated values. Additional data which may be required include variations in air flow rate through the cake during each dewatering period and chemical and physical data for those tests involving cake washing.

It is difficult to plan a filtration leaf test program until one test has been run. In the case of a bottom-feed test, the first run is normally started with the intention of using a 30-s cake formation time. However, if the filtrate rate is very high, it is usually wise to terminate the run at the end of 15 s. Should the filtrate rate be very low, the initial form period should be extended to at least 1 min. If cake washing is to be employed, it is useful to apply a quantity of wash water to measure its rate of passage through the cake. The results of this first run will give the experimenter an approximation of cake formation rate, cake washing rate, and the type of cake discharge that must be used. The

FILTRATION LEAF TEST DATA SHEET – VACUUM AND PRESSURE

Company________ Mat'l as Received: Date ________ Test No. ________

Address ________ Solids: ________ % Date Tested ________

________ Analysis ________ By ________

________ Liquid: ________ % Location ________

Filter Type______ Leaf Size______ Ft.^{2} Analysis ________

Used Shim: No______ Yes______ Precoat Forming Liquid ________ Temp. ________ °F/°C

Run No.	Filter Module and/or Precoat Type	Feed Temp., °F/°C	% Solids in Food		Vacuum = in. Hg. Pressure = PSI.				TIME, MIN.					Air Flow		Filtrate		Precoat Penetration	Wash		Cake/Precoat Thickness, In.	Dia. of Shared Area, In.	Cake Weights			
			As Prepared	Back Calculated	Form	Wash	Dry	After Cake Cracks	Form	Dewater	Wash	Dry	To Crack or Gas Breakthrough After Form/Wash	(1)	Temp., °F/°C	ML.	Clarity		ML.	Temp., °F/°C			Tare GMS.	Wet & Tare GMS.	Dry & Tare GMS.	Dish No.

CAKE DISCHARGE		REAGENT TREATMENT	
RUNS	COMMENT	RUNS	COMMENT

REMARKS: (1) Record Basis of Observation in Space Provided.

FIG. 18-109 Sample data sheet.

rest of the leaf test program can then be planned accordingly.

In any leaf test program there is always a question as to what vacuum level should be used. With very porous materials, a vacuum in the range of 0.1 to 0.3 bar (3 to 9 in Hg) should be used, and, except for thermal-drying applications using hot air, the vacuum level should be adjusted to give an air rate in the range of 450 to 900 $m^3/m^2 \cdot h$ (30 to 40 cfm/ft^2) measured at the vacuum.

For materials of moderate to low porosity, a good starting vacuum level is 0.6 to 0.7 bar (18 to 21 in Hg), as the capacity of most vacuum pumps starts to fall off rapidly at vacuum levels higher than 0.67 bar (20 in Hg). Unless there is a critical moisture content which requires the use of higher vacuums, or unless the deposited cake is so impervious that the air rate is extremely low, process economics will favor operation at vacuums below this level. When test work is carried out at an elevation above sea level different than that of the plant, the elevation at the plant should be taken into account when determining the vacuum system capacity for high vacuum levels (>0.5 bar).

Generalized correlations are available for each of the operations which make up the full filter cycle. This means that simulated operating conditions can be varied to obtain a maximum of information without requiring an excessive number of test runs. The minimum number of test runs required for a given feed will, of course, vary with the expertise of the experimenter and the number of operations performed during the filter cycle. If, for example, the operation involves only the dewatering of a slurry which forms a cake of relatively low to moderate porosity, frequently sufficient data can be obtained in as little as six runs. For more difficult tests, more runs are usually advisable, and the novice certainly should make a larger number of runs as there is likely to be more data scatter.

Bottom-Feed Test Procedure The procedure for collecting data using bottom-feed leaf test techniques is as follows:

1. Fit the test leaf with a filter cloth expected to give reasonable results and seal the back of the leaf and side of the dam with silicone or other suitable material.
2. Hand-crimp the hose in back of the test leaf, and then turn on the vacuum pump and regulate the bypass valve on the pump to give the desired vacuum level in the receiver.
3. Agitate the slurry by hand or with a wide spatula to maintain a homogeneous suspension. Immerse the test leaf face downward to approximately one-half the depth of the slurry.
4. Simultaneously start the timer and release the crimped hose to begin cake formation. Maintain agitation during cake formation and move the leaf as may be required to ensure that solids do not settle out in any part of the container. It is not necessary to try to simulate the velocity with which the full-scale unit's filtration surface passes through the slurry in the filter tank.
5. Remove the leaf from the slurry at the end of the cake-formation period and note the time. If the slurry is particularly thick and viscous, the leaf may be gently shaken to remove excess slurry and prevent the dam from scooping up extra material. Maintain the leaf in an upright position (cake surface on top) and elevated so that liquid within the drainage passages may pass to the receiver. Tilt and rotate the leaf to help the filtrate reach the drain outlet. Continue this dewatering period until:
    a. the preselected time has elapsed, or
    b. the cake cracks.
6. If the cake is to be washed, apply a measured quantity of wash fluid and note the time required for free fluid to disappear from the surface of the cake. Pour the wash fluid onto a deflecting baffle, such as a bent spatula, to prevent the cake from being gouged. Washing must begin before cake cracking occurs. In particular, observe that there is no crack along the edge between the cake and the dam.
7. Continue with the various operations in the predetermined sequence.
8. During each of the operations record all pertinent information such as vacuum level, temperature, time required for the cake to crack, filtrate foaming characteristics, air flow rate during the drying periods, etc.
9. At the end of the run, measure and record the filtrate volume (and weight, if appropriate), cake thickness, final cake temperature (if appropriate), wet cake weight, and note the cake discharge characteristics (roll, sticks to media, etc.).
10. For runs involving cake dewatering only, it is usually convenient to dry the total cake sample, if the associated solution contains little or no dissolved solids.
11. When cake washing is involved, it is usually convenient to weigh the wet cake and then repulp it in a known quantity of distilled water or in water at the same pH as the filtrate, if precipitation of solute could occur in distilled water. The resultant slurry is then filtered using a clean dry filter and flask and a sample of the clear liquid analyzed for the reference constituent.

Should the mother liquor contain a significant quantity of dissolved solids, the filter cake should be thoroughly washed (after the sample for analysis has been taken) so that the final dry weight of the cake will represent suspended solids only. The quantity of reference constituent in the final washed cake can be readily calculated from the wet and dry cake weights and the known amount of distilled water used for repulping.

In cake-washing tests, it is important that the feed slurry liquid be analyzed for total dissolved solids and density as well as the reference constituent.

Top-Feed Test Procedure The sequence of operations with a top-feed leaf test is the same as in a bottom-feed test, except that the leaf is not immersed in the slurry. The best method for transferring the slurry to the top-feed leaf is, of course, a function of the characteristics of the slurry. If the particles in the slurry do not settle rapidly, the feed can usually be transferred to the leaf from a beaker. If, however, the particles settle very rapidly, it is virtually impossible to pour the slurry out of a beaker satisfactorily. In this case, the best method is to make use of an Erlenmeyer flask, preferably one made of plastic. The slurry is swirled in the flask until it is completely suspended and then abruptly inverted over the leaf. This technique will ensure that all of the solids are transferred to the leaf.

When the solids involved are coarse and fast settling, the vacuum should be applied an instant after the slurry reaches the surface of the filter medium.

Precoat Test Procedure Precoat filtration tests are run in exactly the same manner as bottom-feed tests except that the leaf must first be precoated with a bed of diatomaceous earth, perlite, or other shaveable inert solids. Some trial and error is involved in selecting a grade of precoat material which will retain the filtered solids to be removed on the surface of the bed without any significant penetration. During this selection process, relatively thin precoat beds of 1 to 2 cm are satisfactory. After a grade has been selected, bench-scale tests should be run using precoat beds of the same thickness as expected on the full-scale unit.

Where the resistance of the precoat bed is significant in comparison to the resistance of the deposited solids, the thickness of the precoat bed effectively controls the filtration rate. In some instances, the resistance of the deposited solids is very large with respect to even a thick precoat bed. In this case, variations in thickness through the life of the precoat bed have relatively little effect on filtration rate. This type of information readily becomes apparent when the filtration rate data are correlated.

The depth of cut involved in precoat filtration is a very important economic factor. There is some disagreement as to the method required to accurately predict the minimum permissible depth of cut. Some investigators maintain that the depth of cut can be evaluated only in a qualitative manner during bench-scale tests by judging whether the process solids remain on the surface of the precoat bed. This being so, they indicate that it is necessary to run a continuous pilot-plant test to determine the minimum permissible depth of cut. The use of a continuous pilot-plant filter is a very desirable approach and will provide accurate information under a variety of operating conditions.

However, it is not always possible to run a pilot-plant test in order to determine the depth of cut. A well-accepted alternative approach makes use of the more sophisticated test leaf illustrated in Fig. 18-110. This test leaf is designed so that the cake and precoat are extruded axially out the open end of the leaf. The top of the retaining wall on this end of the leaf is a machined surface which serves as a support for a sharp discharge knife. This approach permits variable and known depths of cut to be made so that the minimum depth of cut may be determined. Test units are available from Betts Advanced Metal, hompoc, Calif., (805) 735-5130.

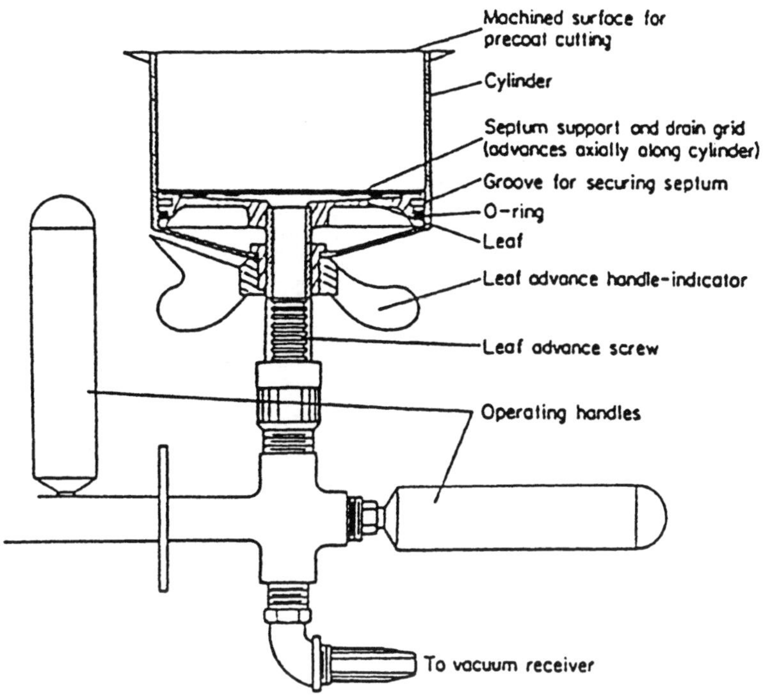

FIG. 18-110 Special test leaf for precoat filtration.

Lacking the above-described actual data, it is possible to estimate precoat consumption by using these values: nonpenetrating solids, 0.06-mm cut/drum revolution (0.0024 in); visible penetration, 0.15- to 0.20-mm cut/drum revolution (0.006 to 0.008 in); precoat bed density, 4.2 kg/m^2·cm of bed depth (2.2 lb/ft^2·in) for diatomaceous earth or 2.1 to 3.0 kg/m^2·cm (1.1 to 1.6 lb/ft^2·in) for perlite.

DATA CORRELATION

[Purchas (ed.), *Solid/Liquid Separation Equipment Scale-Up,* Uplands Press, Croydon, England, 1977.]

The correlations used are based partly on theoretical consideration and partly on empirical observations. The basic filtration data are correlated by application of the classic cake-filtration equation, aided by various simplifying assumptions which are sufficiently valid for many (but not all) situations. Washing and drying correlations are of a more empirical nature but with strong experimental justification. If steam or thermal drying is being examined, additional correlations are required beyond those summarized below; for such applications, it is advisable to consult an equipment manufacturer or refer to published technical papers for guidance.

Dry Cake Weight vs. Thickness It is convenient to convert the test dry cake weight to the weight of dry cake per unit area per cycle (W), and plot these values as a function of cake thickness (Fig. 18-111). Cake weight is measured quite accurately, while cake thickness measurements are subject to some variation. By plotting the data, variations in thickness measurements are averaged. The data usually give a straight line passing through the origin. However, with compressible material, sometimes a slightly curved line best represents the data, since thinner cakes are usually compressed more than thicker cakes.

Dry Solids or Filtrate Rate Filtration rate, expressed either in terms of dry solids or filtrate volume, may be plotted as a function of time on log-log paper. However, it is more convenient to delay the rate calculation until the complete cycle of operations has been defined.

It is most useful to plot either dry cake weight (weight of dry solids/unit area/cycle) or filtrate volume (volume/unit area/cycle) as a function of time on log-log paper. These data should give straight-line plots for constant operating conditions in accordance with Eqs. (18-55) and (18-56). The expected slope of the resultant rate/time plots is +0.50, as in Fig. 18-112. In practice, the vast majority of slopes range from +0.50 to +0.35. Slopes steeper than +0.5 indicate that there is some significant resistance other than that of the cake solids,

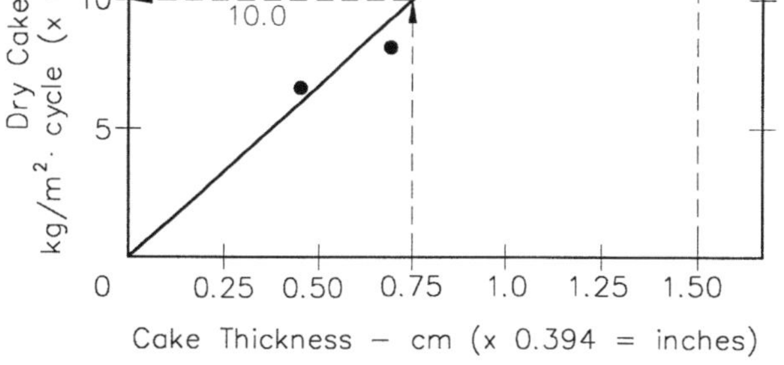

FIG. 18-111 Dry cake weight vs. cake thickness.

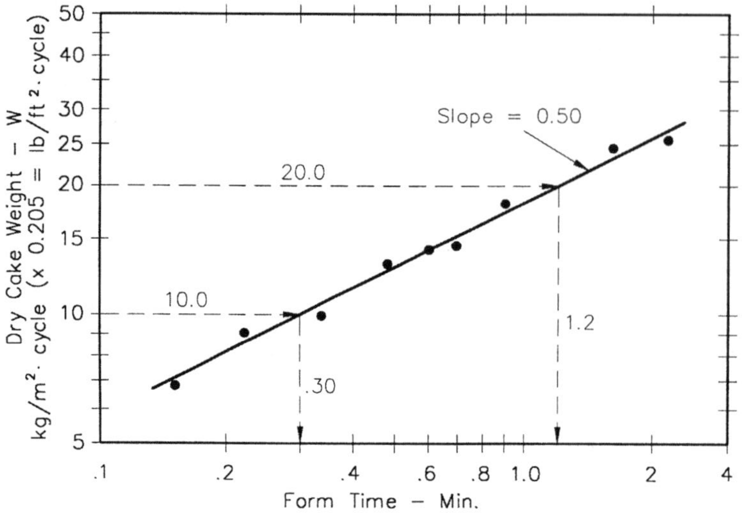

FIG. 18-112 Dry cake weight vs. form time.

such as a hydraulic restriction in the equipment or an exceptionally tight filter cloth.

Data from precoat tests, however, generally produce filtrate curves with much steeper slopes. The precoat bed has a greater resistance than most filter fabrics, and the particles which are separated on a continuous precoat usually form a cake which has a relatively low resistance when compared to that of the precoat bed. Once the thickness of the deposited solids becomes significant, their resistance increases. Thus, at very short form times, the slope of the filtrate curve may be close to 1.0, but as form time increases, the slope of the curve will decrease and will approach +0.5 (Fig. 18-113).

There are some solids, however, which form a less permeable cake, even in very thin layers. With these solids, the resistance of the deposited cake will be very high when compared to that of the precoat bed, and the slope of the filtrate curve will be +0.5 for all values of form time.

Effect of Time on Flocculated Slurries Flocculated slurries usually show significant decreases in filterability with time (Fig. 18-114). The rate of degradation may be established by running a series of repetitive leaf tests at frequent intervals on a flocculated slurry, starting as soon as practical after the addition of the flocculant. If there is little change in the filtration rate, this factor need be given no more consideration. However, it is usually found that there is significant degradation.

When a flocculated feed is added to a filter tank, there is a definite time lag before this material reaches the surface of the filter medium.

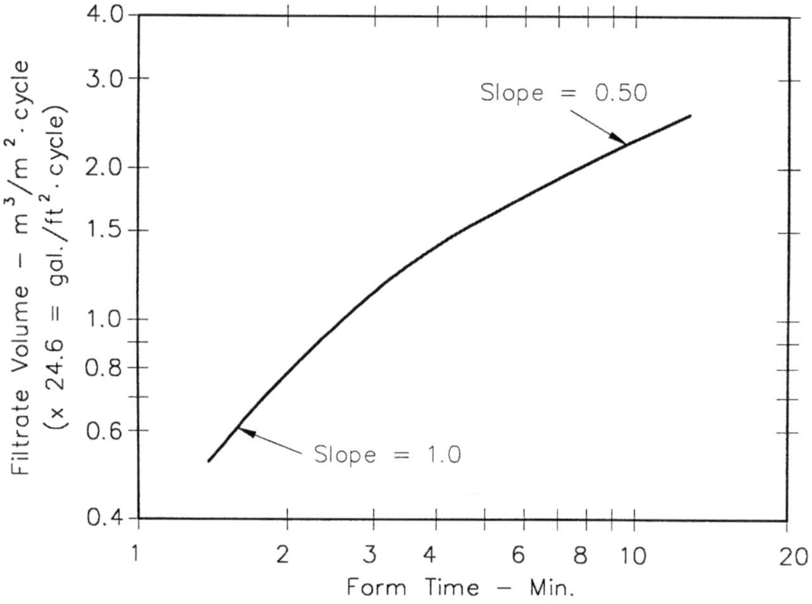

FIG. 18-113 Filtrate volume per cycle vs. form time.

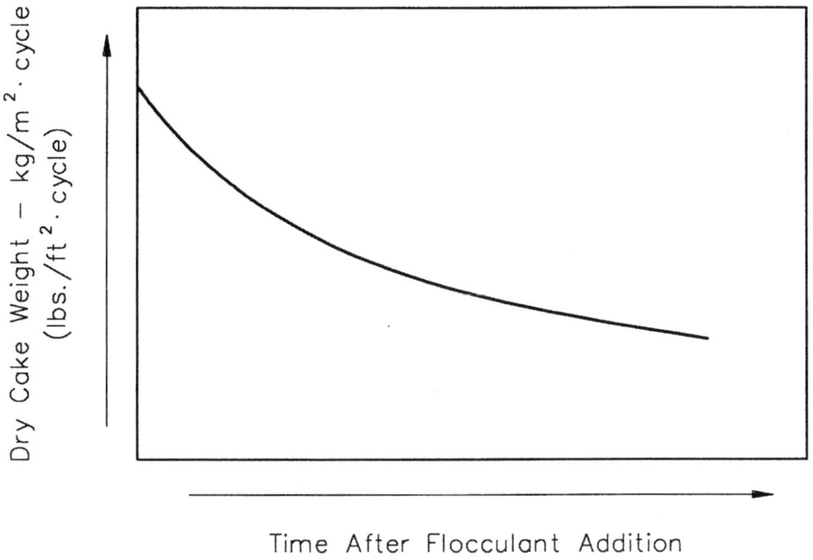

FIG. 18-114 Degradation of flocculation with time.

Since this lag time is not known at the time of testing, a lag time of 8 to 10 minutes should be allowed before starting the first leaf test on a flocculated slurry. Two, or perhaps three, tests can be run before the elapsed time exceeds the probable retention time in the full-scale filter tank. With knowledge of the elapsed time after flocculation and data relating to the rate of degradation, the rates obtained on the leaf test runs can be adjusted to some constant lag time consistent with the anticipated full-scale design.

Cake Moisture Results on a wide variety of materials have shown that the following factor is very useful for correlating cake moisture content data:

$$\text{Correlating factor} = (\text{m}^3/\text{m}^2\cdot\text{h})(P_c/W)(\Theta_D/\mu), \qquad (18\text{-}67)$$

where $\text{m}^3/\text{m}^2\cdot\text{h}$ = air rate through filter cake measured at downstream pressure or vacuum

P_c = pressure drop across cake

W = dry cake weight/unit area/cycle

Θ_D = dry time per cycle

μ = viscosity of liquid phase

For a more rigorous discussion of cake moisture correlation, the reader is referred to an earlier article by Nelson and Dahlstrom [*Chem. Eng. Progress*, **53,** 7, 1957]. Figure 18-115 shows the general shape of the curve obtained when using the cake moisture correlating factor. The value of the correlating factor chosen for design should be somewhere past the knee of the curve. Values at and to the left of the knee are in an unstable range where a small change in operating conditions can result in a relatively large change in cake moisture content.

It is not always necessary to use all of the terms in the correlating factor, and those conditions which are held constant throughout

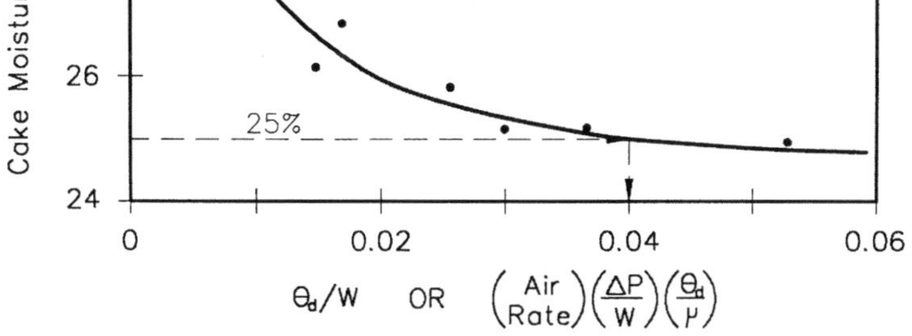

FIG. 18-115 Cake moisture correlation.

the testing may be dropped from the correlating factor. Many times, air rate data are not available and reasonable correlations can be obtained without this information, particularly if the cakes are relatively low in permeability. By dropping these terms, the correlating factor is reduced to the simplified version, Θ_D/W, involving only drying time and cake weight per unit area per revolution. While this is a very convenient factor to use, care must be taken in its application, as it is no longer a generalized factor, and there will be a tendency for the data to produce different curves for changes in operating conditions such as vacuum level and cake thickness.

Cake Washing Wash efficiency data are most conveniently represented by a semilog plot of percent remaining R as a function of wash ratio N as shown in Fig. 18-116. Percent remaining refers to that portion of the solute in the dewatered but unwashed cake which is left in the washed and dewatered cake. Since a cake-washing operation involves the displacement of one volume of liquid by another volume, the removal of solute is related to the ratio of the volume of washing fluid divided by the volume of liquid in the cake. This ratio is defined as wash ratio N.

Practical experience has shown that the most convenient and best means of expressing R is in terms of the solute concentrations in the washed cake liquid, the feed liquid (or unwashed cake liquid), and the cake wash liquid. Furthermore, the wash ratio N may also be expressed either as a volume or weight ratio.

Percent remaining is defined as follows:

$$\frac{R}{100} = \frac{C_2 - C_w}{C_1 - C_w} \tag{18-68}$$

where R = % solute remaining after washing
C_2 = solute concentration in washed cake liquid
C_1 = solute concentration in unwashed cake liquid
C_w = solute concentration in wash liquid

If the cake is washed with solute-free liquid, percent remaining is readily calculated by dividing the solute concentration in the liquid remaining in the washed cake by the solute concentration in the liquid in the original feed.

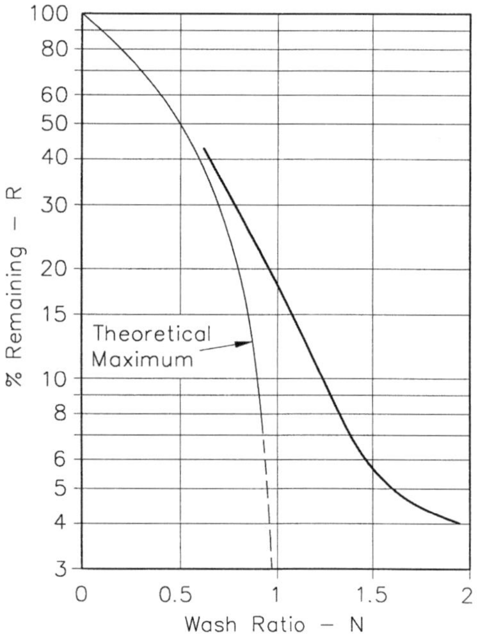

FIG. 18-116 Wash effectiveness.

The residence time of the cake-washing fluid within the cake is relatively short and is not normally considered useful for any kind of leaching operation. Therefore, it is assumed that all of the solute is in solution.

If it were possible to obtain a perfect slug displacement wash, the fraction remaining would be numerically equal to 1 minus the wash ratio. This ideal condition is represented by the maximum theoretical line as shown in Fig. 18-116. Since it represents the best that can be done, no data point should fall to the left of this curve. Most, but not all, cake-washing curves tend to fall along the heavy solid line shown. In the absence of actual data, one may estimate washing results by using this curve.

The quantity of wash water to be used in a given operation is dictated by flowsheet considerations and the required solute content of the washed cake. Generally speaking, the maximum wash water quantity should be equivalent to a wash ratio of 1.5 to 2.5. Where high solute removals are required, it is frequently necessary to use a two-stage filtration system with intermediate repulping. These two stages may involve countercurrent flow of wash water, or fresh wash water may be used on both filters and for the intermediate repulping step.

Wash Time Cake-washing time is the most difficult of the filtration variables to correlate. It is obviously desirable to use one which provides a single curve for all of the data. Filtration theory suggests three possible correlations [Eqs. (18-62) to (18-64)]. These are listed below, beginning with the easiest to use:

1. Wash time vs. WV_w
2. Wash time vs. NW^2
3. Wash time/form time vs. wash volume/form volume

where W = weight dry cake/unit area/cycle
V_w = volume of cake wash/unit area/cycle
N = wash ratio
= volume of wash/volume of liquid in discharged cake

Fortunately, the easiest correlation to use usually gives satisfactory results. This curve is usually a straight line passing through the origin, but frequently falls off as the volume of wash water increases (Fig. 18-117). If for some reason this correlation is not satisfactory, one of the other two should be tried.

Air Rate Air rate through the cake, and thus vacuum pump capacity, can be determined from measurements of the air flow for various lengths of dry time. Figure 18-118 represents instantaneous air rate data. The total volume of gas passing through the cake during a dry period is determined by integrating under the curve.

Note that the air rate at the beginning of each drying or dewatering period within a given cycle starts at zero and then increases with time.

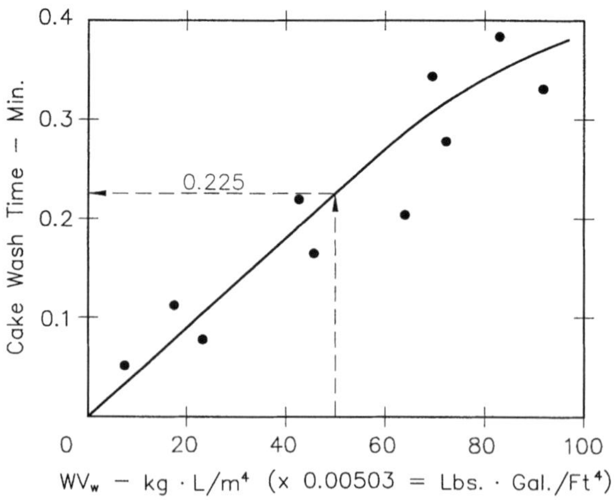

FIG. 18-117 Cake wash time correlation.

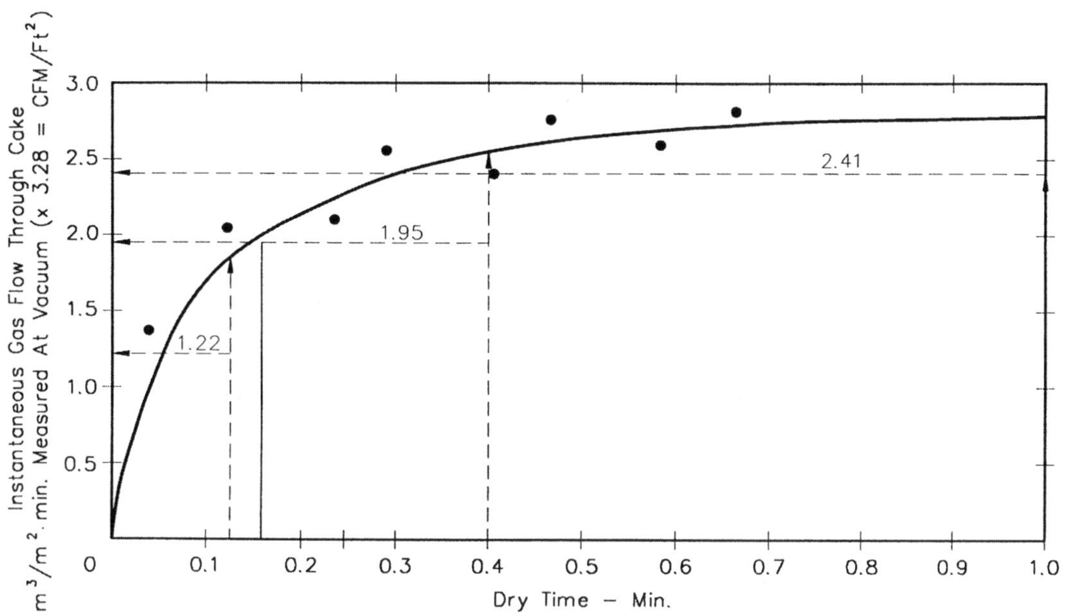

FIG. 18-118 Airflow through cake.

The shape of this curve will be a function of the permeability of the deposited cake.

Vacuum pump capacity is conventionally based on the total cycle and expressed as $m^3/h \cdot m^2$ (cfm/ft^2) of filter area measured at pump inlet conditions. Thus, the gas volumes per unit area passing during each dry period in the cycle are totaled and divided by the cycle time to arrive at the design air rate. Since air rate measurements in the test program are based on pressure drop across the cake and filter medium only, allowance must be made for additional expansion due to pressure drop within the filter and auxiliary piping system in arriving at vacuum pump inlet conditions.

Air rate measurements made during a leaf test program account only for gas flow through the cake. An operating filter has drainage passages that must also be evacuated. The extra air flow may be conveniently accounted for in most cases by multiplying the leaf test rate by 1.10. With horizontal belt filters, one must also add for the leakage that occurs along the sliding seal and, depending upon the type of filter cloth used, edge leakage due to lateral permeability of the cloth. Typically, the total leakage can be significant, amounting to about 35 to 50 $m^3/h \cdot m^2$ (2 to 3 cfm/ft^2).

Adjustment may also be required for differences in altitude between the test site and the commercial installation. In general terms, if the plant elevation is higher, the vacuum pump size must be increased, and conversely.

Darcy's law has been used to derive an expression which reflects not only the effect of a change in elevation, but also provides a means for estimating changes in air rate resulting from changes in vacuum level and cake thickness (or cake weight per unit area). In order for this relationship to hold for changes in vacuum and cake thickness, it must be assumed that both cakes have the same specific resistance.

The generalized equation is as follows:

$$(\text{Air rate})_{b2} = (\text{air rate})_{a2} \frac{(W_a)}{(W_b)} \frac{(P_{b1}^2 - P_{b2}^2)}{(P_{a1}^2 - P_{a2}^2)} \frac{(P_{a2})}{(P_{b2})} \quad (18\text{-}69)$$

where P_1 = inlet absolute pressure
P_2 = outlet absolute pressure
a = base condition
b = revised condition
W = weight of dry cake solids/unit area/cycle

SCALE-UP FACTORS

[Purchas (ed.), *Solid/Liquid Separation Equipment Scale-Up,* Uplands Press, Croydon, England, 1977.]

The overall scale-up factor used to convert a rate calculated from bench-scale data to a design rate for a commercial installation must incorporate separate factors for each of the following:

Scale-up on rate
Scale-up on cake discharge
Scale-up on actual area

Special note should be made that these scale-up factors are *not* safety factors to allow for additional plant capacity at some future date. Their purpose is to account for differences in scale, including such things as minor deviations in the plant slurry from the sample tested, edge effects due to the size of the test equipment, close control of operating conditions during the leaf test program, and long-term medium blinding.

Scale-up on Rate Filtration rates calculated from bench-scale data should be multiplied by a factor of 0.8 for all types of commercial units which do not employ continuous washing of the filter medium and on which there is a possibility of filter-medium blinding. For those units which employ continuous filter-medium washing, belt-type drum and horizontal units, the scale-up factor may be increased to 0.9.

The use of this scale-up factor assumes the following:

Complete cake discharge
Nominal filter area approximately equal to actual area
A representative sample
Suitable choice of filter medium
Operating conditions equal to those used in testing
Normal cloth conditioning during testing and operation

The scale-up factor on rate specifically does not allow for:

Changes in slurry filterability
Changes in feed rate
Changes in operating conditions
Cloth blinding

Where there is any doubt about some of the conditions listed above (except for cake discharge and actual area which should be handled separately) a more conservative scale-up factor should be used.

Scale-up on Cake Discharge When a filter is selected for a particular application, it is intended that the unit be capable of discharging essentially 100 percent of the cake which is formed. There are, however, many applications which are marginal, regardless of the type of discharge mechanism used. In these cases, the experimenter must judge the percent of cake discharge to be expected and factor the design rate accordingly.

Scale-up on Actual Area The nominal area of a filter as used by equipment manufacturers is based upon the overall dimensions of the filtering surface. The fraction of this total area that is active in filtration is a function of the filterability of the material being handled and any special treatment which the surface may receive.

The filtering surface is divided into a number of sections by division strips, radial rods, or some other impervious separator. Material which forms a thin, rather impervious cake will not form across the dividers, and thus the actual area is somewhat less than the nominal. Where relatively thick cakes of at least 1.5 cm are formed, the cake tends to form across the dividers due to cross-drainage in both the filter cake and the filter medium. In this case, the effective area is relatively close to the nominal area.

For most applications, the actual area of a drum filter will generally be no less than 94 to 97 percent of the nominal area, depending upon the size and number of sections. This variation is generally not accounted for separately and is assumed to be taken care of in the scale-up factor on filtration rate.

There are, however, certain special applications where the filter medium around the edge of the section may be deliberately blinded by painting in order to improve cake discharge. This technique is most frequently used on disc filters, with the result that the actual area may be only 75 to 85 percent of the nominal area. This is a significant deviation from the nominal area and must be considered separately.

Overall Scale-up Factor The final design filtration rate is determined by multiplying the bench-scale filtration rate by each of the scale-up factors discussed above. While this approach may seem to be ultraconservative, one must realize that the experimenter maintains careful control over the various steps during the filter cycle while running a bench-scale test, whereas a commercial filter operates with a minimum of attendance and at average conditions which are chosen to provide a satisfactory result in a production context.

FULL-SCALE FILTER PERFORMANCE EVALUATION

The correlations which have been presented for the evaluation of bench-scale data are the same correlations which should be used to evaluate the performance of a commercial installation.

A few random samples taken from a commercial installation most probably will not provide enough insight to determine that the filter is performing as expected. However, by making use of reasonable variations in the most important parameters, the desired correlations can be developed. Bench-scale tests should be run on representative feed samples taken at the same time test runs are made on the commercial unit. The bench-scale tests can be varied over a much wider range to provide a sound basis for both the location and shape of the appropriate correlation. A comparison of these results with the data taken from the commercial installation provides a good measure for efficiency of the commercial unit and a basis for identifying problem areas on the full-scale unit.

FILTER SIZING EXAMPLES

[Purchas (ed.), *Solid/Liquid Separation Equipment Scale-Up,* Uplands Press, Croydon, England, 1977.]

The examples which follow show how data from the correlations just presented and a knowledge of the physical characteristics of a particular filter are used to determine a filtration cycle and, subsequently, the size of the filter itself. The three examples which follow involve a disc, a drum belt, and a horizontal belt filter.

Example 5: Sizing a Disc Filter Equipment physical factors, selected from Table 18-9: Maximum effective submergence = 28%; maximum portion of filter cycle available for dewatering = 45%. (High submergence versions require trunnion seals, and their use is limited to specific applications.)

Scale-up factors: On rate = 0.8; On area = 0.8; On discharge = 0.9. (Scale-up on discharge may be increased to 0.97 if based on previous experience or to 0.95 if the total filter area is based on the measured effective area of the disc.)

Objective: Determine the filter size and vacuum system capacity required to dewater 15 mtph (metric tons per hour) of dry solids and produce a cake containing an average moisture content of 25 wt %.

Calculation procedure:

1. Choose cake thickness = 1.5 cm, slightly thicker than minimum value listed in Table 18-8.
2. From Fig. 18-111, W = 20.0 kg dry cake/(m^2 × rev.).
3. From Fig. 18-112, form time, ${}_f$ = 1.20 min.
4. Use simplified moisture content correlating factor in Fig. 18-115.
   Choose ${}_d/W$ = 0.04 at avg. moisture content of 25 wt %.
   Dry time = ${}_d$ = 0.04 × 20.0 = 0.80 min.
5. Calculate cycle time CT both on the basis of form time and dry time to determine which is controlling:
   CT_{form} = 1.20/.28 = 4.29 mpr (min/rev.).
   CT_{dry} = 0.80/.45 = 1.78 mpr.
   Therefore, cake formation rate is controlling and a cycle time of 4.29 mpr must be used.
6. Overall scale-up factor based on the factors presented previously = 0.8 × 0.8 × 0.9 = 0.58
7. Design filtration rate = (20.0/4.29)(60 × 0.58) = 162 kg/h × m^2
8. Area required to filter 15 metric tons of dry solids per hour = 15 × 1000/162 = 92.6 m^2. The practical choice would then be the nearest commercial size of filter corresponding to this calculated area.
9. Dry time = 45% of CT = 0.45 × 4.29 = 1.93 min. This is a much longer dry time than required. Therefore, reduce the dry time to 1.00 min by proper bridging in the filter valve.
10. From Fig. 18-118, the average gas flow rate during the 1.00 min drying period was found by graphical integration to be 2.41 m^3/m^2 × min.
11. Total volume of air flowing during dry period = 2.41 × 1.00 = 2.41 m^3 per m^2 per cycle. Add 10% to allow for evacuation of drainage passages. Total flow = 2.65 m^3/m^2 × cycle.
12. Required vacuum pump capacity = 2.65/4.29 = 0.62 m^3/min × m^2 of total filter area. Allow for pressure drop within system when specifying the vacuum pump. See next example.

Example 6: Sizing a Drum Belt Filter with Washing Equipment physical factors, selected from Table 18-9: Maximum effective submergence = 30%; max. apparent subm. = 35%; max. arc for washing = 29%; portion of cycle under vacuum = 75%.

Scale-up factors: On rate = 0.9; On area = 1.0; On discharge = 1.0.

Process data: Sp. gr. of feed liquid = 1.0; TDS (total dissolved solids) in feed liquid = 4.0 wt %; fresh water used for washing; vacuum level = 18 in Hg; final cake liquid content = 25 wt %.

TABLE 18-9 Typical Equipment Factors for Cycle Design

	% of cycle					
	Submergence		Total under active vac. or pres.	Max. for washing	Max. for dewatering only	Req'd. for cake discharge
Filter type	Apparent	Max. effective				
Drum						
Standard scraper	35	30	80	29	50–60	20
Roll discharge	35	30	80	29	50–60	20
Belt	35	30	75	29	45–50	25
Coil or string	35	30	75	29	45–50	25
Precoat	35,55	35,55	93	30	60,40	5
Horizontal belt	As req'd.	As req'd.	Lengthen as req'd.	As req'd.	As req'd.	0
Horizontal table	As req'd.	As req'd.	80	As req'd.	As req'd.	20
Tilting pan	As req'd.	As req'd.	75	As req'd.	As req'd.	25
Disc	35	28	75	None	45	25

Objective: Determine the filter size and vacuum capacity required to dewater and wash 15 mtph of dry solids while producing a final washed cake with a moisture content of 25 wt % and containing 0.10 wt % TDS based on dry cake solids.

Calculation procedure:

1. Choose cake thickness = 0.75 cm, slightly thicker than the minimum in Table 18-8.
2. From Fig. 18-111, $W = 10$ kg/m² × cycle.
3. From Fig. 18-112, form time = 0.30 min.
4. From Fig. 18-115, $_d/W = 0.04$ for 25 wt % residual moisture.
5. Dry time = $_d = W \times 0.04 = 10.0 \times 0.04 = 0.40$ min.
6. Determine required wash quantity:
   Calculated TDS concentration in washed cake liquor:
   Liquid in final cake = 10 × 0.25/0.75 = 3.33 kg/m² × cycle.
   TDS in dry washed solids = 10 × 0.001/0.999 = 0.010 kg/m² × cycle.
   TDS in final washed cake liquor = (0.010/3.33)100 = 0.300 wt %.
   Percent remaining, $R = ((C_2 - C_w)/(C_1 - C_w))100$.
   Since $C_w = 0$,
   Required percent remaining, $R = (C_2/C_1)100 = (0.300/4.00)100 = 7.5\%$.
   From Fig. 18-116, required wash ratio $N = 1.35$.
   For design, add 10% → $N = 1.35 \times 1.1 = 1.49$.
   Wash vol. = $V_w = 1.49 \times 3.33/1.00 = 4.96$ L/m² × cycle.
7. Determine wash time:
   $WV_w = 10.0 \times 4.96 = 49.6$ kgL/m⁴.
   From Fig. 18-117, wash time = $_w = 0.225$ min.
8. Summary of minimum times for each operation:
   Form (step 3) = 0.30 min.
   Wash (step 6) = 0.225 min.
   Final dry (step 5) = 0.40 min.
9. Maximum washing arc = horizontal centerline to 15° past top dead center, or 29% of total cycle. Minimum percent of cycle between end of form and earliest start of wash = area between horizontal centerline and maximum apparent submergence = (50% − 35%)/2 = 7.5%.
10. Maximum percentage of cycle for wash + final dry = 75 − 30 − 7.5 = 37.5%.
11. Determine cycle time based on the rate-controlling operation:
    a. $CT_{form} = 0.30/.30 = 1.00$ mpr.
    b. $CT_{wash} = 0.225/.29 = 0.77$ mpr.
    c. $CT_{wash+dry} = (0.225 + 0.40)/.375 = 1.67$ mpr.

Therefore, the cake wash + final dry rate is controlling and a cycle time of 1.67 mpr must be used.

12. Since (*c*) is larger than (*a*) in the previous step, too thick a cake will be formed and it will not wash or dry adequately unless the effective submergence is artificially restricted to yield the design cake thickness. This may be accomplished by proper bridge-block adjustment or by vacuum regulation within the form zone of the filter valve.
13. The required washing arc of (0.225/1.67)360 = 48.5° is assumed to start at the horizontal center line. Careful control of the wash sprays will be required to minimize runback into the slurry in the vat.
14. Overall scale-up factor = 0.9 × 1.0 × 1.0 = 0.9.
15. Design filtration rate = (10.0/1.67)(60 × 0.9).
    = 323.3 kg/h × m².
16. Total filter area required = 15 × 1000/323.3 = 46.4 m².

Nearest commercial size for a single unit could be a 10 ft dia. × 16 ft long with a total area of 502 ft² = 46.7 m².

17. Determine required vacuum capacity:
    Initial dry time = 1.67 × 0.075 = 0.125 min.
    Calculate gas vol. through cake using data from Fig. 18-118:
    Initial dry = 0.125 × 1.22 = 0.153 m³/m² × rev.
    Final dry = 0.40 × 1.95 = 0.780 m³/m² × rev.
    Total, including 10% for evacuation of drainage passages = 0.933 × 1.10 = 1.03 m³/m² × rev.
    Air rate based on total cycle = 1.03/1.67 = 0.62 m³/min × m² measured at 18 in Hg vacuum.
    If pressure drop through system = 1.0 in Hg and barometric pressure = 30 in Hg, design air rate = 0.62 × 12/11 = 0.68 m³/min × m² measured at 19 in Hg vacuum.

Horizontal Belt Filter Since the total cycle of a horizontal belt filter occurs on a single, long horizontal surface, there is no restriction with respect to the relative portions of the cycle. Otherwise, scale-up procedures are similar.

BATCH FILTRATION

Since most batch-type filters operate under pressure rather than vacuum, the following discussion will apply primarily to pressure filtration and the various types of pressure filters.

To use Eq. (18-54) one must know the pattern of the filtration process, i.e., the variation of the flow rate and pressure with time. Generally the pumping mechanism determines the filtration flow characteristics and serves as a basis for the following three categories* [Tiller and Crump, *Chem. Eng. Prog.*, **73**(10), 65 (1977)]:

1. *Constant-pressure filtration.* The actuating mechanism is compressed gas maintained at a constant pressure.
2. *Constant-rate filtration.* Positive-displacement pumps of various types are employed.
3. *Variable-pressure, variable-rate filtration.* The use of a centrifugal pump results in this pattern: the discharge rate decreases with increasing back pressure.

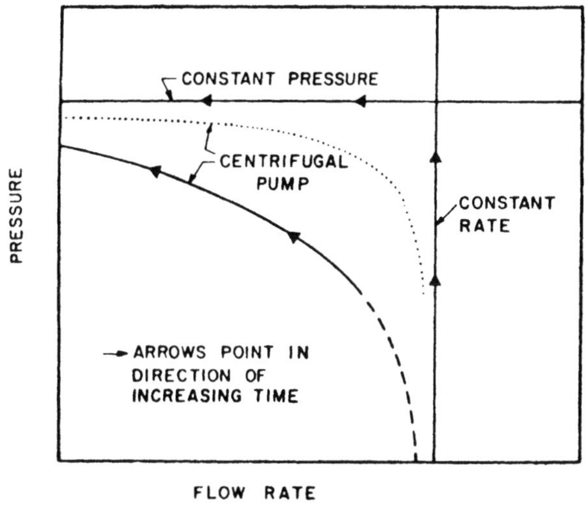

FIG. 18-119 Typical filtration cycles. [*Tiller and Crump,* Chem. Eng. Prog. *73(10), 72(1977), by permission.*]

Flow rate and pressure behavior for the three types of filtration are shown in Fig. 18-119. Depending on the characteristics of the centrifugal pump, widely differing curves may be encountered, as suggested by the figure.

Constant-Pressure Filtration For constant-pressure filtration Eq. (18-54) can be integrated to give the following relationships between total time and filtrate measurements:

$$\frac{\theta}{V/A} = \frac{\mu\alpha}{2P}\frac{W}{A} + \frac{\mu r}{P} \tag{18-70}$$

$$\frac{\theta}{V/A} = \frac{\mu\alpha w}{2P}\frac{V}{A} + \frac{\mu r}{P} \tag{18-71}$$

$$\frac{\theta}{V/A} = \frac{\mu\alpha\rho c}{2P(1-mc)}\frac{V}{A} + \frac{\mu r}{P} \tag{18-72}$$

For a given constant-pressure filtration, these may be simplified to

$$\frac{\theta}{V/A} = K_p\frac{W}{A} + C = K_p'\frac{V}{A} + C \tag{18-73}$$

where K_p, K_p', and C are constants for the conditions employed. It should be noted that K_p, K_p', and C depend on filtering pressure not only in the obvious explicit way but also in the implicit sense that α, m, and r are generally dependent on P.

Constant-Rate Filtration For substantially incompressible cakes, Eq. (18-54) may be integrated for a constant rate of slurry feed to the filter to give the following equations, in which filter-medium resistance is treated as the equivalent constant-pressure component to be deducted from the rising total pressure drop to give the variable pressure through the filter cake [Ruth, *Ind. Eng. Chem.*, **27,** 717 (1935)]:

* A combination of category 2 followed by category 1 as parts of the same filtration cycle is considered by some as a fourth category. For a method of combining the constant-rate and constant-pressure equations for such a cycle, see Brown, loc. cit.

$$\frac{\theta}{V/A} = \frac{1}{\text{rate per unit area}} = \frac{\mu\alpha}{P - P_1}\frac{W}{A} \tag{18-74}$$

which may also be written

$$\frac{\theta}{V/A} = \frac{\mu\alpha w}{P - P_1}\frac{V}{A} = \frac{\mu\alpha\rho c}{(P - P_1)(1 - mc)}\frac{V}{A} \tag{18-75}$$

In these equations P_1 is the pressure drop through the filter medium.

$$P_1 = \mu r(V/A\theta)$$

For a given constant-rate run, the equations may be simplified to

$$V/A = P/K_r + C' \tag{18-76}$$

where K_r and C' are constants for the given conditions.

Variable-Pressure, Variable-Rate Filtration The pattern of this category complicates the use of the basic rate equation. The method of Tiller and Crump (loc. cit.) can be used to integrate the equation when the characteristic curve of the feed pump is available.

In the filtration of small amounts of fine particles from liquid by means of bulky filter media (such as absorbent cotton or felt) it has been found that the preceding equations based upon the resistance of a cake of solids do not hold, since no cake is formed. For these cases, in which filtration takes place on the surface or within the interstices of a medium, analogous equations have been developed [Hermans and Bredée, *J. Soc. Chem. Ind.,* **55T,** 1 (1936)]. These are usefully summarized, for both constant-pressure and constant-rate conditions, by Grace [*Am. Inst. Chem. Eng. J.,* **2,** 323 (1956)]. These equations often apply to the clarification of such materials as sugar solutions, viscose and other spinning solutions, and film-casting dopes.

If a **constant-pressure test** is run on a slurry, care being taken that not only the pressure but also the temperature and the solid content remain constant throughout the run and that time readings begin at the exact start of filtration, one can observe values of filtrate volume or weight and corresponding elapsed time. With the use of the known filtering area, values of $\theta/(V/A)$ can be calculated for various values of V/A which, when plotted with $\theta/(V/A)$ as the ordinate and V/A as the abscissa (Fig. 18-120*a*), result in a straight line having the slope $\mu\alpha w/2P$ and an intercept on the vertical axis of $\mu r/P$. Since μ, w, and P are known, α and r can be calculated from

$$\alpha = 2P/\mu w \times (\text{slope})$$

and

$$r = P/\mu \times (\text{vertical intercept})$$

The effect of the change of any variable not affecting α or r can now be estimated. It should be remembered that α and r usually depend on P and may be affected by w.

The symbol α represents the average specific cake resistance, which is a constant for the particular cake in its immediate condition. In the usual range of operating conditions it is related to the pressure by the expression

$$\alpha = \alpha' P^s \tag{18-77}$$

where α' is a constant determined largely by the size of the particles forming the cake; s is the cake compressibility, varying from 0 for rigid, incompressible cakes, such as fine sand and diatomite, to 1.0 for very highly compressible cakes. For most industrial slurries, s lies between 0.1 and 0.8. The symbol r represents the resistance of unit area of filter medium but includes other losses (besides those across the cake and the medium) in the system across which P is the pressure drop.

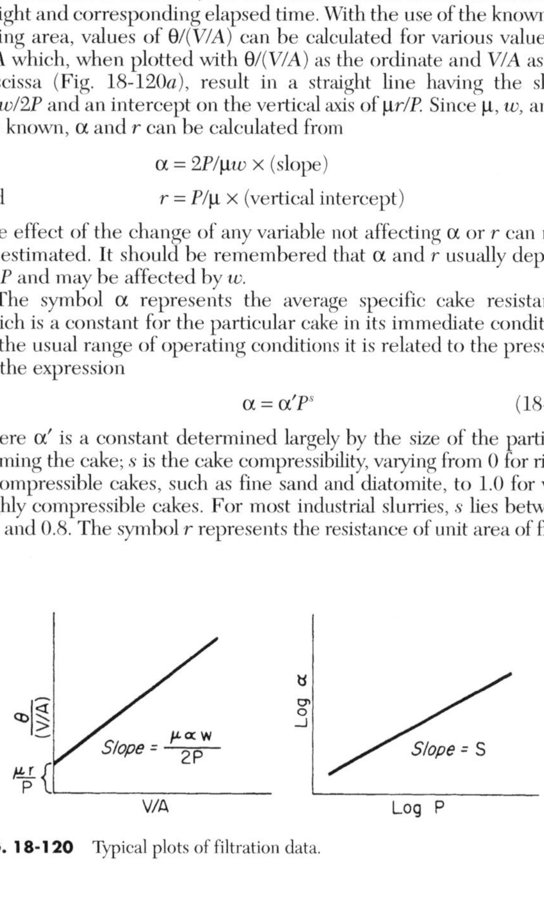

FIG. 18-120 Typical plots of filtration data.

It should be noted also that the intercept is difficult to determine accurately because of large potential experimental error in observing the time of the start of filtration and the time-volume correspondence during the first moments when the filtration rate is high. The value of r calculated from the intercept may vary appreciably from test to test, and will almost always be different from the value measured with clean medium in a permeability test.

To determine the effect of a change in pressure, it is necessary to run tests at three or more pressures, preferably spanning the range of interest. Plotting α or r against P on log-log paper (or log α or log r against the log P on cartesian coordinates) results in an approximate straight line (Fig. 18-120*b*) from which one may estimate values of α or r at interpolated or reasonably extrapolated magnitudes of P. The slope of the line is the index of a power relationship between α and P or r and P.

Not uncommonly r is found to be only slightly dependent on pressure. When this is true and especially when the filter-medium resistance is, as it should be, relatively small, an average value may be used for all pressures.

It is advisable to start a constant-pressure filtration test, like a comparable plant operation, at a low pressure, and smoothly increase the pressure to the desired operating level. In such cases, time and filtrate-quantity data should not be taken until the constant operating pressure is realized. The value of r calculated from the extrapolated intercept then reflects the resistance of both the filter medium and that part of the cake deposited during the pressure-buildup period. When only the total mass of dry cake is measured for the total cycle time, as is usually true in vacuum leaf tests, at least three runs of different lengths should be made to permit a reliable plot of θ/V against W. If rectification of the resulting three points is dubious, additional runs should be made.

Pressure Tests

Leaf Tests A bomb filter is used for small-scale leaf tests to simulate the performance of pressure-leaf (leaf-in-shell) filters. The equipment used is a small [50.8- by 50.8-mm (2- by 2-in)] leaf, covered with appropriate filter medium, suspended in a cell large enough to contain sufficient slurry to form the desired cake (Fig. 18-121). The slurry may be agitated gently, for example, by an air sparger.

Although incremental time and filtrate volume may be taken during a cake-forming cycle at a selected pressure to permit a plot like Fig. 18-120*a* from a single run, it may be more satisfactory to make several successive quick runs at the same pressure but for different lengths of time, recording only the terminal values of filtrate volume, time, and cake mass. Operation of the commercial unit should be kept in mind when the test cycles are planned. Displacement washing and air blowing of the cake should be tried if appropriate. Wet discharge can be simulated by opening the cell and playing a jet of water on the cake; dry discharge, by applying a gentle air blast to the filtrate-discharge tube. Tests at several pressures must be conducted to determine the compressibility of the cake solids.

Plate-and-Frame Tests These tests should be conducted if the use of a filter press in the plant is anticipated; at least a few confirming tests are advisable after preliminary leaf tests, unless the slurry is very rapidly filtering. A laboratory-size filter press consisting of two plates and a single frame may be used. It will permit the observation of solids-settling, cake-packing, and washing behavior, which may be quite different for a frame than for a leaf.

Compression-Permeability Tests Instead of model leaf tests, compression-permeability experiments may be substituted with advantage for appreciably compressible solids. As in the case of constant-rate filtration, a single run provides data equivalent to those obtained from a series of constant-pressure runs, but it avoids the data-treatment complexity of constant-rate tests.

The equipment consists of a cylindrical cell with a permeable bottom and an open top, into which is fitted a close-clearance, hollow, cylindrical piston with a permeable bottom. Slurry is poured into the cell, and a cake is formed by applying gentle vacuum to the filtrate discharge line. The cell is then filled with filtrate, and the counterweighted piston is allowed to descend to the cake level. Successive

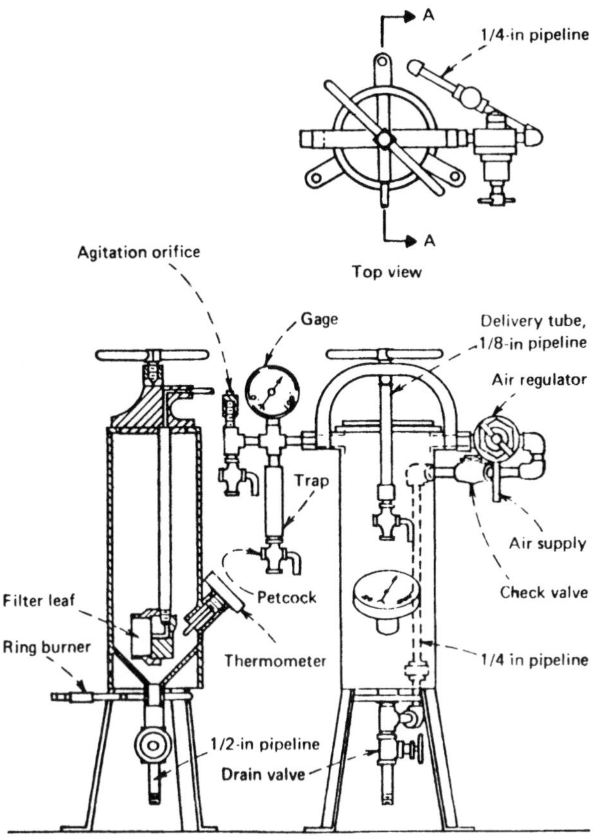

FIG. 18-121 Bomb filter for small-scale pressure filtration tests. [*Silverblatt et al.,* Chem. Eng., ***81****(9), 132 (1974), by permission.*]

increments of mechanical stress are applied to the solids, at each of which the permeability of the cake is determined by passing filtrate through the piston under low head.

The experimental procedure and method of treatment of compression-permeability data have been explained by Grace [*Chem. Eng. Prog.,* **49,** 303, 427 (1953)], who showed that the values of α measured in such a cell and in a pressure filter were the same, and by Tiller [*Filtr. Sep.,* **12,** 386 (1975)].

Scaling Up Test Results The results of small-scale tests are determined as dry weight of solids or volume of filtrate per unit of area per cycle. This quantity multiplied by the number of cycles per day permits the calculation of either the filter area required for a stipulated daily capacity or the daily capacity of a specified plant filter. The scaled-up filtration area should be increased by 25 percent as a factor of uncertainty. In the calculation of cycle length, proper account must be made of the downtime of a batch filter.

FILTER MEDIA

All filters require a filter medium to retain solids, whether the filter is for cake filtration or for filter-medium or depth filtration. Specification of a medium is based on retention of some minimum particle size at good removal efficiency and on acceptable life of the medium in the environment of the filter. The selection of the type of filter medium is often the most important decision in success of the operation. For cake filtration, medium selection involves an optimization of the following factors:

1. Ability to bridge solids across its pores quickly after the feed is started (i.e., minimum propensity to bleed)
2. Low rate of entrapment of solids within its interstices (i.e., minimum propensity to blind)
3. Minimum resistance to filtrate flow (i.e., high production rate)
4. Resistance to chemical attack
5. Sufficient strength to support the filtering pressure
6. Acceptable resistance to mechanical wear
7. Ability to discharge cake easily and cleanly
8. Ability to conform mechanically to the kind of filter with which it will be used
9. Minimum cost

For filter-medium filtration, attributes 3, 4, 5, 8, and 9 of the preceding list apply and must have added to them (*a*) ability to retain the solids required, (*b*) freedom from discharge of lint or other adulterant into the filtrate, and (*c*) ability to plug slowly (i.e., long life).

Filter-medium selection embraces many types of construction: fabrics of woven fibers, felts, and nonwoven fibers, porous or sintered solids, polymer membranes, or particulate solids in the form of a permeable bed. Media of all types are available in a wide choice of materials.

Fabrics of Woven Fibers For cake filtration these fabrics are the most common type of medium. A wide variety of materials are available; some popular examples are listed in Table 18-10, with ratings for chemical and temperature resistance. In addition to the material of the fibers, a number of construction characteristics describe the filter cloth: (1) weave, (2) style number, (3) weight, (4) count, (5) ply, and (6) yarn number. Of the many types of weaves available, only four are extensively used as filter media: plain (square) weave, twill, chain weave, and satin.

All these weaves may be made from any textile fiber, natural or synthetic. They may be woven from spun staple yarns, multifilament continuous yarns, or monofilament yarns. The performance of the filter cloth depends on the weave and the type of yarn.

A recently developed medium known as a double weave incorporates different yarns in warp and fill in order to combine the specific advantages of each type. An example of this is Style 99FS, made by Madison Filtration, in which multifilament warp yarns provide good cake release properties and spun staple fill yarns contribute to greater retentivity.

Metal Fabrics or Screens These are available in several types of weave in nickel, copper, brass, bronze, aluminum, steel, stainless steel, Monel, and other alloys. In the plain weave, 400 mesh is the closest wire spacing available, thus limiting use to coarse crystalline slurries, pulps, and the like. The "Dutch weaves" employing relatively large, widely spaced, straight warp wires and relatively small crimped filling wires can be woven much more closely, providing a good medium for filtering fine crystals and pulps. This type of weave tends to plug readily when soft or amorphous particles are filtered and makes the use of filter aid desirable. Good corrosion and high temperature resistance of properly selected metals makes filtrations with metal media desirable for long-life applications. This is attractive for handling toxic materials in closed filters to which minimum exposure by maintenance personnel is desirable.

Pressed Felts and Cotton Batting These materials are used to filter gelatinous particles from paints, spinning solutions, and other viscous liquids. Filtration occurs by deposition of the particles in and on the fibers throughout the mat.

Nonwoven media consist of web or sheet structures which are composed primarily of fibers or filaments bonded together by thermal, chemical, or mechanical (such as needlepunching) means. Needled felts are the most commonly used nonwoven fabric for liquid filtration. Additional strength often is provided by including a scrim of woven fabric encapsulated within the nonwoven material. The surface of the medium can be calendered to improve particle retention and assist in filter cake release. Weights range from 270 to 2700 gm/m^2 (8 to 80 oz/yd^2). Because of their good retentivity, high strength, moderate cost, and resistance to blinding, nonwoven media have found wide acceptance in filter press use, particularly in mineral concentrate filtration applications. They are used frequently on horizontal belt filters where their dimensional stability reduces or eliminates wrinkling and biasing problems often encountered with woven belts.

Filter Papers These papers come in a wide range of permeability, thickness, and strength. As a class of material, they have low strength, however, and require a perforated backup plate for support.

TABLE 18-10 Characteristics of Filter-Fabric Materials*

Generic name and description	Breaking tenacity, g/denier	Abrasion resistance	Resistance to acids	Resistance to alkalies	Resistance to oxidizing agents	Resistance to solvents	Specific gravity	Maximum operating temperature, °F†
Acetate—cellulose acetate. When not less than 92% of the hydroxyl groups are acetylated, "triacetate" may be used as a generic description.	1.2–1.5	G	F	P	G	G	1.33	210
Acrylic—any long-chain synthetic polymer composed of at least 85% by weight of acrylonitrile units.	2.0–4.8	G	G	F	G	E	1.18	300
Glass—fiber-forming substance is glass.	3.0–7.2	P	E	P	E	E	2.54	600
Metallic—composed of metal, metal-coated plastic, plastic-coated metal, or a core completely covered by metal.	—	G						
Modacrylic—fiber-forming substance is any long-chain synthetic polymer composed of less than 85% but at least 35% by weight of acrylonitrile units.	2.5–3.0	G	G	G	G	G	1.30	180
Nylon—any long-chain synthetic polyamide having recurring amide groups as an integral part of the polymer chain.	3.8–9.2	E	F–P	G	F–P	G	1.14	225
Polyester—any long-chain synthetic polymer composed of at least 85% by weight of an ester of a dihydric alcohol and terephthalic acid (p—HOOC—C_6H_4—COOH).	2.2–7.8	E–G	G	G–F	G	G	1.38	300
Polyethylene—long-chain synthetic polymer composed of at least 85% weight of ethylene.	1.0–7.0	G	G	G	F	G	0.92	165‡
Polypropylene—long-chain synthetic polymer composed of at least 85% by weight of propylene.	4.8–8.5	G	E	E	G	G	0.90	250§
Cotton—natural fibers.	3.3–6.4	G	P	F	G	E–G	1.55	210
Fluorocarbon—long-chain synthetic polymer composed of tetrafluoroethylene units.	1.0–2.0	F	E	E	E	G	2.30	550¶

*Adapted from Mais, *Chem. Eng.*, **78**(4), 51 (1971). Symbols have the following meaning: E = excellent, G = good, F = fair, P = poor.
†°C = (°F − 32)/1.8; K = (°F + 459.7)/1.8.
‡Low-density polymer. Up to 230°F for high-density.
§Heat-set fabric; otherwise lower.
¶Requires ventilation because of release of toxic gases above 400°F.

Rigid Porous Media These are available in sheets or plates and tubes. Materials used include sintered stainless steel and other metals, graphite, aluminum oxide, silica, porcelain, and some plastics—a gamut that allows a wide range of chemical and temperature resistance. Most applications are for clarification.

Polymer Membranes These are used in filtration applications for fine-particle separations such as microfiltration and ultrafiltration (clarification involving the removal of 1-μm and smaller particles). The membranes are made from a variety of materials, the commonest being cellulose acetates and polyamides. Membrane filtration, discussed in Sec. 22, has been well covered by Porter (in Schweitzer, op. cit., sec. 2.1).

Media made from woven or nonwoven fabrics coated with a polymeric film, such as Primapor, and Primapor II made by Madison Filtration, Gore-Tex, made by W. L. Gore and Associates, and Tetratex, made by Donaldson Company, combine the high retentivity characteristics of a membrane with the strength and durability of a thick filter cloth. These media are used on both continuous and batch filters where excellent filtrate clarity is required.

Granular Beds of Particulate Solids Beds of solids like sand or coal are used as filter media to clarify water or chemical solutions containing small quantities of suspended particles. Filter-grade grains of desired particle size can be purchased. Frequently beds will be constructed of layers of different materials and different particle sizes.

Various types of filter media and the materials of which they are constructed are surveyed extensively by Purchas (*Industrial Filtration of Liquids,* CRC Press, Cleveland, 1967, chap. 3), and characterizing measurements (e.g., pore size, permeability) are reviewed in detail by Rushton and Griffiths (in Orr, op. cit., chap. 3). Briefer summaries of classification of media and of practical criteria for the selection of a filter medium are presented by Shoemaker (op. cit., p. 26) and Purchas [*Filtr. Sep.,* **17,** 253, 372 (1980)].

FILTER AIDS

Use of filter aids is a technique frequently applied for filtrations in which problems of slow filtration rate, rapid medium blinding, or unsatisfactory filtrate clarity arise. Filter aids are granular or fibrous solids capable of forming a highly permeable filter cake in which very fine solids or slimy, deformable flocs may be trapped. Application of filter aids may allow the use of a much more permeable filter medium than the clarification would require to produce filtrate of the same quality by depth filtration.

Filter aids should have low bulk density to minimize settling and aid good distribution on a filter-medium surface that may not be horizontal. They should also be porous and capable of forming a porous cake to minimize flow resistance, and they must be chemically inert to the filtrate. These characteristics are all found in the two most popular commercial filter aids: diatomaceous earth (also called diatomite), which is an almost pure silica prepared from deposits of diatom skeletons; and expanded perlite, particles of "puffed" lava that are principally aluminum alkali silicate. Cellulosic fibers (ground wood pulp) are sometimes used when siliceous materials cannot be used but are much more compressible. The use of other less effective aids (e.g., carbon and gypsum) may be justified in special cases. Sometimes a combination of carbon and diatomaceous earth permits adsorption in addition to filter-aid performance. Various other materials, such as salt, fine sand, starch, and precipitated calcium carbonate, are employed in specific industries where they represent either waste material or inexpensive alternatives to conventional filter aids.

Diatomaceous Earth Filter aids of diatomaceous earth have a dry bulk density of 128 to 320 kg/m^3 (8 to 20 lb/ft^3), contain particles mostly smaller than 50 μm, and produce a cake with porosity in the range of 0.9 (volume of voids/total filter-cake volume). The high porosity (compared with a porosity of 0.38 for randomly packed uniform spheres and 0.2 to 0.3 for a typical filter cake) is indicative of its filter-aid ability. Different methods of processing the crude diatomite result in a series of filter aids having a wide range of permeability.

Perlite Perlite filter aids are somewhat lower in bulk density (48 to 96 kg/m^3, or 3 to 6 lb/ft^3) than diatomaceous silica and contain a higher fraction of particles in the 50- to 150-μm range. Perlite is also available in a number of grades of differing permeability and cost, the grades being roughly comparable to those of diatomaceous earth. Diatomaceous earth will withstand slightly more extreme pH levels than perlite, and it is said to be somewhat less compressible.

Filter aids are used in two ways: (1) as a precoat and (2) mixed with the slurry as a "body feed." Precoat filtration, employing a thin layer of about 0.5 to 1.0 kg/m^2 (0.1 to 0.2 lb/ft^2) deposited on the filter medium prior to beginning feed to the filter, is in wide use to protect the filter medium from fouling by trapping solids before they reach the medium. It also provides a finer matrix to trap fine solids and assure filtrate clarity. Body-feed application is the continuous addition of filter aid to the filter feed to increase the porosity of the cake. The amount of addition must be determined by trial, but in general, the quantity added should at least equal the amount of solids to be removed. For solids loadings greater than 1000 ppm this may become a significant cost factor. An acceptable alternative might be to use a rotary vacuum precoat filter [Smith, *Chem. Eng.*, **83**(4), 84 (1976)]. Further details of filter-aid filtration are set forth by Cain (in Schweitzer, op. cit., sec. 4.2) and Hutto [*Am. Inst. Chem. Eng. Symp. Ser.*, **73**(171), 50 (1977)]. Figure 18-122 shows a flow sheet indicating arrangements for both precoat and body-feed applications. Most filter aid is used on a one-time basis, although some techniques have been demonstrated to reuse precoat filter aid on vertical-tube pressure filters.

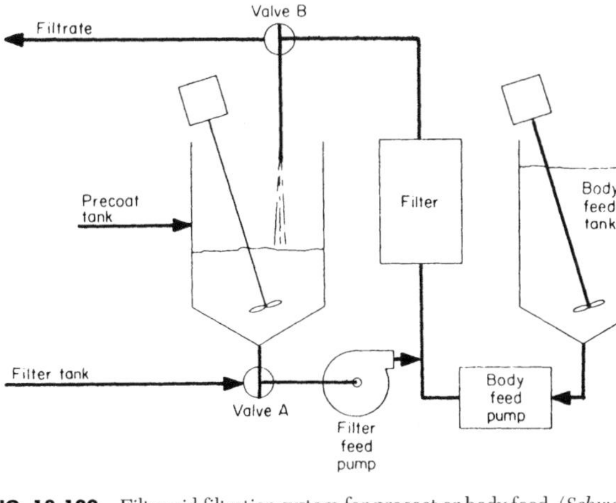

FIG. 18-122 Filter-aid filtration system for precoat or body feed. (*Schweitzer,* Handbook of Separation Techniques for Chemical Engineers, *p. 4-12. Copyright 1979 by McGraw-Hill, Inc. and used with permission.*)

FILTRATION EQUIPMENT

Cake Filters Filters that accumulate appreciable visible quantities of solids on the surface of a filter medium are called cake filters. The slurry feed may have a solids concentration from about 1 percent to greater than 40 percent. The filter medium on which the cake forms is relatively open to minimize flow resistance, since once the cake forms, it becomes the effective filter medium. The initial filtrate therefore may contain unacceptable solids concentration until the cake is formed. This situation may be made tolerable by recycling the filtrate until acceptable clarity is obtained or by using a downstream polishing filter (clarifying type).

Cake filters are used when the desired product of the operation is the solids, the filtrate, or both. When the filtrate is the product, the degree of removal from the cake by washing or blowing with air or gas becomes an economic optimization. When the cake is the desired product, the incentive is to obtain the desired degree of cake purity by washing, blowing, and sometimes mechanical expression of residual liquid.

Implicit in cake filtration is the removal and handling of solids, since the cake is usually relatively dry and compacted. Cakes can be sticky and difficult to handle; therefore, the ability of a filter to discharge the cake cleanly is an important equipment-selection criterion.

In the operational sense, some filters are batch devices, whereas others are continuous. This difference provides the principal basis for classifying cake filters in the discussion that follows. The driving force by which the filter functions—hydrostatic head ("gravity"), pressure imposed by a pump or a gas blanket, or atmospheric pressure ("vacuum")—will be used as a secondary criterion.

Batch Cake Filters

Nutsche Filters A nutsche is one of the simplest batch filters. It is a tank with a false bottom, perforated or porous, which may either support a filter medium or act as the filter medium. The slurry is fed into the filter vessel, and separation occurs by gravity flow, gas pressure, vacuum, or a combination of these forces. The term "nutsche" comes from the German term for sucking, and vacuum is the common operating mode.

The design of most nutsche filters is very simple, and they are often fabricated by the user at low cost. The filter is very frequently used in laboratory, pilot-plant, or small-plant operation. For large-scale processing, however, the excessive floor area encumbered per unit of filtration area and the awkwardness of cake removal are strong deterrents. For small-scale operations, cake is manually removed. For large-scale applications, cake may be further processed by reslurrying or redissolving; or it may be removed manually (by shovel) or by mechanical discharge arrangements such as a movable filter medium belt.

Thorough displacement washing is possible in a nutsche if the wash solvent is added before the cake begins to be exposed to air displacement of filtrate. If washing needs to be more effective, an agitator can be provided in the nutsche vessel to reslurry the cake to allow adequate diffusion of solute from the solids.

Horizontal Plate Filter The horizontal multiple-plate pressure filter consists of a number of horizontal circular drainage plates and guides placed in a stack in a cylindrical shell (Fig. 18-123). In normal practice the filtering pressure is limited to 345 kPa (50 psig), although special filters have been designed for shell pressures of 2.1 MPa (300 psig) or higher.

Filter Press The filter press, one of the most frequently used filters in the early years of the chemical industry, is still widely employed. Often referred to generically (in error) as the plate-and-frame filter, it has probably over 100 design variations. Two basic popular designs are the flush-plate, or plate-and-frame, design and the recessed-plate press. Both are available in a wide range of materials: metals, coated metals, plastics, or wood.

Plate-and-frame press. This press is an alternate assembly of plates covered on both sides with a filter medium, usually a cloth, and hollow frames that provide space for cake accumulation during filtration. The frames have feed and wash manifold ports, while the plates have filtrate drainage ports. The plates and frames usually are

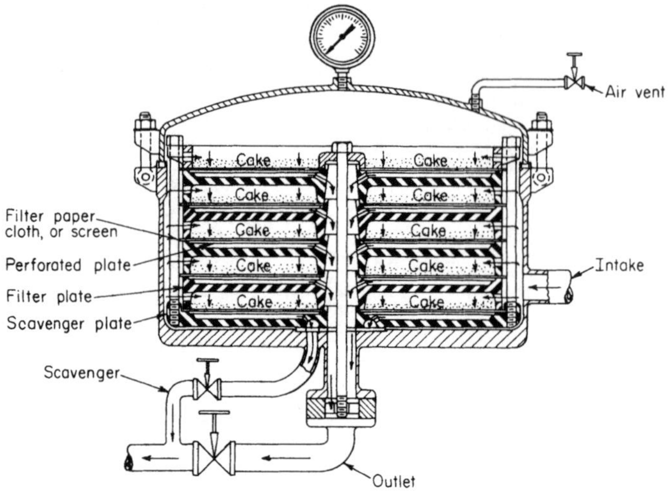

FIG. 18-123 Elevation section of a Sparkler horizontal plate filter. (*Sparkler Filters, Inc.*)

rectangular, although circles and other shapes also are used (Fig. 18-124). They are hung on a pair of horizontal support bars and pressed together during filtration to form a watertight closure between two end plates, one of which is stationary. The press may be closed manually, hydraulically, or by a motor drive. Several feed and filtrate discharge arrangements are possible. In the most popular, the feed and discharge of the several elements of the press are manifolded via some of the holes that are in the four corners of each plate and frame (and filter cloth) to form continuous longitudinal channels from the stationary end plate to the other end of the press. Alternatively, the filtrate may be drained from each plate by an individual valve and spigot (for open discharge) or tubing (for closed). Top feed to and bottom discharge from the chambers provide maximum recovery of filtrate and maximum mean cake dryness. This arrangement is especially suitable for heavy fast-settling solids. For most slurries, bottom feed and top filtrate discharge allow quick air displacement and produce a more uniform cake.

Two wash techniques are used in plate-and-frame filter presses, illustrated in Fig. 18-125. In simple washing, the wash liquor follows the same path as the filtrate. If the cake is not extremely uniform and highly permeable, this type of washing is ineffective in a well-filled press. A better technique is thorough washing, in which the wash is introduced to the faces of alternate plates (with their discharge channels valved off). The wash passes through the entire cake and exits through the faces of the other plates. This improved technique requires a special design and the assembly of the plates in proper order. Thorough washing should be used only when the frames are well filled, since an incomplete fill of cake will allow cake collapse during the wash entry. The remainder of the wash flow will bypass through cracks or channels opened in the cake.

Filter presses are made in plate sizes from 10 by 10 cm (4 by 4 in) to 2.4 by 2.4 m (94 by 94 in). Frame thickness ranges from 0.3 to 20 cm (0.125 to 8 in). Operating pressures up to 689 kPa (100 psig) are common, with some presses designed for 6.9 MPa (1000 psig). Some metal units have cored plates for steam or refrigerant. Maximum pressure for wood or plastic frames is 410 to 480 kPa (60 to 70 psig).

The filter press has the advantage of simplicity, low capital cost, flexibility, and ability to operate at high pressure in either a cake-filter or a clarifying-filter application. Floor-space and headroom needs per unit of filter area are small, and capacity can be adjusted by adding or removing plates and frames. Filter presses are cleaned easily, and the filter medium is easily replaced. With proper operation a denser, drier cake compared with that of most other filters is obtained.

There are several serious disadvantages, including imperfect washing due to variable cake density, relatively short filter-cloth life due to the mechanical wear of emptying and cleaning the press (often involving scraping the cloth), and high labor requirements. Presses frequently drip or leak and thereby create housekeeping problems, but the biggest problem arises from the requirement to open the filter for cake discharge. The operator is thus exposed routinely to the contents of the filter, and this is becoming an increasingly severe disadvantage as more and more materials once believed safe are given restricted exposure limits.

Recessed-plate filter press. This press is similar to the plate-and-frame press in appearance but consists only of plates (Fig. 18-126). Both faces of each plate are hollowed to form a chamber for cake accumulation between adjacent plates. This design has the advantage of about half as many joints as a plate-and-frame press, making a tight closure more certain. Figure 18-127 shows some of the features of one type of recessed-plate filter which has a gasket to further minimize leaks. Air can be introduced behind the cloth on both sides of each plate to assist cake removal.

Some interesting variations of standard designs include the ability to roll the filter to change from a bottom to a top inlet or outlet and the

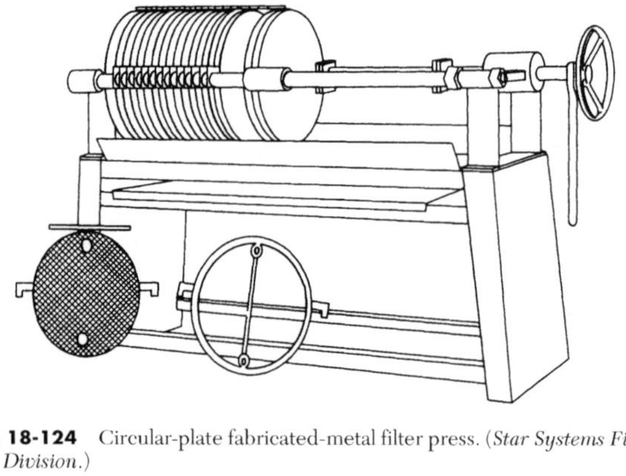

FIG. 18-124 Circular-plate fabricated-metal filter press. (*Star Systems Filtration Division.*)

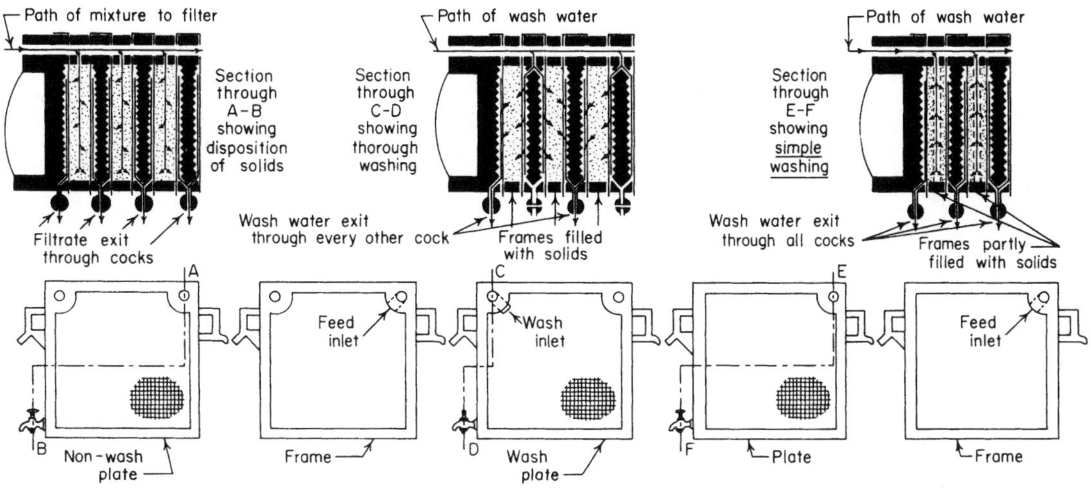

FIG. 18-125 Filling and washing flow patterns in a filter press. (*D. R. Sperry & Co.*)

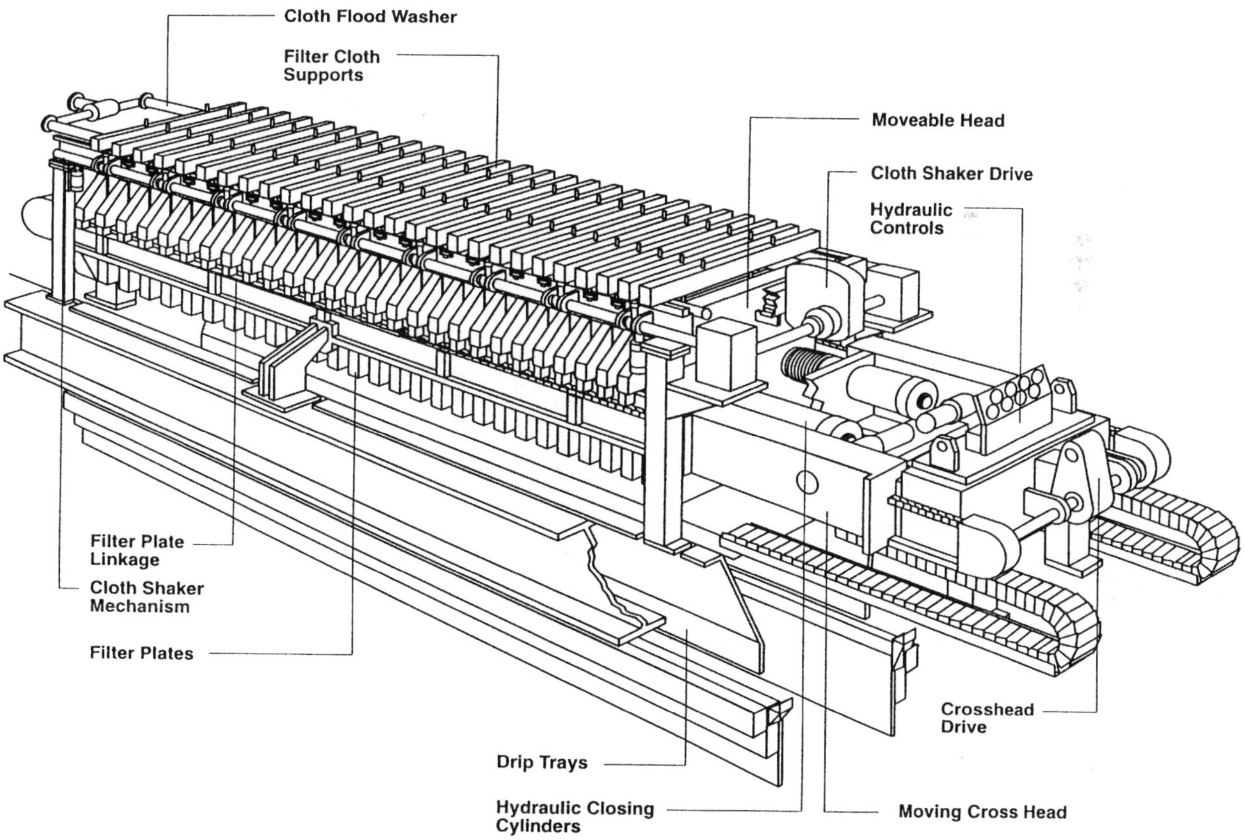

FIG. 18-126 Automated recessed-plate filter press used in mineral applications. (*Dorr-Oliver EIMCO.*)

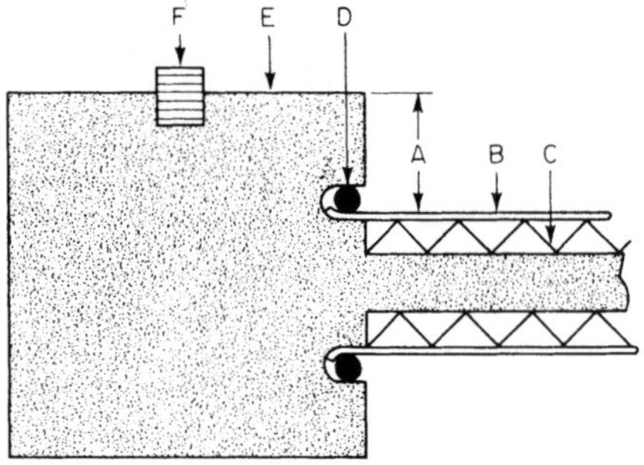

FIG. 18-127 Section detail of a caulked-gasketed-recessed filter plate: (*a*) cake recess; (*b*) filter cloth; (*c*) drainage surface of plate; (*d*) caulking strip; (*e*) plate joint; (*f*) sealing gasket. (*Dorr-Oliver EIMCO*)

ability to add blank dividers to convert a press to a multistage press for further clarification of the filtrate or to do two separate filtrations simultaneously in the same press. Some designs have rubber membranes between plates which can be expanded when filtration is finished to squeeze out additional moisture. Some designs feature automated opening and cake-discharge operations to reduce labor requirements. Examples of this type of pressure filter include Larox, Vertipress, and Oberlin.

Internal Cake Tube Filters or Liquid Bag Filters This type of filter, such as manufactured by Industrial Filter and Pump Mfg. Co., Rosedale, Illinois, and many others, utilizes one or more perforated tubes supported by a tube sheet or by the lip of the pressure vessel. A cylindrical filter bag sealed at one end is inserted into the perforated tube. The open end of the filter bag generally has a flange or special seal ring to prevent leakage.

Slurry under pressure is admitted to the chamber between the head of the shell and the tube sheet, whence it enters and fills the tubes. Filtration occurs as the filtrate passes radially outward through the filter medium and the wall of each tube into the shell and on out the filtrate discharge line, depositing cake on the medium. The filtration cycle is ended when the tubes have filled with cake or when the media have become plugged. The cake can be washed (if it has not been allowed to fill the tubes completely) and air-blown. The filter has a removable head to provide easy access to the tube sheet and mouth of the tubes; thus "sausages" of cake can be removed by taking out the filter bags or each tube and bag assembly together. The tubes themselves are easily removed for inspection and cleaning.

The advantages of the tubular filter are that it uses an easily replaced filter medium, its filtration cycle can be interrupted and the shell can be emptied of prefilt at any time without loss of the cake, the cake is readily recoverable in dry form, and the inside of the filter is conveniently accessible. There is also no unfiltered heel. Disadvantages are the necessity and attendant labor requirements of emptying by hand and replacing the filter media and the tendency for heavy solids to settle out in the header chamber. Applications are as a scavenger filter to remove fines not removed in a prior-filtration stage with a different kind of equipment, to handle the runoff from other filters, and in semiworks and small-plant operations in which the filter's size, versatility, and cleanliness recommend it.

External-Cake Tubular Filters Several filter designs are available with vertical tubes supported by a filtrate-chamber tube sheet in a vertical cylindrical vessel (Fig. 18-128). The tubes may be made of wire cloth; porous ceramic, carbon, plastic, or metal; or closely wound wire. The tubes may have a filter cloth on the outside. Frequently a filter-aid precoat will be applied to the tubes. The prefilt slurry is fed near the bottom of the vertical vessel. The filtrate passes from the outside to the inside of the tubes and into a filtrate chamber at the top or the bottom of the vessel. The solids form a cake on the outside of the tubes with the filter area actually increasing as the cake builds up, partially compensating for the increased flow resistance of the thicker cake. The filtration cycle continues until the differential pressure reaches a specified level, or until about 25 mm (1 in) of cake thickness is obtained.

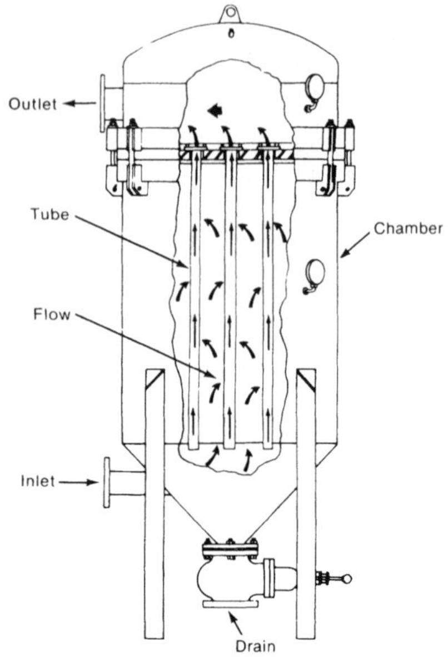

FIG. 18-128 Top-outlet tubular filter. (*Industrial Filter & Pump Mfg. Co.*)

Cake-discharge methods are the chief distinguishing feature among the various designs. That of the Industrial Filter & Pump Hydra-Shoc, for example, removes cake from the tubes by filtrate backflushing assisted by the "shocking" action of a compressed-gas pocket formed in the filtrate chamber at the top of the vertical vessel. Closing the filtrate outlet valve while continuing to feed the filter causes compression of the gas volume trapped in the dome of the vessel until, at the desired gas pressure, quick-acting valves stop the feed and open a bottom drain. The compressed gas rapidly expands, forcing a rush of filtrate back across the filter medium and dislodging the cake, which drains out the bottom with the flush liquid. Of course, this technique may be used only when wet-cake discharge is permitted.

Dry cake discharge can be achieved with a Fundabac candle-type filter manufactured by DrM, Dr. Müller, AG, of Switzerland. This filter uses a candle made up of six small-diameter tubes around a central filtrate delivery tube. This design allows the filter cloth to be flexed outward upon blowback, easily achieving an effective dry cake discharge (Fig. 18-129).

Pressure Leaf Filters Sometimes called tank filters, they consist of flat filtering elements (leaves) supported in a pressure shell. The leaves are circular, arc-sided, or rectangular, and they have filtering surfaces on both faces. The shell is a cylindrical or conical tank. Its axis may be horizontal or vertical, and the filter type is described by its shell axis orientation.

A filter leaf consists of a heavy screen or grooved plate over which a filter medium of woven fabric or fine wire cloth may be fitted. Textile fabrics are more commonly used for chemical service and are usually applied as bags that may be sewed, zippered, stapled, or snapped. Wire-screen cloth is frequently used for filter-aid filtrations, particularly if a precoat is applied. It may be attached by welding, riveting, bolting, or caulking or by the clamped engagement of two 180° bends in the wire cloth under tension, as in Multi Metal's Rim-Lok leaf. Leaves may also be of all-plastic construction. The filter medium, regardless of material, should be as taut as possible to minimize sagging when it is loaded with a cake; excessive sag can cause cake cracking or dropping. Leaves may be supported at top, bottom, or center and may discharge filtrate from any of these locations. Figure 18-130 shows the elevation section of a precoated bottom-support wire leaf.

Pressure leaf filters are operated batchwise. The shell is locked, and the prefilt slurry is admitted from a pressure source. The slurry enters in such a way as to minimize settling of the suspended solids. The shell

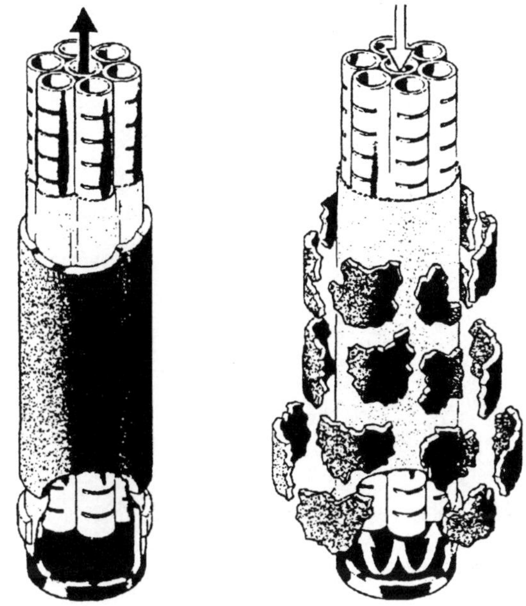

FIG. 18-129 Cake formation and discharge with the Fundabac filter element. (*DrM, Dr. Müller AG, Switzerland.*)

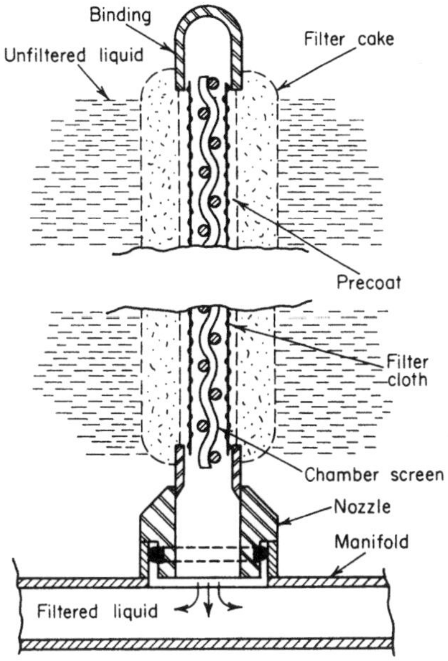

FIG. 18-130 Section of precoated wire filter leaf.

is filled, and filtration occurs on the leaf surfaces, the filtrate discharging through an individual delivery line or into an internal manifold, as the filter design dictates. Filtration is allowed to proceed only until a cake of the desired thickness has formed, since to overfill will cause cake consolidation with consequent difficulty in washing and discharge. The decision of when to end the filtering cycle is largely a matter of experience, guided roughly by the rate in a constant-pressure filter or pressure drop in a constant-rate filter. This judgment may be supplanted by the use of a detector which "feels" the thickness of cake on a representative leaf.

If the cake is to be washed, the slurry heel can be blown from the filter and wash liquor can be introduced to refill the shell. If the cake tends to crack during air blowing, it may be necessary to displace the slurry heel with wash gradually so as never to allow the cake to dry. Upon the completion of filtration and washing, the cake is discharged by one of several methods, depending on the shell and leaf configuration.

Horizontal pressure leaf filters. In these filters the leaves may be rectangular leaves which run parallel to the axis and are of varying sizes since they form chords of the shell; or they may be circular or square elements parallel to the head of the shell, and all of the same dimension. The leaves may be supported in the shell from an independent rack, individually from the shell, or from a filtrate manifold. Horizontal filters are particularly suited to dry-cake discharge.

Most of the currently available commercial horizontal pressure filters have leaves parallel to the shell head. Cake discharge may be wet or dry; it can be accomplished by sluicing with liquid sprays, vibration of the leaves, or leaf rotation against a knife, wire, or brush. If a wet-cake discharge is allowable, the filters will probably be sluiced with high-pressure liquid. If the filter has a top or bottom filtrate manifold, the leaves are usually in a fixed position, and the spray header is rotated to contact all filter surfaces. If the filtrate header is center-mounted, the leaves are generally rotated at about 3 r/min and the spray header is fixed. Some units may be wet-cake-discharged by mechanical vibration of the leaves with the filter filled with liquid. Dry-cake discharge normally will be accomplished by vibration if leaves are top- or bottom-manifolded and by rotation of the leaves against a cutting knife, wire, or brush if they are center-manifolded.

In many designs the filter is opened for cake discharge, and the leaf assembly is separated from the shell by moving one or the other on rails (Fig. 18-131). For processes involving toxic or flammable materials, a closed filter system can be maintained by sloping the bottom of the horizontal cylinder to the drain nozzle for wet discharge or by using a screw conveyor in the bottom of the shell for dry discharge.

Vertical pressure leaf filters. These filters have vertical, parallel, rectangular leaves mounted in an upright cylindrical pressure tank. The leaves usually are of such different widths as to allow them to conform to the curvature of the tank and to fill it without waste space. The leaves often rest on a filtrate manifold, the connection being sealed by an O ring, so that they can be lifted individually from the top of the filter for inspection and repair. A scavenger leaf frequently is installed in the bottom of the shell to allow virtually complete filtration of the slurry heel at the end of a cycle.

Vertical filters are not convenient for the removal of dry cake, although they can be used in this service if they have a bottom that can be retracted to permit the cake to fall into a bin or hopper below. They are adapted rather to wet-solids discharge, a process that may be assisted by leaf vibration, air or steam sparging of a filter full of water, sluicing from fixed, oscillating, or traveling nozzles, and blowback. They are made by many companies, and they enjoy their widest use for filter-aid precoat filtration.

Advantages and uses. The advantages of pressure leaf filters are their considerable flexibility (up to the permissible maximum, cakes of various thickness can be formed successfully), their low labor charges, particularly when the cake may be sluiced off or the dry cake discharged cleanly by blowback, the basic simplicity of many of the designs, and their adaptability to quite effective displacement washing. Their disadvantages are the requirement of exceptionally intelligent and watchful supervision to avoid cake consolidation or dropping, their inability to form as dry a cake as a filter press, their tendency to classify vertically during filtration and to form misshapen nonuniform cakes unless the leaves rotate, and the restriction of most models to 610 kPa (75 psig) or less.

Pressure leaf filters are used to separate much the same kinds of slurries as are filter presses and are used much more extensively than filter presses for filter-aid filtrations. They should be seriously considered whenever uniformity of production permits long-time operation

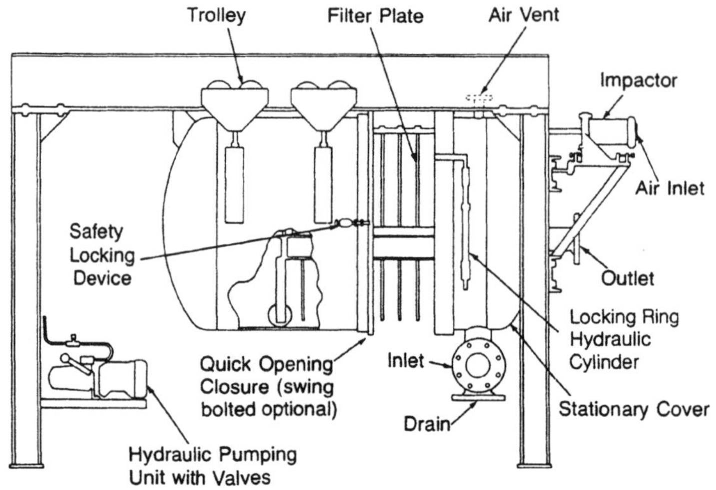

FIG. 18-131 Horizontal-tank pressure leaf filter designed for dry cake discharge. (*Sparkler Filter, Inc.*)

under essentially constant filtration conditions, when thorough washing with a minimum of liquor is desired, or when vapors or fumes make closed construction desirable. Under such conditions, if the filter medium does not require frequent changing, they may show a considerable advantage in cycle and labor economy over a filter press, which has a lower initial cost, and advantages of economy and flexibility over continuous vacuum filters, which have a higher first cost.

Pressure leaf filters are available with filtering areas of 930 cm^2 (1 ft^2) (laboratory size) up to about 440 m^2 (4734 ft^2) for vertical filters and 158 m^2 (1700 ft^2) for horizontal ones. Leaf spacings range from 5 to 15 cm (2 to 6 in) but are seldom less than 7.5 cm (3 in) since 1.3 to 2.5 cm (0.5 to 1 in) should be left open between surfaces.

Centrifugal-Discharge Filter Horizontal top-surface filter plates may be mounted on a hollow motor-connected shaft that serves both as a filtrate-discharge manifold and as a drive shaft to permit centrifugal removal of the cake. An example is the Funda filter (marketed in the United States by Steri Technologies), illustrated schematically in Fig. 18-132. The filtering surface may be a textile fabric or a wire screen, and the use of a precoat is optional. The Funda filter is driven from the top, leaving the bottom unobstructed for inlet and drainage lines; a somewhat similar machine that employs a bottom drive, providing a lower center of mass and ground-level access to the drive system, is the German-made **Schenk filter** (marketed in the United States by Pall Seitz Schenk Filtersystems).

During filtration, the vessel that coaxially contains the assembly of filter plates is filled with prefilt under pressure, the filtrate passes through the plates and out the hollow shaft, and cake is formed on the top surfaces of the plates. After filtration, the vessel is drained, or the heel may be filtered by recirculation through a cascade ring at the top of the filter. The cake may be washed—or it may be extracted, steamed, air-blown, or dried by hot gas. It is discharged, wet or dry, by rotation of the shaft at sufficiently high speed to sling away the solids. If flushing is permitted, the discharge is assisted by a backwash of appropriate liquid.

The operating advantages of the centrifugal-discharge filter are those of a horizontal-plate filter and, further, its ability to discharge cake without being opened. It is characterized by low labor demand, easy adaptability to automatic control, and amenability to the processing of hazardous, noxious, or sterile materials. Its disadvantages are its complexity and maintenance (stuffing boxes, high-speed drive) and its cost. The Funda filter is made in sizes that cover the filtering area range of 1 to 50 m^3 (11 to 537 ft^2). The largest Schenk filter provides 100 m^2 (1075 ft^2) of area.

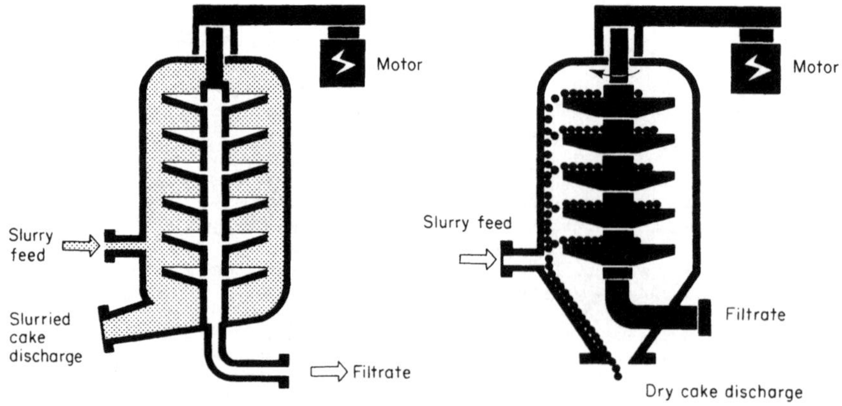

FIG. 18-132 Schematic of a centrifugal-discharge filter. (*Steri Technologies.*)

Continuous Cake Filters Continuous cake filters are applicable when cake formation is fairly rapid, as in situations in which slurry flow is greater than about 5 L/min (1 to 2 gal/min), slurry concentration is greater than 1 percent, and particles are greater than 0.5 μm in diameter. Liquid viscosity below 0.1 Pa·s (100 cP) is usually required for maintaining rapid liquid flow through the cake. Some designs of continuous filters can compromise some of these guidelines by sacrificial use of filter aid when the cake is not the desired product.

Rotary Drum Filters The rotary drum filter is the most widely used of the continuous filters. There are many design variations, including operation as either a pressure filter or a vacuum filter. The major difference between designs is in the technique for cake discharge, to be discussed later. All the alternatives are characterized by a horizontal-axis drum covered on the cylindrical portion by filter medium over a grid support structure to allow drainage to manifolds. Basic materials of construction may be metals or plastics. Sizes (in terms of filter areas) range from 0.37 to 186 m^2 (4 to 2000 ft^2).

All drum filters (except the single-compartment filter) utilize a rotary-valve arrangement in the drum-axis support trunnion to facilitate removal of filtrate and wash liquid and to allow introduction of air or gas for cake blowback if needed. The valve controls the relative duration of each cycle as well as providing "dead" portions of the cycle through the use of bridge blocks. A typical valve design is shown in Fig. 18-133. Internal piping manifolds connect the valve with various sections of the drum.

Most drum filters are fed by operating the drum with about 35 percent of its circumference submerged in a slurry trough, although submergence can be set for any desired amount between zero and almost total. Some units contain an oscillating rake agitator in the trough to aid solids suspension. Others use propellers, paddles, or no agitator.

Slurries of free-filtering solids that are difficult to suspend are sometimes filtered on a top-feed drum filter or filter-dryer. An example application is in the production of table salt. An alternative for slurries of extremely coarse, dense solids is the internal drum filter. In the chemical-process industry both top-feed and internal drums (which are described briefly by Emmett in Schweitzer, op. cit., p. 4-41) have largely been displaced by the horizontal vacuum filter (q.v.).

Most drum filters operate at a rotation speed in the range of 0.1 to 10 r/min. Variable-speed drives are usually provided to allow adjustment for changing cake-formation and drainage rates.

Drum filters commonly are classified according to the feeding arrangement and the cake-discharge technique. They are so treated in this subsection. The characteristics of the slurry and the filter cake usually dictate the cake-discharge method.

Scraper-Discharge Filter The filter medium is usually caulked into grooves in the drum grid, with cake removal facilitated by a scraper blade just prior to the resubmergence of the drum (Fig. 18-134). The scraper serves mainly as a deflector to direct the cake, dislodged by an air blowback, into the discharge chute, since actual contact with the medium would cause rapid wear. In some cases the filter medium is held by circumferentially wound wires spaced 50 mm (2 in) apart, and a flexible scraper blade may rest lightly against the wire winding. A taut wire in place of the scraper blade may be used in some applications in which physical dislodging of sticky, cohesive cakes is needed.

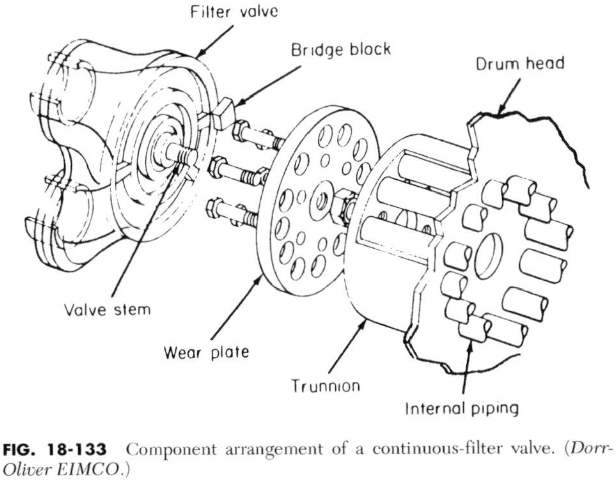

FIG. 18-133 Component arrangement of a continuous-filter valve. (*Dorr-Oliver EIMCO.*)

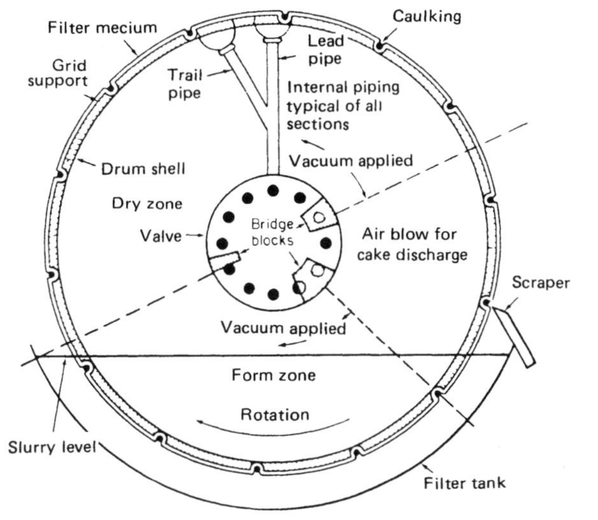

FIG. 18-134 Schematic of a rotary-drum vacuum filter with scraper discharge, showing operating zones. (*Schweitzer,* Handbook of Separation Techniques for Chemical Engineers, *p. 4-38. Copyright 1979 by McGraw-Hill, Inc. and used with permission.*)

For a given slurry, the maximum filtration rate is determined by the minimum cake thickness which can be removed—the thinner the cake, the less the flow resistance and the higher the rate. The minimum thickness is about 6 mm (0.25 in) for relatively rigid or cohesive cakes of materials such as mineral concentrates or coarse precipitates like gypsum or calcium citrate. Solids that form friable cakes composed of less cohesive materials such as salts or coal will usually require a cake thickness of 13 mm (0.5 in) or more. Filter cakes composed of fine precipitates such as pigments and magnesium hydroxide, which often produce cakes that crack or adhere to the medium, usually need a thickness of at least 10 mm (0.38 in).

String-Discharge Filter A system of endless strings or wires spaced about 13 mm (0.5 in) apart pass around the filter drum but are separated tangentially from the drum at the point of cake discharge, lifting the cake off as they leave contact with the drum. The strings return to the drum surface guided by two rollers, the cake separating from the strings as they pass over the rollers. If it has the required body, a thinner cake (5 mm or about 3⁄16 in) than can be handled by drum filters is feasible, allowing more difficult materials to be filtered. This is done at the expense of greater dead area on the drum. Success depends on the ability of the cake to be removed with the strings and must be determined experimentally. Applications are mainly in the starch and pharmaceutical industries, with some in the metallurgical field.

Removable-Medium Filters Some drum filters provide for the filter medium to be removed and reapplied as the drum rotates. This feature permits the complete discharge of thin or sticky cake and provides the regenerative washing of the medium to reduce blinding. Higher filtration rates are possible because of the thinner cake and clean medium, but this is compromised by a less pure filtrate than normally produced by a nonremovable medium.

Belt-discharge filter. This is a drum filter carrying a fabric that is removed, passed over rollers, washed, and returned to the drum. Figure 18-135 shows the path of the medium while it is off the drum.

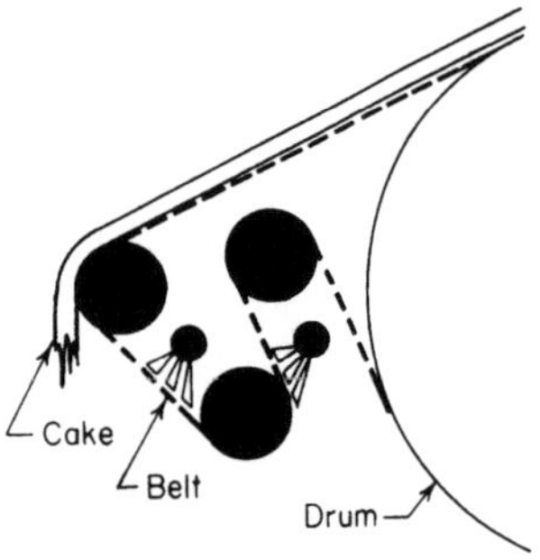

FIG. 18-135 Cake discharge and medium washing on an EIMCO belt filter. (*Dorr-Oliver EIMCO.*)

A special aligning device keeps the medium wrinkle-free and in proper line during its travel. Thin cakes of difficult solids which may be slightly soluble are good applications. When acceptable, a sluice discharge makes cakes as thin as 1.5 to 2 mm (about 1⁄16 in) feasible. Several manufacturers offer belt-discharge filters.

Coilfilter. The **Coilfilter** (Komline-Sanderson Engineering Corp.) is a drum filter with a medium consisting of one or two layers of stainless-steel helically coiled springs, about 10 mm (0.4 in) in diameter, placed in a corduroy pattern around the drum. The springs follow the drum during filtration with cake forming the coils. They are separated from the drum to discharge the cake and undergo washing; if two layers are used, the coils of each layer are further separated from those of the other, passing over different sets of rolls. The use of stainless steel in spring form provides a relatively permanent medium that is readily cleaned by washing and flexing. Filtrate clarity is poorer than with most other media, and a relatively large vacuum pump is needed to handle greater air leakage than is characteristic of fabric media. Material forming a slimy, matlike cake (e.g., raw sewage) is the typical application.

Roll-Discharge Filters A roll in close proximity to the drum at the point of cake discharge rotates in the opposite direction at a peripheral speed equal to or slightly faster than that of the drum (Fig. 18-136). If the cake on the drum is adequately tacky and cohesive for this discharge technique, it adheres to cake on the smaller roll and separates from the drum. A blade or taut wire removes the material from the discharge roll. This design is especially good for thin, sticky cakes. If necessary, a slight air blow may be provided to help release the cake from the drum. Typical cake thickness is 1 to 10 mm (0.04 to 0.4 in).

Single-Compartment Drum Filter

Bird-Young filter. This filter (Bird Machine Co.) differs from most drum filters in that the drum is not compartmented and there is no internal piping or rotary valve. The entire inside of the drum is subjected to vacuum, with its surface perforated to pass the filtrate. Cake is discharged by an air blowback applied through a "shoe" that covers a narrow discharge zone on the inside surface of the drum to interrupt the vacuum, as illustrated in Fig. 18-137. The internal drum surface must be machined to provide close clearance of the shoe to avoid leakage. The filter is designed for high filtration rates with thin cakes. Rotation speeds to 40 r/min are possible with cakes typically 3 to 6 mm (0.12 to 0.24 in) thick. Filter sizes range from 930 cm^2 to 19 m^2 (1 to 207 ft^2) with 93 percent of the area active. The slurry is fed into a conical feed tank designed to prevent solids from settling without the use of mechanical agitators. The proper liquid level is maintained by overflow, and submergence ranges from 5 to 70 percent of the drum circumference.

The perforated drum cylinder is divided into sections about 50 to 60 mm (2 to 2.5 in) wide. The filter medium is positioned into tubes between the sections and locked into place by round rods. No caulking, wires, or other fasteners are needed.

Wash sprays may be applied to the cake, with collection troughs or pans inserted inside the drum to keep the wash separate from the filtrate. Filtrate is removed from the lower section of the drum by a pipe passing through the trunnions.

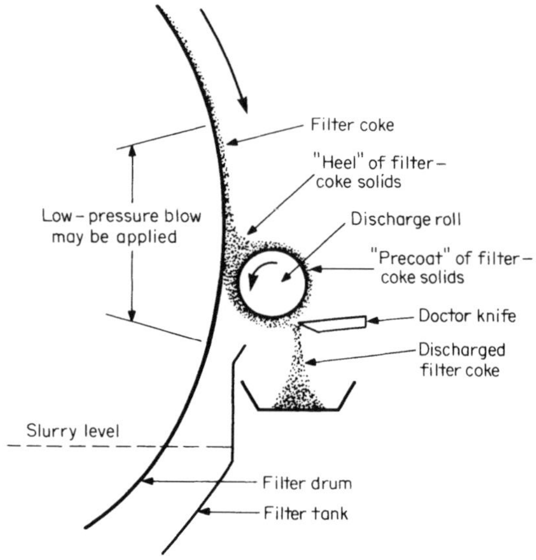

FIG. 18-136 Operating principles of a roll-discharge mechanism. (*Schweitzer,* Handbook of Separation Techniques for Chemical Engineers, *p. 4-40. Copyright 1979 by McGraw-Hill, Inc. and used with permission.*)

The major advantages of the Bird-Young filter are its ability to handle thin cakes and operate at high speeds, its washing effectiveness, and its low internal resistance to air and filtrate flow. An additional advantage is the possibility of construction as a pressure filter with up to 1.14-MPa (150-psig) operating pressure to handle volatile liquids. The chief disadvantages are its high cost and the limited flexibility imposed by not having an adjustable rotary valve. Best applications are on free-draining nonblinding materials such as paper pulp or crystallized salts.

Continuous Pressure Filters These filters consist of conventional drum or disc filters totally enclosed in pressure vessels. Filtration takes place with the vessel pressurized up to 6 bar and the filtrate discharging either at atmospheric pressure or into a receiver maintained at a suitable backpressure. Cake discharge is facilitated through a dual valve and lock-hopper arrangement in order to maintain vessel pressure. Alternatively, the discharged filter cake can be reslurried within the filter or in an adjoining pressure vessel and removed through a control valve.

One variation in design, the Ceramec, offered by Outokumpu Mintec, employs "gasless" ceramic media instead of traditional filter fabrics, relying partly on capillary action to achieve low moistures. This results in a significant drop in power consumption by greatly reducing the compressed air requirements.

Continuous Precoat Filters These filters may be operated as either pressure or vacuum filters, although vacuum operation is the prevailing one. The filters are really not continuous but have an extremely long batch cycle (1 to 10 days). Applications are for continuous clarification of liquids from slurries containing 50 to 5000 ppm of solids when only very thin unacceptable cakes would form on other filters and where "perfect" clarity is required.

Construction is similar to that of other drum filters, except that vacuum is applied to the entire rotation. Before feeding slurry a precoat layer of filter aid or other suitable solids, 75 to 125 mm (3 to 5 in) thick, is applied. The feed slurry is introduced and trapped in the outer surface of the precoat, where it is removed by a progressively advancing doctor knife which trims a thin layer of solids plus precoat (Fig. 18-138). The blade advances 0.05 to 0.2 mm (0.002 to 0.008 in) per revolution of the drum. When the precoat has been cut to a predefined minimum thickness, the filter is taken out of service, washed, and freshly precoated. This turnaround time may be 1 to 3 h.

Disc Filters A disc filter is a vacuum filter consisting of a number of vertical discs attached at intervals on a continuously rotating

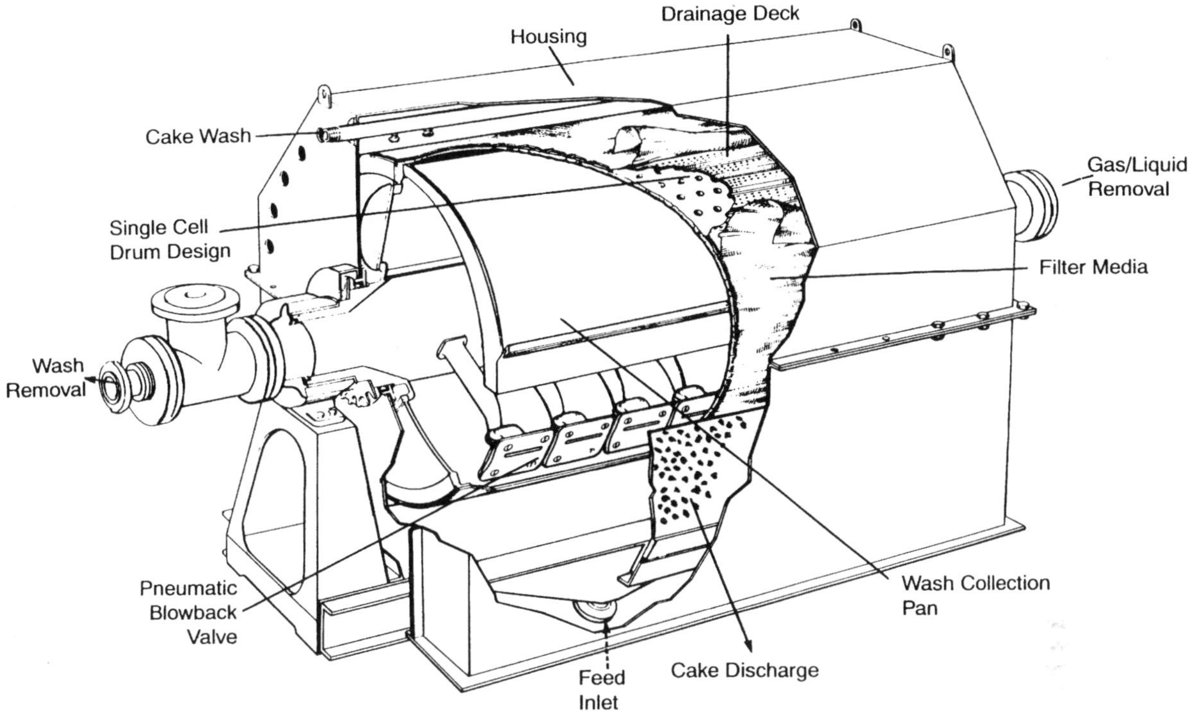

FIG. 18-137 Cutaway of the single-compartment drum filter. (*Andritz Bird.*)

horizontal hollow central shaft (Fig. 18-139). Rotation is by a gear drive. Each disc consists of 10 to 30 sectors of metal, plastic, or wood, ribbed on both sides to support a filter cloth and provide drainage via an outlet nipple into the central shaft. Each sector may be replaced individually. The filter medium is usually a cloth bag slipped over the sectors and sealed to the discharge nipple. For some heavy-duty applications on ores, stainless-steel screens may be used.

The discs are typically 30 to 50 percent submerged in a troughlike vessel containing the slurry. Another horizontal shaft running beneath the discs may contain agitator paddles to maintain suspension of the solids, as in the EIMCO **Agidisc filter.** In some designs, feed is distributed through nozzles below each disc. Vacuum is supplied to the sectors as they rotate into the liquid to allow cake formation. Vacuum is maintained as the sectors emerge from the liquid and are exposed to air. Wash may be applied with sprays, but most applications are for dewatering only. As the sectors rotate to the discharge point, the vacuum is cut off, and a slight air blast is used to loosen the cake. This allows scraper blades to direct the cake into discharge chutes positioned between the discs. Vacuum and air blowback is controlled by an automatic valve as in rotary-drum filters.

Of all continuous filters, the vacuum disc is the lowest in cost per unit area of filter when mild steel, cast iron, or similar materials of construction may be used. It provides a large filtering area with minimum floor space, and it is used mostly in high-tonnage dewatering applications in sizes up to about 300 m^2 (3300 ft^2) of filter area.

The main disadvantages are the inadaptability to have effective wash and the difficulty of totally enclosing the filter for hazardous-material operations.

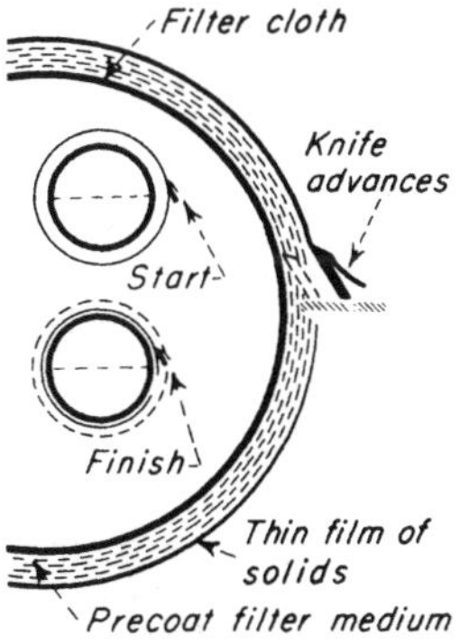

FIG. 18-138 Operating method of a vacuum precoat filter. (*Dorr-Oliver EIMCO.*)

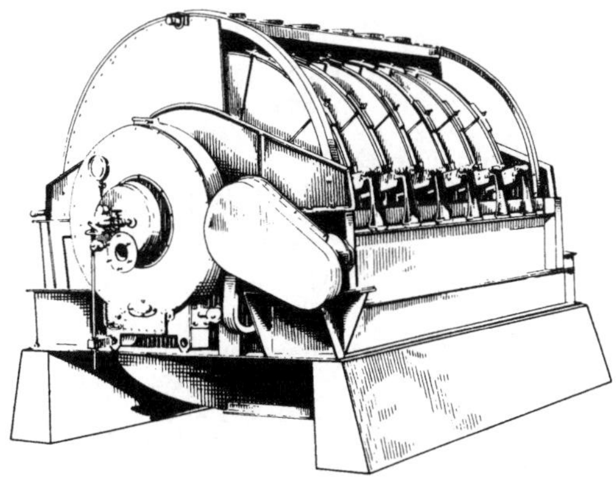

FIG. 18-139 Rotary disc filter. (*Dorr-Oliver EIMCO.*)

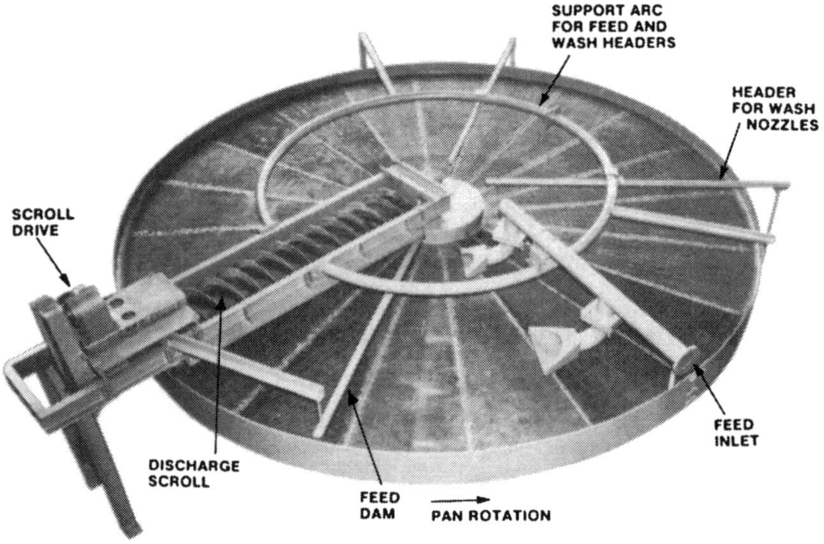

FIG. 18-140 Continuous horizontal vacuum table filter. (*Dorr-Oliver EIMCO.*)

Horizontal Vacuum Filters These filters are generally classified into two broad classes: rotary circular and belt-type units. Regardless of geometry, they have similar advantages and limitations. They provide flexibility of choice of cake thickness, washing time, and drying cycle. They effectively handle heavy, dense solids, allow flooding of the cake with wash liquor, and are easily designed for true countercurrent leaching or washing. The disadvantages are they are more expensive to build than drum or disc filters, they use a large amount of floor space per filter area, and they are difficult to enclose for hazardous applications.

Horizontal-Table, Scroll-Discharge, and Pan Filters These are all basically revolving annular tables with the top surface a filter medium (Fig. 18-140). The table is divided into sectors, each of which is a separate compartment. Vacuum is applied through a drainage chamber beneath the table that leads to a large rotary valve. Slurry is fed at one point, and cake is removed after completing more than three-fourths of the circle, by a horizontal scroll conveyor which elevates the cake over the rim of the filter. A clearance of about 10 mm (0.4 in) is maintained between the scroll and the filter medium to prevent damage to the medium. Residual cake on the medium may be loosened by an air blow from below or with high-velocity liquid sprays from above. This residual cake is a disadvantage peculiar to this type of filter. With material that can cause blinding, frequent shutdowns for thorough cleaning may be needed. Unit sizes range from about 0.9 to 9 m (3 to 30 ft) in diameter, with about 80 percent of the surface available for filtration.

Tilting-Pan Filter This is a modification of the table or pan filter in which each of the sectors is an individual pan pivoted on a radial axis to allow its inversion for cake discharge, usually assisted by an air blast. Filter-cake thicknesses of 50 to 100 mm (2 to 4 in) are common. Most applications involve free-draining inorganic-salt dewatering. In addition to the advantages and disadvantages common to all horizontal continuous filters, tilting-pan filters have the relative advantages of complete wash containment per sector, good cake discharge, filter-medium washing, and feasibility of construction in very large sizes, up to about 25 m (80 ft) in diameter, with about 75 percent of the area usable. Relative disadvantages are high capital cost (especially in smaller sizes) and mechanical complexity leading to higher maintenance costs.

Horizontal-Belt Filter This filter consists of a slotted or perforated elastomer drainage belt driven as a conveyor belt carrying a filter fabric belt (Fig. 18-141). Both belts are supported by and pass across a lubricated support deck. A vacuum pan, aligned with the slots in the elastomer belt, forms a continuous vacuum surface which may include multiple zones for cake formation, washing and final dewatering. Several manufacturers provide horizontal-belt filters, the major differences among which lie in the construction of the drainage belt, the method of retaining the slurry/cake on the belt, and the method of maintaining the alignment of the filter medium. The filters are rated according to the available active filtration area. *Indexing horizontal-belt filters* do away with the elastomer drainage belt of the original design in favor of large drainage pans directly beneath the filter medium. Either the pans or the filter medium is indexed to provide a pseudo-continuous filtration operation. The applied vacuum is cycled with the indexing operation to minimize wear to the sliding surfaces. As a result the indexing filter must be de-rated for the indexing cycle. The indexing horizontal-belt filter avoids the problem of process compatibility with the elastomer drainage belt. The major differences among the indexing machines of several manufacturers lie in the method of indexing and the method of cycling the applied vacuum.

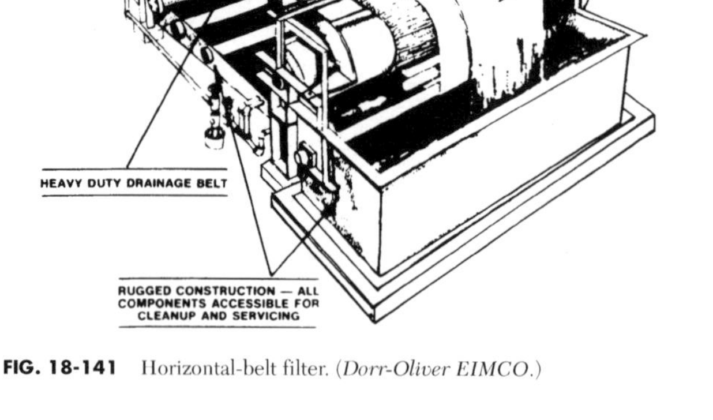

FIG. 18-141 Horizontal-belt filter. (*Dorr-Oliver EIMCO.*)

The method of feeding, washing, dewatering and discharging is essentially the same with all horizontal-belt filters. Slurry is fed at one end by overflow weirs or a fantail chute; wash liquor, if required, is applied by sprays or weirs at one or more locations as the formed cake moves along the filter. Wiping dams and separations in the drainage pan(s) provide controlled wash application. The cake is discharged as the filter-medium belt passes over the end pulley after separation from the drainage surface. Separating the filter-medium from the drainage surface allows thorough spray cleaning of the filter-medium belt. The duration of the filtration cycle is controlled by belt speed which may be as high as 1 m/s (3.3 ft/s) and is typically variable. The minimum possible cake thickness, at a given solids loading, which can be effectively discharged limits the belt speed from a process point of view. The maximum cake thickness is dependent on the method used to retain the slurry during cake formation and can be 100 to 150 mm (4 to 6 in) with fast draining materials.

Some of the advantages of horizontal-belt filters are the precise control of the filtration cycle including the capability for countercurrent washing of the cake, effective cake discharge and thorough cleaning of the filter medium belt. The horizontal-belt filter's primary disadvantage is that at least half of the filtration medium is always idle during the return loop. This contributes to a significantly higher capital cost which can be two to four times that of a drum or disc filter with equal area. Horizontal-belt filters with active filtration area ranging from 0.18 m^2 to 150 m^2 (2 ft^2 to 1600 ft^2) on a single machine have been installed.

Additional equipment is sometimes integrated with horizontal-belt filters to further dewater the cake through expression. The addition of such equipment shouldn't be confused with expression equipment that utilizes filter medium belts. Belt type expression equipment is described later in the "Expression" subsection.

Filter Thickeners Thickeners are devices which remove a portion of the liquid from a slurry to increase the concentration of solids in suspension. Thickening is done to prepare a dilute slurry for more economical filtration or to change the consistency or concentration of the slurry for process reasons. The commonest method of thickening is by gravity sedimentation, discussed earlier in this section. Occasions may arise, however, in which a filter may be called upon for thickening service. Many of the filters previously discussed as cake filters can be operated as thickeners: the filter press with special plates containing flow channels that keep velocity high enough to prevent cake buildup, cycled tube or candle filters with the cake discharge into the filter tank, and continuous leaf filters which use rotating elements adjacent to the filtering surfaces to limit filter cake buildup. Examples of these filters include the Shriver Thickener, the Industrial Hydra-Shoc Filter employing Back Pulse Technology, the DrM, Dr. Müller AG, Contibac Thickener, and the Ingersoll Rand Continuous Pressure Filter.

Clarifying Filters Clarifying filters are used to separate liquid mixtures which contain only very small quantities of solids. When the solids are finely divided enough to be observed only as a haze, the filter which removes them is sometimes called a polishing filter. The prefilt slurry generally contains no more than 0.10 percent solids, the size of which may vary widely (0.01 to 100 μm). The filter usually produces no visible cake, sometimes because the amount of solids removed is so small, sometimes because the particles are removed by being entrapped within rather than upon the filter medium. Compared with cake filters, clarifying filters are of minor importance to pure chemical-process work, their greatest use being in the fields of beverage and water polishing, pharmaceutical filtration, fuel- and lubricating-oil clarification, electroplating-solution conditioning, and dry-cleaning-solvent recovery. They are essential, however, to the processes of fiber spinning and film extrusion; the spinning solution or dope must be free of particles above a certain size to maintain product quality and to prevent the clogging of spinnerets.

Most cake filters can be so operated as to function as clarifiers, although not necessarily with efficiency. On the other hand, a number of clarifying filters which can be used for no purpose other than clarifying or straining have been developed. In general, clarifying filters are less expensive than cake filters. Clarifying filters may be classified as disc and plate presses, cartridge clarifiers, precoat pressure filters, deep-bed filters, and miscellaneous types. Membrane filters constitute a special class of plate presses and cartridge filters. Simple strainers sometimes are used as clarifiers of liquids containing very large particles. Because they more closely resemble wet screens than filters and because they have little primary process application, they are not discussed here.

Disc Filters and Plate Presses Filters employing asbestos-pulp discs, cakes of cotton fibers (filtermasse), or sheets of paper or other media are used widely for the polishing of beverages, plating solutions, and other low-viscosity liquids containing small quantities of suspended matter. The term **disc filter** is applied to assemblies of pulp discs made of asbestos and cellulose fibers and sealed into a pressure case. The discs may be preassembled into a self-supporting unit (Fig. 18-142), or each disc may rest on an individual screen or plate against which it is sealed as the filter is closed (Fig. 18-143). The liquid flows through the discs, and into a central or peripheral discharge manifold. Flow rates are on the order of 122 L/(min·m^2) [3 gal/(min·ft^2)], and the operating pressure does not normally exceed 345 kPa (50 psig) (usually it is less). Disc filters are almost always operated as pressure filters. Individual units deliver up to 378 L/min (6000 gal/h) of low-viscosity liquid.

Disc-and-plate assemblies somewhat resemble horizontal-plate pressure filters, which, in fact, may be used for polishing. In one design (**Sparkler VR filter**) both sides of each plate are used as filtering surfaces, having paper or other media clamped against them.

Pulp filters. These filters employ one or more packs of filter-masse (cellulose fibers compressed to a compact cylinder) stacked into a pressure case. The packs are sometimes supported in individual trays which provide drainage channels and sometimes rest on one another with a loose spacer plate between each two packs and with a drainage screen buried in the center of each pack. The liquid being clarified flows under a pressure of 345 kPa (50 psig) or less through the pulp packs and into a drainage manifold. Flow rates are somewhat less than for disc filters, on the order of 20 L/(min·m^2) [0.5 gal/(min·ft^2)]. Pulp filters are used chiefly to polish beverages. The filtermasse may be washed in special washers and re-formed into new cakes.

Plate presses. Sometimes called sheet filters, these are assemblies of plates, sheets of filter media, and sometimes screens or frames. They are essentially modified filter presses with practically no cake-holding capacity. A press may consist of many plates or of a single filter sheet between two plates, the plates may be rectangular or circular, and the sheets may lie in a horizontal or vertical plane. The operation is similar to that of a filter press, and the flow rates are about the same as for disc filters. The operating pressure usually does not exceed 138 kPa (20 psig). The presses are used most frequently for low-viscosity liquids, but an ordinary filter press with thin frames is commonly used as a clarifier for 100-Pa·s (1000-P) rayon-spinning solution. Here the filtration pressure may be 6900 kPa (1000 psig).

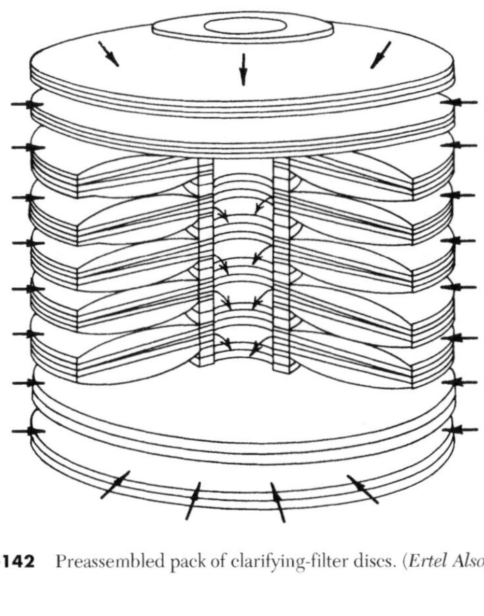

FIG. 18-142 Preassembled pack of clarifying-filter discs. (*Ertel Alsop.*)

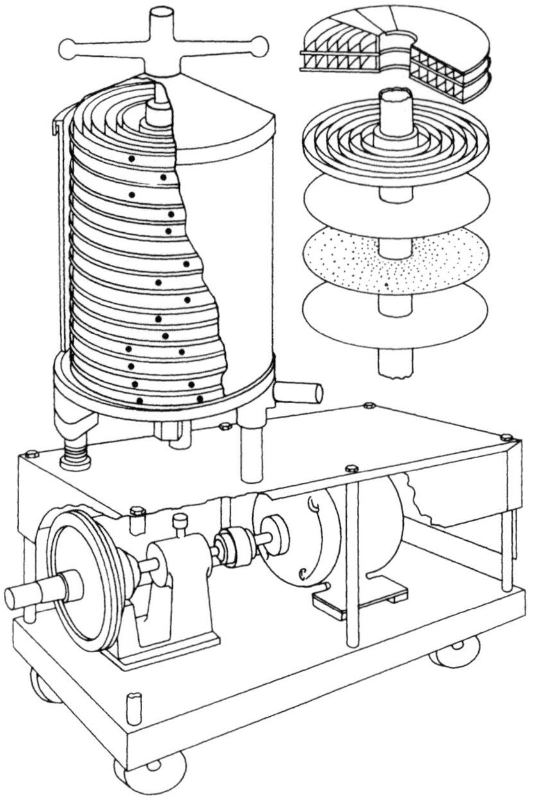

FIG. 18-143 Disc-and-plate clarifying-filter assembly. (*Ertel Alsop.*)

Disc, pulp, and sheet filters accomplish extreme clarification. Not infrequently their mission is complete removal of particles above a stipulated cut size, which may be much less than 1 µm. They operate over a particle-size range of four to five orders of magnitude, contrasting with two orders of magnitude for most other filters. It is not surprising, therefore, that they involve a variety of kinds and grades of filter media, often in successive stages. In addition to packs or discs of cellulosic, polymeric, or asbestos fiber, sheets of pulp, paper, asbestos, carded fiber, woven fabrics, and porous cellophane or polymer are employed. Sandwich-pack composites of several materials have been used for viscous-dope filtration.

The use of asbestos has been greatly diminished because of its identification with health hazards. There have been proposed replacement materials such as the **Zeta Plus** filter media from the AMF Cuno Division, consisting of a composite of cellulose and inorganic filter aids that have a positive charge and provide an electrokinetic attraction to hold colloids (usually negatively charged). These media therefore provide both mechanical straining and electrokinetic adsorption.

Cartridge Clarifiers Cartridge clarifiers are units which consist of or use one or more replaceable or renewable cartridges containing the active filter element. The unit usually is placed in a line carrying the liquid to be clarified; clarification thus occurs while the liquid is in transit.

Mechanical or edge filters. These consist of stacks of discs separated to precise intervals by spacer plates, or a wire wound on a cage in grooves of a precise pitch, or a combination of the two. The liquid to be filtered flows radially between the discs, wires, or layers of paper, and particles larger than the spacing are screened out. Edge filters can remove particles down to 0.001 in (25 µm) but more often have a minimum spacing of twice this value. They have small solids-retaining capacity and hence must be cleaned often to avoid plugging. Continuous cleaning is provided in some filters. For example, the Cuno **Auto-Klean,** a wire-wound unit, employs a slowly rotating scraper that fits into the interdisc slots to comb away accumulated solids. In either case, the dislodged solids fall into a sump that may be drained at intervals.

Micronic clarifiers. The greatest number of cartridge clarifiers are of the micronic class, with elements of fiber, resin-impregnated filter paper, porous stone, or porous stainless steel of controlled porosity. Other rustless metals are also available. The elements may be chosen to remove particles larger than a fraction of a micrometer, although many are made to pass 10-µm solids and smaller. By proper choice of multiple-cylinder cartridges or multiple cartridges in parallel any desired flow rate can be obtained at a reasonable pressure drop, often less than 138 kPa (20 psig).

When the pressure rises to the permissible maximum, the cartridge must be opened and the element replaced. Micronic elements of the fiber type cannot be cleaned and are so priced that they can be discarded or the filter medium replaced economically. Stone elements usually must be cleaned, a process best accomplished by the manufacturer of the porous ceramic or in accordance with the manufacturer's directions. The user can clean stainless-steel elements by chemical treatment.

Flexibility. Cartridge filters are flexible: cartridges of different ratings and materials of construction can be interchanged, permitting ready accommodation to shifting conditions. They have the disadvantage of very limited solid-handling capability so that the maximum solid concentrations in the feed are limited to about 0.01 percent solids. The biggest limitation for modern process-plant operation is the need to open the filter to replace cartridges, which makes their use for the processing of hazardous materials undesirable. Some manufacturers—for example, the Hydraulic Research Division of Textron Inc. and the Fluid Dynamics Division of Brunswick Corp.—have designed cartridges of bonded metal fibers that can be back-flushed or chemically cleaned without opening the unit. These filters, which can operate at temperatures to 482°C (900°F) and at pressures of 33 MPa (325 atm) or greater, are particularly useful for filtering polymers.

Granular Media Filters Many types of granular media filters are used for clarification, operating either as gravity or pressure filters. Gravity filters rely on a difference in elevation between inlet and outlet to provide the driving force necessary to force the liquid through the granular media. Pressure filters employ enclosed vessels operating at relatively low pressure differentials, in the order of 50 to 70 kPa (7 to 10 psig), which may function in either an upflow or a downflow mode.

The media may be a single material, such as sand, but more often will consist of two or even three layers of different materials, such as anthracite coal in the top layer and sand in the lower one. Solids are captured throughout the bed depth, rather than on the surface, and the gradient in void size provides substantially more solids-holding capacity. The anthracite layer, typically employing 1-mm grain size, serves as a roughing filter and also provides a flocculating action which helps the finer sand, ~0.5-mm particle size, to serve as an effective polishing zone. Media depths vary, but 0.7 to 1.0 m is typical of a dual media installation. Deeper beds of up to 2.5 m (8 ft) are employed in some cases involving special applications where greater solids-holding capacity is desired.

Filtrate is collected in the underdrain system, which may be as simple as a network of perforated pipes covered by graded gravel or a complex structure with slotted nozzles or conduits that will retain the finest sand media while maintaining high flow rates. This latter design allows the use of both air and liquid for the backwashing and cleaning operations.

Backwashing usually is carried out when a limiting pressure drop is reached and before the bed becomes nearly filled with solids, which would lead to a deterioration in filtrate clarity. Cleaning the media is greatly aided by the use of an air scour which helps break loose the trapped solids and provides efficient removal of this material in the subsequent backflushing step. The filtration action tends to agglomerate the filtered solids and, as a result, these generally will settle out readily from the backwash fluid. If the filter is handling a clarifier overflow, usually it is possible to discharge the backwash liquid into the clarifier without risk of these solids returning to the filter. Filter media consumption is low, with normal replacement usually being less than 5 percent per year.

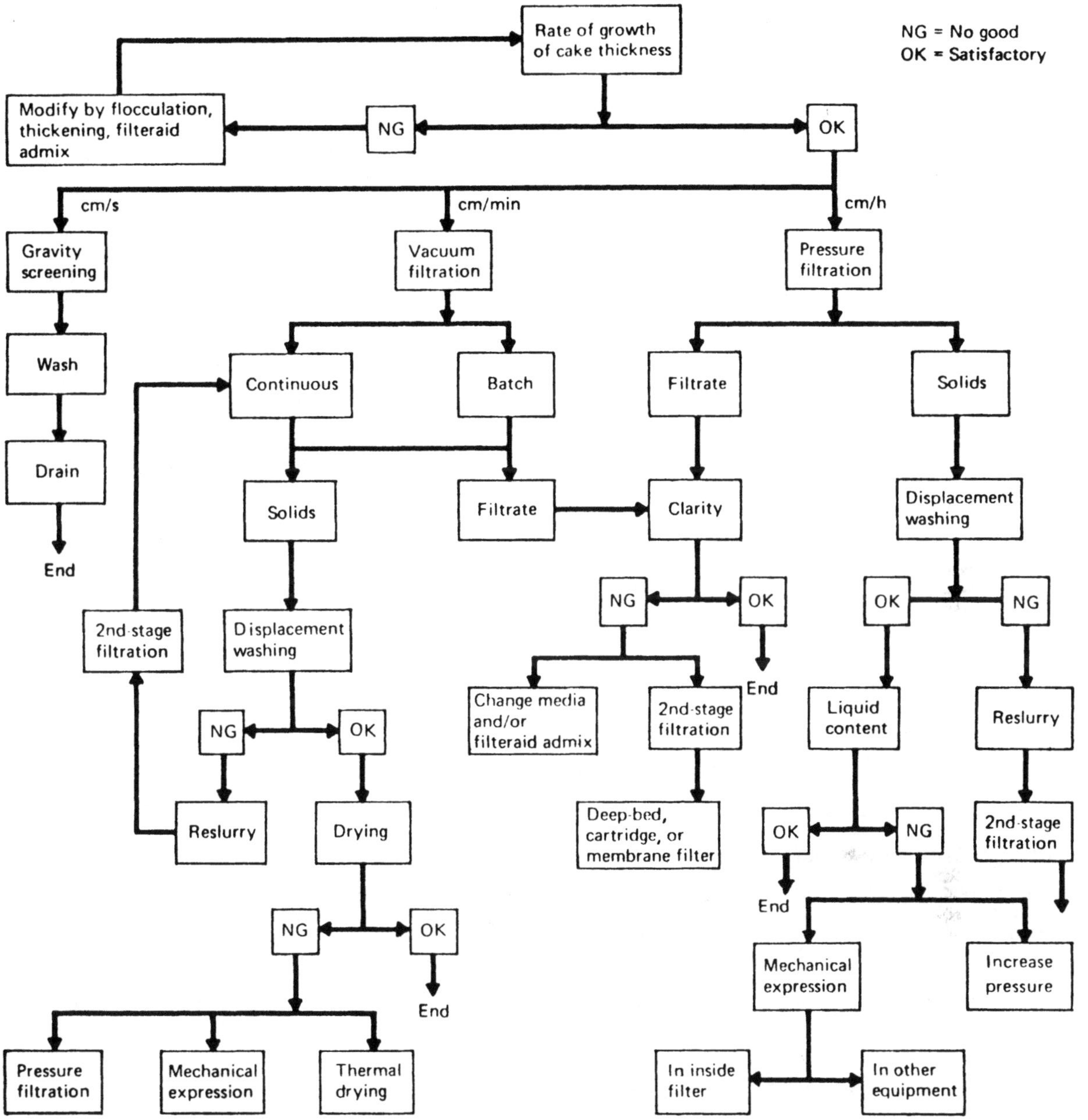

FIG. 18-144 Decision pattern for solving a filtration problem. [*Tiller,* Chem. Eng., ***81***(9), *118 (1974), by permission.*]

These filters are best applied on relatively dilute suspensions, <150 mg/L suspended solids, allowing operation at relatively high rates, 7.5 to 15.0 $m^3/m^2/h$ (3 to 6 gpm/ft^2). Solids capture will range from 90 to 98 percent in a well-designed system. Typical operating cycles range from 8 to 24 h of filtration (and up to 48 h in municipal water treatment), followed by a backwash interval of 15 to 30 min. Applications are principally in municipal and wastewater treatment, but granular media filters also have been employed in industrial uses such as pulp and paper plant inlet water treatment; removal of oil, grease, and scale from steelmaking process wastewater; and clarification of electrolyte in copper electrowinning operations.

United States Filter Corp. Maxi-Flo Filter. The **Maxi-Flo Filter** is an example of the upflow closed-vessel design. Filtration rates to 0.0081 $m^3/(m^2 \cdot s)$ [12 $gal/(ft^2 \cdot min)$] and filter cross-section areas up to 10.5 m^2 (113 ft^2) are possible. Deep-bed filtration has been reviewed by Tien and Payatakes [*Am. Inst. Chem. Eng. J.,* **25,** 737 (1970)] and by Oulman and Baumann [*Am. Inst. Chem. Eng. Symp. Ser.,* **73**(171), 76 (1977)].

Dyna Sand Filter. A filter that avoids batch backwashing for cleaning, the **Dyna Sand Filter** is available from Parkson Corporation. The bed is continuously cleaned and regenerated by recycling solids internally through an air-lift pipe and a sand washer. Thus a constant pressure drop is maintained across the bed, and the need for parallel filters to allow continued on-stream operation, as with conventional designs, is avoided.

Miscellaneous Clarifiers Various types of filters such as cartridge, magnetic, and bag filters are widely used in polishing operations,

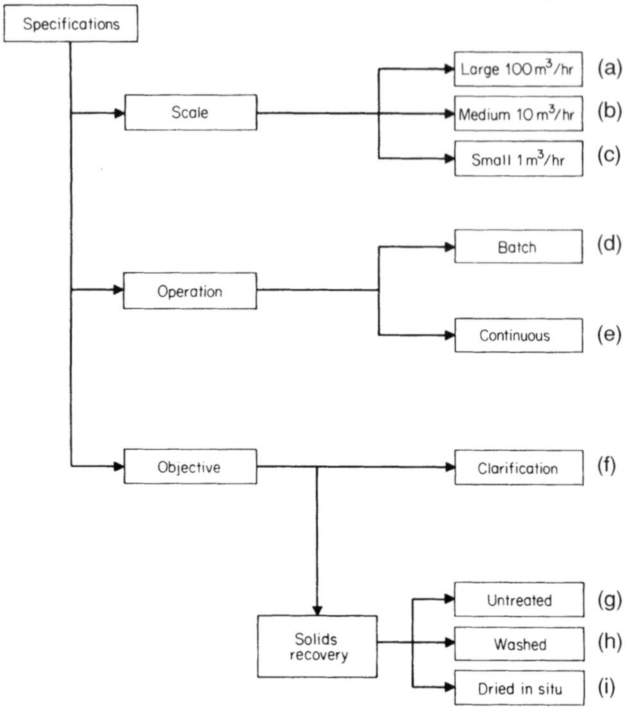

FIG. 18-145 Coding the problem specification. (*Purchas,* Solid/Liquid Separation Equipment Scale-Up, *Uplands Press, Croydon, England, 1977, p. 10, by permission.*)

generally to remove trace amounts of suspended solids remaining from prior unit operations. A thorough discussion of cartridge and felt strainer bag filters is available in Schweitzer, op. cit., Section 4.6 (Nickolaus) and Section 4.7 (Wrotnowski).

SELECTION OF FILTRATION EQUIPMENT

If a process developer who must provide the mechanical separation of solids from a liquid has cleared the first decision hurdle by determining that filtration is the way to get the job done (see the final subsection of Sec. 18, "Selection of a Solids-Liquid Separator")—or that it must remain in the running until some of the details of equipment choice have been settled—choosing the right filter and right filtration conditions may still be difficult. Much as in the broader determination of which unit operation to employ, the selection of filtration equipment involves the balancing of process specifications and objectives against capabilities and characteristics of the various equipment choices (including filter media) available. The important process-related factors are slurry character, production throughput, process conditions, performance requirements, and permissible materials of construction. The important equipment-related factors are type of cycle (batch or continuous), driving force, production rates of the largest and smallest units, separation sharpness, washing capability, dependability, feasible materials of construction, and cost. The estimated cost must account for installed cost, equipment life, operating labor, maintenance, replacement filter media, and costs associated with product-yield loss (if any). In between the process and equipment factors are considerations of slurry preconditioning and use of filter aids.

Slurry characteristics determine whether a clarifying or a cake filter is appropriate; and if the latter, they determine the rate of formation and nature of the cake. They affect the choice of driving force and cycle as well as specific design of machine.

There are no absolute selection techniques available to come up with the "best" choice since there are so many factors involved, many of them difficult to make quantitative and, not uncommonly, some contradictory in their demands. However, there are some published general suggestions to guide the thinking of the engineer who faces the selection of filtration equipment. Figure 18-144 is a decision tree designed by Tiller [*Chem. Eng.*, **81**(9), 118 (1974)] to show the steps to be followed in solving a filtration problem. It is erected on the premise that rate of cake formation is the most important guide to equipment selection. A filter-selection process proposed by Purchas (op. cit., pp. 10–14) employs additional criteria and is based on a combination of process specifications and the results of simple tests. The application is coded by use of Figs. 18-145, 18-146, and 18-147, and the resulting codes are matched against Table 18-11 to identify possible filters. Information needed for Fig. 18-148 can be obtained by observing the settling of a slurry sample (Purchas suggests 1 L) in a graduated cylinder. Filter-cake-growth rate (Fig. 18-148) is determined by small-scale leaf or funnel tests as described earlier.

Almost all types of continuous filters can be adapted for cake washing. The effectiveness of washing is a function of the number of wash displacements applied, and this, in turn, is influenced by the ratio of wash time to cake-formation time. Countercurrent washing, particularly with three or more stages, is usually limited to horizontal filters, although a two-stage countercurrent wash sometimes can be applied on a drum filter handling freely filtering material, such as crystallized

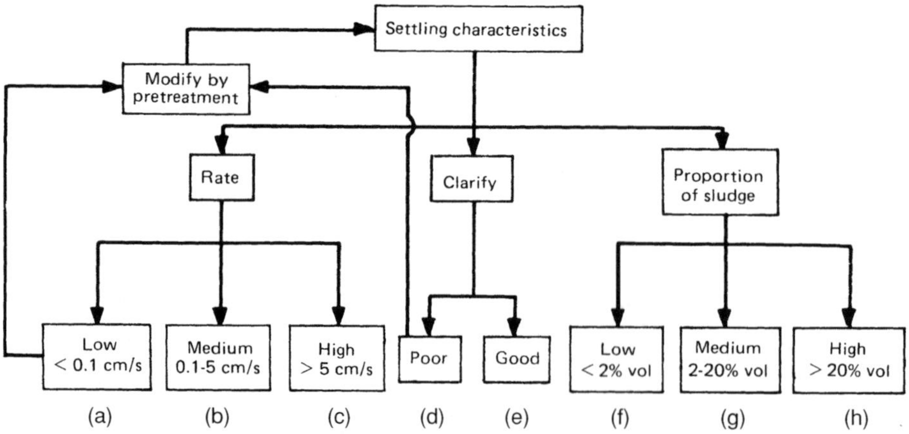

FIG. 18-146 Coding the settling characteristics of a slurry. (*Purchas,* Solid/Liquid Separation Equipment Scale-Up, *Uplands Press, Croydon, England, 1977, p. 11, by permission.*)

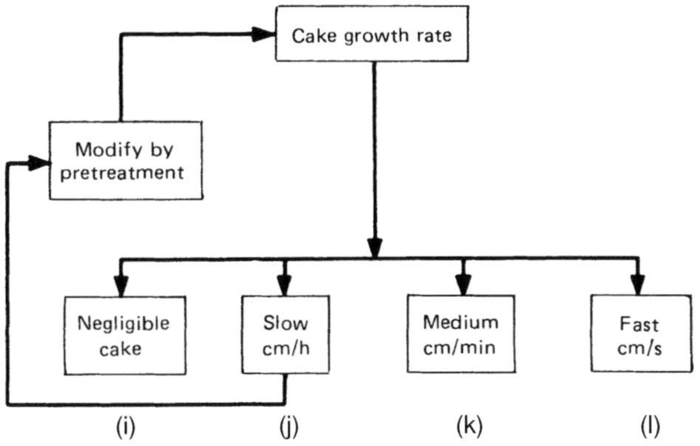

FIG. 18-147 Coding the filtration characteristics of a slurry. (*Purchas,* Solid/Liquid Separation Equipment Scale-Up, *Uplands Press, Croydon, England, 1977, p. 12, by permission.*)

salts. Cake washing on batch filters is commonly done, although, generally, a greater number of wash displacements may be required in order to achieve the same degree of washing obtainable on a continuous filter.

Continuous filters are most attractive when the process application is a steady-state continuous one, but the rate at which cake forms and the magnitude of production rate are sometimes overriding factors. A rotary vacuum filter, for example, is a dubious choice if a 3-mm (0.12-in) cake will not form under normal vacuum in less than 5 min and if less than 1.4 m^3/h (50 ft^3/h) of wet cake is produced. Upper production-rate limits to the practicality of batch units are harder to establish, but any operation above 5.7 m^3/h (200 ft^3/h) of wet cake should be considered for continuous filtration if it is at all feasible. Again, however, other factors such as the desire for flexibility or the need for high pressure may dictate batch equipment.

For estimating filtration rate (therefore, operating pressure and size of the filter), washing characteristics, and other important features,

TABLE 18-11 Classification of Filters according to Duty and Slurry-Separation Characteristics*

Type of equipment	Suitable for duty specification†	Required slurry-separation characteristics‡	
		Slurry-settling characteristics	Slurry-filtering characteristics
Deep-bed filters	a or b	A	T
	e	D	
	f	F	
Cartridges	b or c	A or B	
	d	D or E	
	f	F	
Batch filters			
Pressure vessel with vertical elements	a, b, or c	A or B	I or J
	d	D or E	
	f, g, h, or i	F or G	
Pressure vessel with horizontal elements	b or c	A or B	J or K
	d	D or E	
	g or h	F or G	
Filter presses	a, b, or c	A (or B)	I or J
	d	D or E	
	f, g, h, or i	F, G, or H	
Variable-volume filters	a, b, or c	A (or B)	J or K
	d or e	D or E	
	g (or h)	G or H	
Continuous filters			
Bottom-fed drum or belt drum	a, b, or c	A or B	I, J, K, or L
	e	D or E	
	f, g, h, or i	F, G, or H	
Top-fed drum	a, b, or c	C	L
	e	E	
	g, i (or h)	G or H	
Disc	a, b, or c	A or B	J or K
	e	D or E	
	g	G or H	
Horizontal belt, pan, or table	a, b, or c	A, B, or C	J, K, or L
	d or e	D or E	
	g or h	F, G, or H	

*Adapted from Purchas, *Solid/Liquid Separation Equipment Scale-Up,* Uplands Press, Croydon, England, 1977, p. 13, by permission.
†Symbols are identified in Fig. 18-135.
‡Symbols are identified in Figs. 18-136 and 18-137.

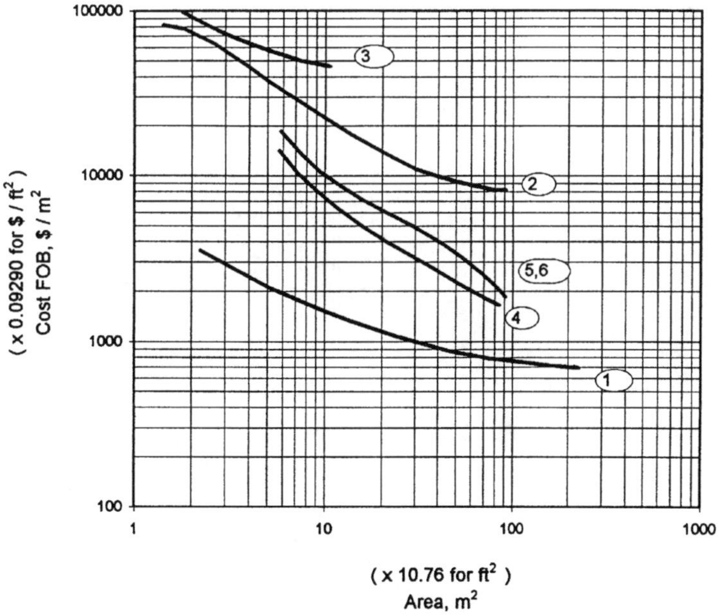

FIG. 18-148 Price of filters installed, FOB point of manufacture. (*EIMCO Process Equipment Co.*)

small-scale tests such as the leaf or pressure bomb tests described earlier are usually essential. In the conduct and interpretation of such tests, and for advice on labor requirements, maintenance schedule, and selection of accessory equipment the assistance of a dependable equipment vendor is advisable.

FILTER PRICES

As indicated, one of the factors affecting the selection of a filter is total cost of carrying out the separation with the selected machine. An important component of this cost item is the installed cost of the filter, which starts with the purchase price.

From a survey of early 1982, prices of a number of widely used types of process filter were collated by Hall and coworkers [*Chem. Eng.*, **89**(7), 80 (1982)]. These data are drawn together in Fig. 18-148, updated to 1995 prices. They have a claimed accuracy of ±10 percent, but they should be used confidently only with study-level cost estimations (±25 percent) at best. Cost of delivery to the plant can be approximated as 3 percent of the FOB price [Pikulik and Diaz, *Chem. Eng.*, **84**(21), 106 (1977)].

The cost of the filter station includes not only the installed cost of the filter itself but also that of all the accessories dedicated to the filtration operation. Examples are feed pumps and storage facilities, precoat tanks, vacuum systems (often a major cost factor for a vacuum filter station), and compressed-air systems. The delivered cost of the accessories plus the cost of installation of filter and accessories generally is of the same order of magnitude as the delivered filter cost and commonly is several times as large. Installation costs, of course, must be estimated with reference to local labor costs and site-specific considerations.

The relatively high prices of pulp and paper filters reflect the construction features that accommodate the very high hydraulic capacity that is required. The absence of data for some common types of filters, in particular the filter press, is explained by Hall as due to the complex variety of individual features and materials of construction. For information about missing filters and for firmer estimates for those types presented, vendors should be consulted. In all cases of serious interest, consultation should take place early in the evaluation procedure so that it can yield timely advice on testing, selection, and price.

CENTRIFUGES

GENERAL REFERENCES: Ambler, in McKetta, *Encyclopedia of Chemical Processing and Design*, vol. 7, Marcel Dekker, New York, 1978; also in Schweitzer, *Handbook of Separation Techniques for Chemical Engineers*, McGraw-Hill, New York, 1979, sec. 4. Ambler and Keith, in Perry and Weissberger, *Separation and Purification Techniques of Chemistry*, 3d ed., vol. 12, Wiley, New York, 1978. Flood, Porter, and Rennie, *Chem. Eng.*, **73**(13), 190 (1966). Greenspan, *J. of Fluid Mech.*, **127**(9), 91 (1983). Hultsch and Wilkesmann, in Purchas, *Solid/Liquid Separation Equipment Scale-Up*, Uplands Press, Croydon, England, 2d ed., 1986, chap. 12. Gerl, Stadager, and Stahl, *Chemical Eng. Progress*, **91**, 48–54, (May 1995). Leung, *Chem. Eng.* (1990). Leung, *Fluid-Particle Sep. J.*, **5**(1), 44 (1992). Leung, *10th Pittsburgh Coal Conf. Proceed.* (1993). Leung and Shapiro, *Filtration and Separation Journal*, Sept. and Oct. 1996. Leung and Shapiro, U.S. Patents 5,520,605 (May 28, 1996), 5,380,266 (Jan. 10, 1995), and 5,401,423 (March 28 1995). Mayer and Stahl, *Aufbereitungs-Technik*, **11**, 619 (1988). Moyers, *Chem. Eng.*, **73**(13), 182 (1966). Records, in Purchas, op. cit., chap. 6. Smith, *Ind. Eng. Chem.*, **43,** 439 (1961). Sullivan and Erikson, ibid., p. 434. Svarovsky, in *Solid-Liquid Separation*, 3d ed., Butterworths, 1990, chap. 7. Tiller, *AICHE J.*, **33**(1), (1987). Zeitsch, in Svarovsky, op. cit., chap. 14. Dr. Andreas Karolis, *The Technology of Solid-Bowl Scroll Centrifuges.* McGillicuddy, *Chem. Process. Mag.*, **59** (12), 54–59 (Dec. 1996).

Nomenclature

Symbol	Definition	SI units	U.S. customary units
a	Acceleration	m/s^2	ft/s^2
B_o	Bond number	Dimensionless	Dimensionless
b	Basket axial length	m	ft
C_f	Frictional coefficient	Dimensionless	Dimensionless
D	Bowl/basket diameter	m	in
d	Particle diameter	m	ft
Ek	Ekman number	Dimensionless	Dimensionless
F	Cumulative fraction	Dimensionless	Dimensionless
G	Centrifugal gravity	m/s^2	ft/s^2
g	Earth gravity	m/s^2	ft/s^2
h	Cake height	m	ft
K	Cake permeability	m^2	ft^2
L	Length	m	ft
M	Mass	kg	lb
m	Bulk mass rate	kg/s	lb/h
N_c	Capillary number	Dimensionless	Dimensionless
P	Power	kw	hp
Q	Flow rate	L/s	gpm
R_m	Filter media resistance	m^{-1}	ft^{-1}
Ro	Rossby number	Dimensionless	Dimensionless
r	Radius	m	ft
Rec	Solids recovery	Dimensionless	Dimensionless
S	Liquid saturation in cake (=volume of liquid/ volume cake void)	Dimensionless	Dimensionless

Nomenclature *(Concluded)*

Symbol	Definition	SI units	U.S. customary units
sg	Specific gravity	Dimensionless	Dimensionless
t	Time	s	s
t_d	Time	Dimensionless	Dimensionless
u	Velocity	m/s	ft/s
$V\theta$	Circumferential velocity	m/s	ft/s
V_c	Bulk cake volume	m^3	ft^3
V	Velocity	m/s	ft/s
W	Weight fraction of solids	Dimensionless	Dimensionless
Y	Yield	Dimensionless	Dimensionless
Z	Capture efficiency	Dimensionless	Dimensionless
	Greek symbols		
ε	Cake void volume fraction	Dimensionless	Dimensionless
ε_s	Cake solids volume fraction	Dimensionless	Dimensionless
μ	Liquid viscosity	Pa·s	P
ρ_s	Solid density	kg/m^3	lb/ft^3
ρ_L	Liquid density	kg/m^3	lb/ft^3
σ	Surface tension	N/m	lbf/ft
σ_h	Hoop stress	Pa	psi
θ	Angle	Radian	degree
Σ	Scale-up factor (equivalent sedimentation area)	m^2	ft^2
ξ	Time	Dimensionless	Dimensionless
ϕ	Feed solids volume fraction	Dimensionless	Dimensionless
τ_y	Yield stress	Pa	psi
Δ	Differential speed	1/s	r/min
Ω	Angular speed	1/s	r/min
	Subscripts		
b	Bowl or basket		
c	Cake		
e	Centrate		
f	Feed		
f	Filtrate		
acc	Acceleration		
p	Pool		
con	Conveyance		
t	Tangential		
L	Liquid		
s	Solid		

INTRODUCTION

Centrifuges for the separation of solids from liquids are of two general types: (1) sedimentation centrifuges, which require a difference in density between the two or three phases present (solid-liquid or liquid-liquid or liquid-liquid-solid or solid-liquid-solid) and (2) filtering centrifuges (for solid-liquid separation), in which the solid phase is retained by the filter medium through which the liquid phase is free to pass. The following discussion is focused on solid-liquid separation for both types of centrifuges; however, a dispersed liquid phase in another continuous liquid phase as used in sedimenting centrifuges exhibits similar behavior as solid in liquid, and therefore the results developed are generally applicable. The use of centrifuges covers a broad range of applications, from separation of fine calcium carbonate particles of less than 10 μm to coarse coal of 0.013 m (½ in).

GENERAL PRINCIPLES

Centripetal and Centrifugal Acceleration A centripetal body force is required to sustain a body of mass moving along a curve trajectory. The force acts perpendicular to the direction of motion and is directed radially inward. The centripetal acceleration, which follows the same direction as the force, is given by the kinematic relationship:

$$a = \frac{V_\theta^2}{r} \tag{18-78}$$

where V_θ is the tangential velocity at a given point on the trajectory and r is the radius of curvature at that point. This analysis holds for the motion of a body in an inertial reference frame, for example, a stationary laboratory. It is most desirable to consider the process in a centrifuge, and the dynamics associated with such, in a noninertial reference frame such as in a frame rotating at the same angular speed as the centrifuge. Here, additional forces and accelerations arise, some of which are absent in the inertial frame. Analogous to centripetal acceleration, an observer in the rotating frame experiences a centrifugal acceleration directed radially outward from the axis of rotation with magnitude:

$$a = \Omega^2 r \tag{18-79}$$

where Ω is the angular speed of the rotating frame and r is the radius from the axis of rotation.

Solid-Body Rotation When a body of fluid rotates in a solid-body mode, the tangential or circumferential velocity is linearly proportional to radius:

$$V_\theta = \Omega r \tag{18-80}$$

as with a system of particles in a rigid body. Under this condition, the magnitude of the centripetal acceleration, Eq. (18-78), equals that of the centrifugal acceleration, Eq. (18-79), despite the fact that these accelerations are considered in two different reference frames. Hereafter, the rotating frame attached to the centrifuge is adopted. Therefore, centrifugal acceleration is exclusively used.

G-Level Centrifugal acceleration G is measured in multiples of earth gravity g:

$$\frac{G}{g} = \frac{\Omega^2 r}{g} \tag{18-81}$$

With the speed of the centrifuge Ω in r/min and D the diameter of the bowl,

$$\frac{G}{g} = 0.000559\Omega^2 D, \; D(m) \tag{18-82}$$

With D in inches, the constant in Eq. (18-82) is 0.0000142. G can be as low as $100g$ for slow-speed, large basket units to as much as $10{,}000g$ for high-speed, small decanter centrifuges and $15{,}000g$ for disc centrifuges. Because G is usually very much greater than g, the effect due to earth's gravity is negligible. In analytical ultracentrifuges used to process small samples, G can be as much as $500{,}000g$ to effectively separate two phases with very small density difference.

Coriolis Acceleration The Coriolis acceleration arises in a rotating frame, which has no parallel in an inertial frame. When a body moves at a linear velocity u in a rotating frame with angular speed Ω, it experiences a Coriolis acceleration with magnitude:

$$a = 2\Omega u \tag{18-83}$$

The Coriolis vector lies in the same plane as the velocity vector and is perpendicular to the rotation vector. If the rotation of the reference frame is anticlockwise, then the Coriolis acceleration is directed 90° clockwise from the velocity vector, and vice versa when the frame rotates clockwise. The Coriolis acceleration distorts the trajectory of the body as it moves rectilinearly in the rotating frame.

Effect of Fluid Viscosity and Inertia The dynamic effect of viscosity on a rotating liquid slurry as found in a sedimenting centrifuge is confined in very thin fluid layers, known as **Ekman layers.** These layers are adjacent to rotating surfaces which are perpendicular to the axis of rotation, such as bowl heads, flanges, and conveyor blades, etc. The thickness of the Ekman layer δ is of the order

$$\delta = \sqrt{\frac{\mu/\rho}{\Omega}} \tag{18-84}$$

where μ/ρ is the kinematic viscosity of the liquid. For example with water at room temperature, μ/ρ is 1×10^{-6} m²/s and for a surface rotating at $\Omega = 3000$ r/min, δ is 0.05 mm! These layers are very thin; nevertheless, they are responsible for transfer of angular momentum between the rotating surfaces to the fluid during acceleration and deceleration. They worked together with the larger-scale inviscid bulk flow transferring momentum in a rather complicated way. This is demonstrated by the teacup example in which the content of the cup is brought to speed when it is stirred and it is brought to a halt after undergoing solid-body rotation. The viscous effect is characterized by the dimensionless Ekman number:

$$\text{Ek} = \frac{\mu/\rho}{\Omega L^2} \tag{18-85}$$

where L is a characteristics length. It measures the scale of the viscous effect to that of the bulk flow.

The effect of fluid inertia manifests during abrupt change in velocity of the fluid mass. It is quantified by the Rossby number:

$$\text{Ro} = \frac{u}{\Omega L} \tag{18-86}$$

Typically, Ro is small to the order of 1 with the high end of the range showing possible effect due to inertia, whereas the Ek number is usually very small, 10^{-6} or smaller. Therefore, the viscous effect is confined to thin boundary layers with thickness $\text{Ek}^{1/2}L$.

Sedimenting and Filtering Centrifuges Under centrifugal force, the solid phase assumed to be denser than the liquid phase settles out to the bowl wall—sedimentation. Concurrently, the lighter, more buoyant liquid phase is displaced toward the smaller diameter—flotation. This is illustrated in Fig. 18-149*a*. Some centrifuges run with an air core, i.e., with free surface, whereas others run with slurry filled to the center hub or even to the axis in which pressure can be sustained.

In a sedimenting centrifuge, the separation can be in the form of **clarification,** wherein solids are separated from the liquid phase in which clarity of the liquid phase is of prime concern. For biological sludge, polymers are used to agglomerate fine solids to facilitate clarification. Separation can also be in the form of **classification** and degritting at which separation is effected by means of particle size and density. Typically, the finer solids (such as kaolin) of smaller size and/or density in the feed slurry are separated in the centrate stream as product (for example 90 percent of particles less than 1 μm, etc.), whereas the larger and/or denser solids are captured in cake as reject. Furthermore, separation can be in the form of **thickening,** where solids settle under centrifugal force to form a stream with concentrated solids. In

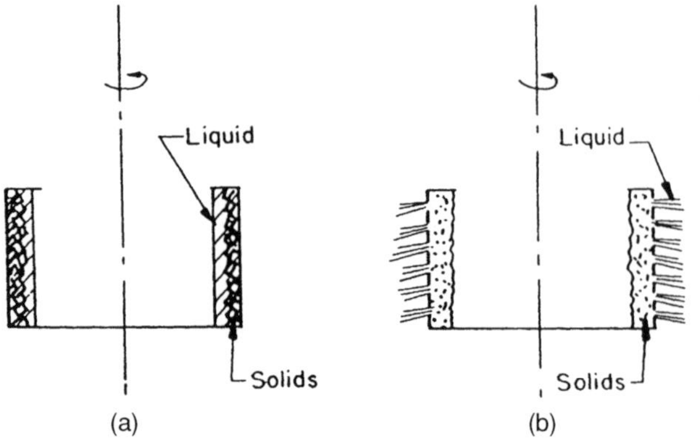

FIG. 18-149 Principles of centrifugal separation and filtration: (*a*) sedimentation in rotating imperforate bowl; (*b*) filtration in rotating perforate basket.

dewatering or **deliquoring,** the objective is to produce dry cake with high solids consistency by centrifugation.

In a filtering centrifuge, separating solids from liquid does not require a density difference between the two phases. Should a density difference exist between the two phases, sedimentation is usually at a much more rapid rate compared to filtration. In both cases, the solid and liquid phases move toward the bowl under centrifugal force. The solids are retained by the filter medium, while the liquid flows through the cake solids and the filter. This is illustrated in Fig. 18-149*b*.

Performance Criteria Separation of a given solid-liquid slurry is usually measured by the purity of the separated liquid phase in the centrate (or liquid effluent) in sedimenting mode or filtrate in filtering mode, and the separated solids in the cake. In addition, there are other important considerations. Generally, a selected subset of the following criteria are used, depending on the objectives of the process:

Cake dryness or moisture content
Total solids recovery
Polymer dosage
Size recovery and yield
Volumetric and solids throughput
Solid purity and wash ratio
Power consumption

Cake Dryness In dewatering, usually the cake needs to be as dry as possible. Cake dryness is commonly measured by the solids fraction by weight W or by volume ε_s. The moisture content is measured by the complement of W or ε_s. The volume fraction of the pores and void in the wet cake is measured by the cake porosity $\varepsilon (= 1 - \varepsilon_s)$; whereas the volume fraction of the liquid in the pores of the cake is measured by the saturation S. For well-defined solids in the cake with solid density (bone dry) ρ_s and liquid density ρ_L, and given that the cake volume V_c and the mass of solids in the cake w_s are known, the cake porosity is determined by

$$\varepsilon = 1 - \frac{w_s}{\rho_s V_c} \tag{18-87}$$

For undersaturated cake with $S < 1$, saturation can be inferred from the weight fraction of solids and the porosity of the cake, together with the solid and liquid densities:

$$S = \left(\frac{1-W}{W}\right)\left(\frac{1-\varepsilon}{\varepsilon}\right)\frac{\rho_s}{\rho_L} \tag{18-88}$$

When the cake is saturated $S = 1$, the cake porosity can be determined from Eq. (18-89) as

$$\varepsilon = \left(1 + \frac{\rho_L}{\rho_s}\frac{W}{1-W}\right)^{-1} \tag{18-89}$$

Cake dewatering by compression and rearrangement of the solids in the cake matrix reduce ε, yet the cake is still saturated with $S = 1$. (Assuming cake solids are ideal spheres of uniform size, the maximum packing, in rhombohedral arrangement, is such that $\varepsilon_s = 74$ percent or $\varepsilon = 26$ percent.) Drainage of liquid within the cake by centrifugation further reduces S to be less than 1. There is a lower limit on S which is determined by the cake height, dewatering time, centrifugal force as compared to the capillary and surface forces, as well as the surface roughness and porosity of the particles.

Total Solids Recovery In clarification, the clarity of the effluent is measured indirectly by the total solids recovered in the cake as

$$\text{Rec} = \frac{m_c W_c}{m_f W_f} \tag{18-90}$$

where subscripts c and f denote, respectively, the cake and the feed. m is the bulk mass flow rate in kg/s (lb/h).

Under steady state, the mass balance on both solids and liquid yield, respectively:

$$m_f W_f = m_c W_c + m_e W_e \tag{18-91}$$

$$m_f = m_c + m_e \tag{18-92}$$

From the above, it follows that

$$\text{Rec} = \frac{1 - (W_e/W_f)}{1 - (W_e/W_c)} \tag{18-93}$$

where subscript e represents liquid centrate. Stringent requirements on centrate quality or capture of valuable solid product often require the recovery to exceed 90 percent and, in some cases, 99+ percent. In such cases, the centrate solids are typically measured in ppm.

Polymer Dosage Cationic and anionic polymers have been commonly used to coagulate and flocculate fine particles in the slurry. This is especially pertinent to biological materials such as are found in wastewater treatment. In the latter, cationic polymers are often used to neutralize the negative-charge ions left on the surface of the colloidal particles. Polymer dosage is measured by kg of dry polymer/1000 kg of dry solids cake (lbm of dry polymer/ton of dry solids cake). With liquid polymers, the equivalent (active) dry solid polymer is used to calculate the dosage. There is a minimum polymer dosage to agglomerate and capture the fines in the cake. Overdose can be undesirable to recovery and cake dryness. The range of optimal dosage is dictated by the type of solids in the slurry, slurry physical properties such as pH, ionic strength, etc., and the operating condition and characteristics of the centrifuge. It is known that flocculated particles or flocs obtained from certain polymers may be more sensitive to shear than others, especially during feed acceleration in the centrifuges. A more gentle feed accelerator is beneficial for this type of polymer. Also, polymers can be introduced to the feed at various locations either within or external to the centrifuge.

Size Recovery and Yield Centrifuges have been applied to classify polydispersed fine particles. The size distribution of the particles is quantified by the cumulative weight fraction F less than a given particle size d for both the feed and the centrate streams. It is measured by a particle size counter which operates based on principles such as sedimentation or optical scattering.

In kaolin classification, the product is typically measured with a certain percentage less than a given size (example 90 percent or 95 percent less than 1 or 2 μm). Each combination of percent and size cut represents a condition by which the centrifuge would have to tune to yield the product specification.

The yield Y is defined as the fraction of feed particles of a given size below which they report to the centrate product. Thus,

$$Y = \frac{m_e W_e F_e}{m_f W_f F_f} \tag{18-94}$$

From material balance, the particle size distribution of the feed and centrate, as well as the total solids recovery, determine the yield,

$$Y(d) = \frac{F_e(d)}{F_f(d)}(1 - \text{Rec}) \tag{18-95}$$

The complementary is the cumulative capture efficiency Z (= 1 − Y), which is defined as the feed particles of a given size and smaller which are captured in the cake, which in most dewatering applications is the product stream.

Volumetric and Solids Throughput The maximum volumetric and solids throughput to a centrifuge are dictated by one or several governing factors, the most common ones are the centrate solids, cake dryness, and capacity (torque and power) of drive/gear unit. The solids throughput is also governed by other factors such as solids conveyance and discharge mechanisms for continuous and batch centrifuges. The settling rate, as may be significantly reduced by increasing feed solids concentration, also becomes crucial to solids throughput, especially if it has to meet a certain specification on centrate quality.

Solid Purity and Wash Ratio Cake washing in a centrifuge is used to remove dissolved impurities on the solids particle surface. It is most effective in filtering centrifuges—typically with a wash ratio of 0.05 to 0.3 kg wash/kg solids in continuous centrifuges, although higher ratios can be achieved with derated capacities to provide sufficient residence time. Batch filtering centrifuges are unlimited in the wash quantity that can be applied. Solid soluble impurities generally cannot be washed out in situ due to insufficient contact time, in which case repulp washing may be more effective. Repulp is often utilized with sedimenting centrifuges where wash is required.

Power Consumption Power is consumed to overcome windage and bearing (and seal) friction, to accelerate feed stream from zero speed to full tangential speed at the pool so as to establish the required *G*-force for separation, and to convey and discharge cake. The power to overcome windage and bearing friction is usually established through tests for a given centrifuge geometry at different rotation speeds. It is proportional to the mass of the centrifuge, to the first power of the speed for the bearing friction, and to the second power of speed for windage. It is also related to the bearing diameter. The seal friction is usually small.

The horsepower for feed acceleration is given by

$$P_{acc} = 5.984(10^{-10})\text{sg}Q(\Omega r_p)^2 \qquad (18\text{-}96)$$

where sg is the specific gravity of the feed slurry, Q the volumetric flow rate of feed in gpm(l/s), Ω the speed in r/min, r_p in meters corresponds to the radius of the pool surface for sedimenting centrifuge, or to the radius of the cake surface for filtering centrifuge. *Note:* To convert horsepower to kilowatts, multiply by 0.746.

The horsepower for cake conveyance for scroll centrifuge is

$$P_{con} = 1.587\ (10^{-5})\ T\Delta \qquad (18\text{-}97)$$

where Δ is the differential speed in r/min (s^{-1}) between the scroll conveyor and the bowl, and T is the conveyance torque in in·lb$_f$ (N·m). For centrifuge where cake is discharged differently, the conveyance power is simply

$$P_{con} = MGC_fV \qquad (18\text{-}98)$$

where M is the mass of the cake, G the centrifugal acceleration, C_f is the coefficient of friction, and V is the cake velocity. Comparing Eqs. (18-97) and (18-98) the conveyance torque is inversely related to the differential speed and directly proportional to the G acceleration, cake velocity, and cake mass.

Stress in the Centrifuge Rotor The stress in the centrifuge rotor is quite complex. Analytical methods, such as the finite element method, are used to analyze the mechanical integrity of a given rotor design. Without getting into an involved analysis, some useful knowledge can be gained from a simple analysis of the hoop stress of a rotating bowl under load. At equilibrium, the tensile hoop stress σ_h of the cylindrical bowl wall with thickness t is balanced by the centrifugal body force due to the mass of the bowl wall with density ρ_m and its contents (cake or slurry or liquid) with equivalent density ρ_L. Consider a circular wall segment with radius r, unit subtended angle, and unit axial length. A force balance requires

$$\sigma_h = \rho_m V_t^2\left[1 + 0.5\frac{\rho_L}{\rho_m}\left(1 - \frac{r_s^2}{r_b^2}\right)\frac{r_b}{t}\right] \qquad (18\text{-}99)$$

$V_t = \Omega r_b$ is the tip speed of the bowl. The term in the bracket is typically of order 1. If the maximum allowed σ_h is designed to be no more than 60 percent of the yield stress of the bowl material, which for steel is about 2.07(10^8) Nm^{-2}(30,000 lb$_f$/in^2). Given that the ρ_m of stainless steel is 7867 kg/m^3 (0.284 lbm/in^3), and there is no liquid load, then $(V_t)_{max} = \{\sigma_h/\rho_m\}^{1/2} = 126$ m/s (412 ft/s). With additional liquid load, $\rho_L = 1000$ kg/m^3 (0.0361 lbm/in^3), $r_b/t = 10$, and further assuming the worst case with liquid filling to the axis, the term in the curly bracket is 1.636. Using Eq. (18-99), $(V_t)_{max} = \{\sigma_h/\rho_m/1.636\}^{1/2} = 98$ m/s (322 ft/s). Indeed, almost all centrifuges are designed with top rim speed about 91 m/s (300 ft/s). With special construction materials for the rotor, such as Duplex Ferritic/Austenitic stainless steel, with higher yield stress, the maximum rim speed under full load can be over 122 m/s (400 ft/s).

G-Force vs. Throughput The *G*-acceleration can be expressed as

$$G = \Omega^2 r_b = \frac{(V_t)^2_{max}}{r_b} \qquad (18\text{-}100)$$

Figure 18-150 shows the range of diameter of commercial centrifuges and the range of maximum G developed in each type. It demonstrates an inverse relationship between G and r_b at $V_t = (V_t)_{max}$, which is constant for a given material. Figure 18-151 shows a log-log plot of G versus Ω for various bowl diameters, [Eq. (18-100)]. Also, the limiting conditions as delineated by $G = \Omega^2 R = \Omega(V_t)_{max}$ with various $(V_t)_{max}$, are superimposed on these curves. These two sets of curves dictate the operable speed and G for a given diameter and a given construction material for the bowl. The throughput capacity of a machine, depending on the process need, is roughly proportional to the nth power of the bowl radius,

$$Q = C_1 r_b^n \qquad (18\text{-}101)$$

where n is normally between 2 to 3, depending on clarification, classification, thickening, or dewatering. Thus,

$$G = c_2(V_t)^2_{max}Q^{1/n} \qquad (18\text{-}102)$$

where c_1 and c_2 are constants. It follows that large centrifuges can deliver high flow rate but separation is at lower *G*-force; vice versa, smaller centrifuges can deliver lower flow rate but separation is at higher *G*-force. Also, using higher-strength material for construction of the rotating assembly permits higher maximum tip speed, thus allowing higher *G*-force for separation at a given feed rate.

Centrifuge bowls are made of almost every machinable alloy of reasonably high strength. Preference is given to those alloys having 1 percent elongation to minimize the risk of cracking at stress-concentration points. Typically the list includes (in increasing cost) rubber-lined carbon steel, SS316L, SS317LMN, duplex SS (SAF2205), Alloy 904L, AL6 XN, Inconel, Hastelloy C22, Hastelloy B, nickel, and titanium Gr. 2. Most coatings (Halar, PTFE, etc.) cause problems with stressed components such as baskets, but can be utilized for static components such as housings. Vertical-basket centrifuges are frequently constructed of carbon steel or stainless steel coated with rubber, neoprene, Penton, or Kynar. Casings and feed, rinse, and discharge lines that are stationary and lightly stressed may be constructed of any suitable rigid corrosion-resistant material. Wear-resistant materials—tungsten and cermaic carbide, hard-facing, and others—are often used to protect the bare metal surfaces in high-wear areas such as the blade tips of the decanter centrifuge.

Critical Speeds In the design of any high-speed rotating machinery, attention must be paid to the phenomenon of critical speed. This is the speed at which the frequency of rotation matches the natural frequency of the rotating part. At this speed, any vibration induced by slight unbalance in the rotor is strongly reinforced, resulting in large deflections, high stresses, and even failure of the equipment. Speeds corresponding to harmonics of the natural frequency are also critical speeds but give relatively small deflections and are much less troublesome than the fundamental frequency. The critical speed of simple shapes may be calculated from the moment of inertia; with complex elements such as a loaded centrifuge bowl, it is best found by tests.

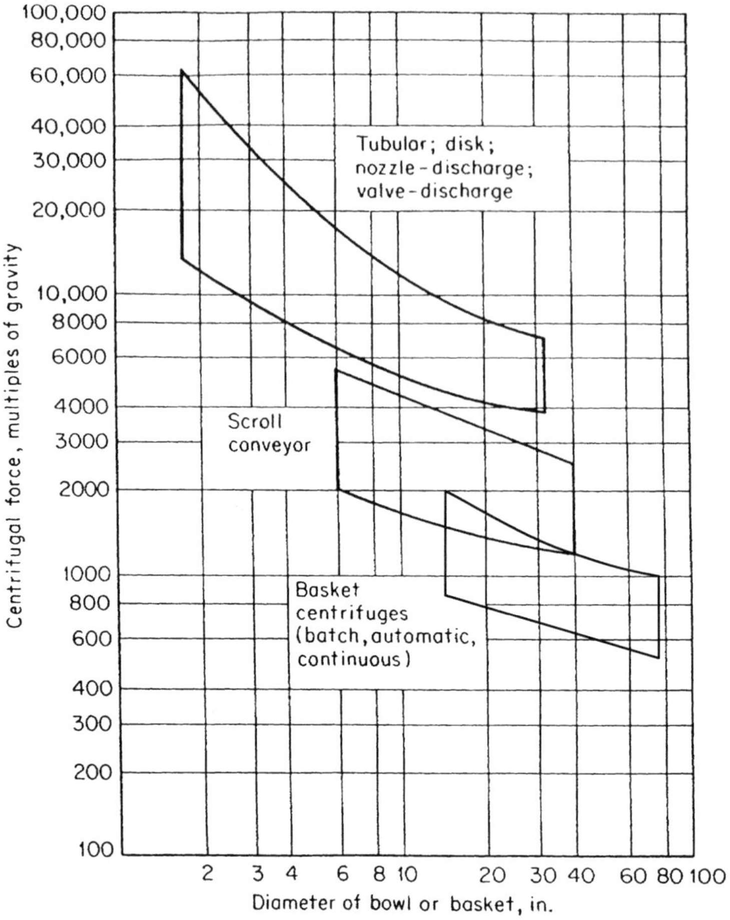

FIG. 18-150 Variation of centrifugal force with diameter in industrial centrifuges.

Nearly all centrifuges operate at speeds well above the primary critical speed and therefore must pass through this speed during acceleration and deceleration. To permit them to do so safely, some degree of damping in their mounting must be provided. This may result from the design of the spindle or driveshaft alone, spring-loading of the spindle bearing nearest the rotor, elastic loading of the suspension, or a combination of these. Smaller and medium-sized centrifuges of the cream-separator and bottle-centrifuge design are frequently mounted on elastic cushions. Horizontal decanter centrifuges are mounted on isolators with dampers to reduce vibration transmitted to the foundation.

SEDIMENTATION CENTRIFUGES

When a spherical particle of diameter d settles in a viscous liquid under earth gravity g, the terminal velocity V_s is determined by the weight of the particle-balancing buoyancy and the viscous drag on the particle in accordance to Stokes' law. In a rotating flow, Stokes' law is modified by the "centrifugal gravity" $G = \Omega^2 r$, thus

$$V_s = \frac{1}{18\mu}\Omega^2 r(\rho_s - \rho_L)d^2 \qquad (18\text{-}103)$$

In order to have good separation or high settling velocity, a combination of the following conditions is generally sufficient:

1. High centrifuge speed
2. Large particle size
3. Large density difference between solid and liquid
4. Large separation radius
5. Low liquor viscosity

Among the five parameters, the settling velocity is very sensitive to change in speed and particle size. It varies as the square of both parameters. The maximum achievable rotational speed of a centrifuge is normally dictated by the stresses exerted by the processing medium on the bowl and the stresses of the bowl on periphery equipment, most notably the drive system, which consists of a gear unit or hydraulic pump. If the particles in the feed slurry are too small to be separated in the existing G-field, coagulation and flocculation by polymers are effective ways to create larger agglomerated particles for settling. Unlike separation under a constant gravitational field, the settling velocity under a centrifugal field increases linearly with the radius. The greater the radius at which the separation takes place in a given centrifuge at a given rotational speed, the better the separation. Sedimentation of particles is favorable in a less viscous liquid. Some processes are run under elevated temperature where liquid viscosity drops to a fraction of its original value at room temperature.

Laboratory Tests

Spin-Tube Tests The objective of the spin-tube test is to check the settleability of solids in a slurry under centrifugation. The clarity of the supernatant liquid and the solid concentration in the sediment can also be evaluated. A small and equal amount of feed is introduced into two diametrically opposite test tubes (typically plastic tubes in a stainless

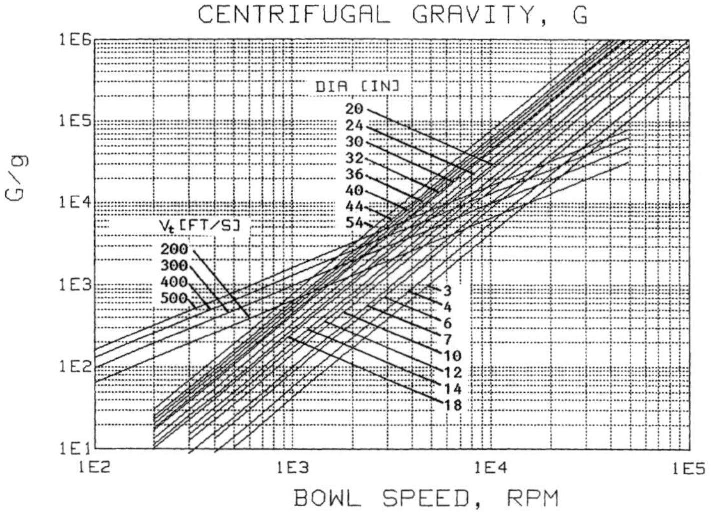

FIG. 18-151 Variation of centrifugal force with r/min.

steel holder) with a volume of 15 to 50 mL. The samples are centrifuged at a given G and for a period of time t. The supernatant liquid is decanted off from the spin tubes from which the clarity (in the form of turbidity or any measurable solids, dissolved and suspended) is measured. The integrity—more precisely, the yield stress—of the cake can be determined approximately by the amount of penetration of a rod into the cake under its weight and accounting for the buoyancy effect due to the wet cake. It is further assumed that the rod does not lean on the sides of the tube. The yield stress τ_y of the centrifuged cake can be determined from:

$$\tau_y = \frac{1}{2}\left(\frac{\rho_r L}{h} - \rho_L\right) g r_d \tag{18-104}$$

where r_d and L are, respectively, the radius and length of the solid circular rod; ρ_r is the density of the rod; and h is the penetration of the rod into the cake. By using rods of various sizes and densities, yield stress which is indicative of cake handling and integrity can be measured for a wide range of conditions.

The solids recovery in the cake can be inferred from measurements using Eq. (18-93). It is shown as a function of G-seconds for different feed solids concentration in Fig. 18-152; see also theory discussed below.

Imperforate Bowl Tests The amount of supernant liquid from spin tubes is usually too small to warrant accurate gravimetric analysis. A fixed amount of slurry is introduced at a controlled rate into a rotating imperforate bowl to simulate a continuous sedimentation centrifuge. The liquid is collected as it overflows the ring weir. The test is

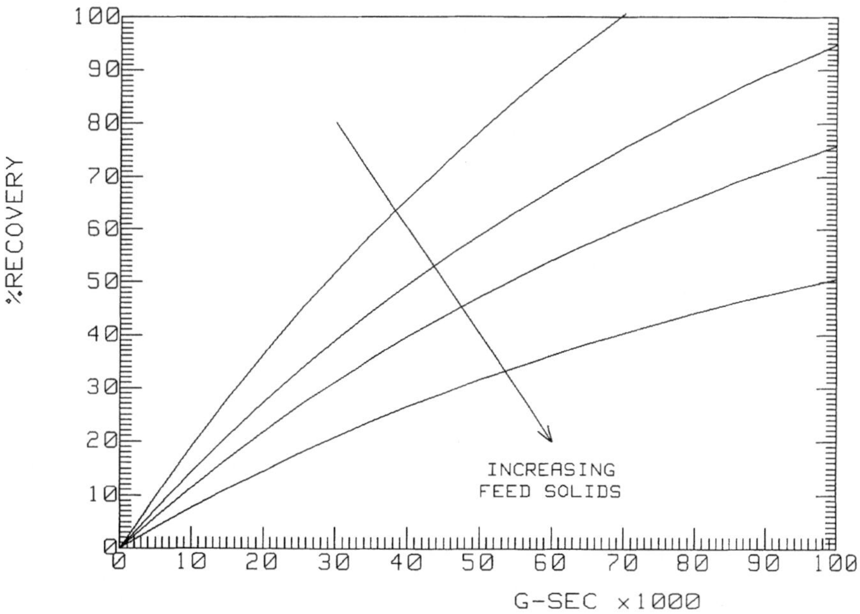

FIG. 18-152 Recovery as a function of G-seconds for centrifugal sedimentation.

stopped when the solids in the bowl build up to a thickness which affects centrate quality. The solid concentration of the centrate is determined similarly to that of the spin tube.

Transient Centrifugation Theory As in gravitational sedimentation, there are three layers which exist during batch settling of a slurry in a centrifuge: a clarified liquid layer closest to the axis, a middle feed slurry layer with suspended solids, and a cake layer adjacent to the bowl wall with concentrated solids. Unlike with constant gravity g, the centrifugal gravity G increases linearly with radius. It is highest near the bowl and is zero at the axis of rotation. Also, the cylindrical surface area through which the particle has to settle increases linearly with radius. Both of these effects give rise to some rather unexpected results.

Consider the simple initial condition $t = 0$ where the solid concentration ϕ_{so} is constant across the entire slurry domain $r_L \le r \le r_b$ where r_L and r_b are, respectively, the radii of the slurry surface and the bowl. At a later time $t > 0$, three layers coexist: the top clarified layer, a middle slurry layer, and a bottom sediment layer. The air-liquid interface remains stationary at radius r_L, while the liquid-slurry interface with radius r_s expands radially outward, with t with r_s given by:

$$\frac{r_s}{r_L} = \sqrt{\frac{\phi_{so}}{\phi_s}} \tag{18-105}$$

Eq. (18-105) can be derived from conservation of angular momentum as applied to the liquid-slurry interface.

Interestingly, the solid concentration in the slurry layer ϕ_s does not remain constant with time as in gravitational sedimentation. Instead, ϕ_s decreases with time uniformly in the entire slurry layer in accordance to:

$$\frac{\phi_s}{\phi_{smax}} = 1 - \left[1 - \left(\frac{\phi_{so}}{\phi_{smax}}\right)\right]e^{2\xi} \tag{18-106}$$

where ξ is a dimensionless time variable:

$$\xi = \left(\frac{V_{go}t}{r_b}\right)\left(\frac{G}{g}\right) \tag{18-107}$$

In Eqs. (18-106) and (18-107), under hindered settling and 1 g, the solids flux $\phi_s V_s$ is assumed to be a linear function of ϕ_s decreasing at a rate of V_{go}. Also, the solids flux is taken to be zero at the "maximum" solids concentration ϕ_{smax}. As $G/g >> 1$, this solids flux behavior based on 1 g is assumed to be ratioed by G/g.

Concurrent with the liquid-slurry interface moving radially outward, the cake layer builds up with the cake-slurry interface moving radially inward, with radial position given by:

$$\frac{r_c}{r_b} = \sqrt{\frac{(\varepsilon_s - \phi_{so})}{(\varepsilon_s - \phi_s)}} \tag{18-108}$$

where ε_s is a constant cake solids concentration. Sedimentation stops when the growing cake-slurry interface meets the decreasing slurry-liquid interface with $r_c = r_s$. This point is reached at $\phi_s = \phi_s^\circ$ and $t = t^\circ$ when

$$\frac{1}{\phi_s^\circ} = \frac{1}{\varepsilon_s} + \left(\frac{r_b}{r_L}\right)^2\left(\frac{1}{\phi_{so}} - \frac{1}{\varepsilon_s}\right) \tag{18-109}$$

$$t^\circ = \frac{1}{2}\left(\frac{gr_b}{GV_{go}}\right)\ln\left(\frac{\phi_{smax} - \phi_s^\circ}{\phi_{smax} - \phi_{so}}\right) \tag{18-110}$$

Example

Calcium carbonate–water slurry

$G/g = 2667$

$V_{go} = 1.31 \times 10^{-6}$ m/s (5.16×10^{-5} in/s)

$\phi_{smax} = 0.26$ (with $\phi_s V_g = 0$)

$\phi_{so} = 0.13$

$r_L = 0.0508$ m (2 in)

$r_b = 0.1016$ m (4 in)

$\xi = 2667\ (1.31 \times 10^{-6})\ (1/0.1016) = 0.0344$

$\varepsilon_s = 0.52$

$t(s)$	ξ	ϕ_s/ϕ_{smax}	r_s (m)	r_c (m)
0.0	0	0.50	0.051	0.102
1.0	0.034	0.46	0.053	0.100
5.0	0.173	0.29	0.067	0.095
7.6	0.261	0.16	0.091	0.091

There are six types of industrial sedimenting centrifuges:

- Tubular-bowl centrifuges
- Multichamber centrifuges
- Skimmer pipe/knife-discharge centrifuges
- Disc centrifuges
- Decanter centrifuges
- Screenbowl centrifuges

The first three types, including the manual-discharge disc, are batch-feed centrifuges, whereas the latter three, including the intermittent and nozzle-discharge discs, are continuous centrifuges.

Tubular-Bowl Centrifuges The tubular-bowl centrifuge is widely employed for purifying used lubricating and other industrial oils and in the food, biochemical, and pharmaceutical industries. Industrial models have bowls 102 to 127 mm (4 to 5 in) in diameter and 762 mm (30 in) long (Table 18-12). It is capable of delivering up to 18,000g. The smallest size, 44 mm × 229 mm (1.75 in × 9 in bowl), is a laboratory model capable of developing up to 65,000g. It is also used for separating difficult-to-separate biological solids with very small density difference, such as cells and virus.

The bowl is suspended from an upper bearing and drive (electric or turbine motor) assembly through a flexible-drive spindle with a loose guide in a controlled damping assembly at the bottom. The unit finds its axis of rotation if it becomes slightly unbalanced due to process load.

The feed slurry is introduced into the lower portion of the bowl through a small orifice. Immediately downstream of the orifice is a distributor and a baffle assembly which distribute and accelerate the feed to circumferential speed. The centrate discharges from the top end of the bowl by overflowing a ring weir. Solids that have sedimented against the bowl wall are removed manually from the centrifuge when the buildup of solids inside the bowl is sufficient to affect the centrate clarity.

The liquid-handling capacity of the tubular-bowl centrifuge varies with use. The low end shown in Table 18-11 corresponds to stripping small bacteria from a culture medium. The high end corresponds to purifying transformer oil and restoring its dielectric value. The solids-handling capacity of this centrifuge is limited to 4.5 kg (10 lb) or less. Typically, the feed stream solids should be less than 1 percent in practice.

Multichamber Centrifuges While the tubular has a high aspect ratio (i.e., length-to-diameter ratio) of 5 to 7, the multichamber centrifuges have aspect ratios of 1 or less. The bowl driven from below consists of a series of short tubular sections of increasing diameter nested to form a continuous tubular passage of stepwise increasing diameter for the liquid flow. The feed is introduced at the center tube and gradually finds its way to tubes with larger diameters. The larger and denser particles settle out in the smaller-diameter tube, while the smaller and lighter particles settle out in the larger-diameter tubes. Classification of particles can be conveniently carried out. Clarification may be significantly improved by spacing especially the outer tubes more closely together to reduce the settling distance, a concept which is fully exploited by the disc-centrifuge design. This also serves to maintain a constant velocity of flow between adjacent tubes. As much as six chambers can be accommodated. The maximum solids-holding capacity is 0.064 m^3 (17 gal). The most common use is for clarifying fruit juices, wort, and beer. For these services it is equipped with a centripetal pump at effluent discharge to minimize foaming and contact with air.

Knife-Discharge Centrifugal Clarifiers Knife-discharge centrifuges with solid instead of perforated bowls are used as sedimenting centrifuges. The liquid flow is usually continuous until the settled solids start to interfere with the effluent liquid. The feed enters the

TABLE 18-12 Specifications and Performance Characteristics of Typical Sedimenting Centrifuges

Type	Bowl diameter	Speed, r/min	Maximum centrifugal force × gravity	Throughput		Typical motor size, hp
				Liquid, gal/min	Solids, tons/h	
Tubular	1.75	50,000*	62,400	0.05–0.25		*
	4.125	15,000	13,200	0.1–10		2
	5	15,000	15,900	0.2–20		3
Disc	7	12,000	14,300	0.1–10		⅓
	13	7,500	10,400	5–50		6
	24	4,000	5,500	20–200		7½
Nozzle discharge	10	10,000	14,200	10–40	0.1–1	20
	16	6,250	8,900	25–150	0.4–4	40
	27	4,200	6,750	40–400	1–11	125
	30	3,300	4,600	40–400	1–11	125
Scroll conveyor	6	8,000	5,500	To 20	0.03–0.25	5
	14	4,000	3,180	To 75	0.5–1.5	20
	18	3,500	3,130	To 100	1–3	50
	24	3,000	3,070	To 250	2.5–12	125
	30	2,700	3,105	To 350	3–15	200
	36	2,250	2,590	To 600	10–25	300
	44	1,600	1,600	To 700	10–25	400
	54	1,000	770	To 750	20–60	250
Knife discharge	20	1,800	920	†	1.0‡	20
	36	1,200	740	†	4.1‡	30
	68	900	780	†	20.5‡	40

*Turbine drive, 100 lb/h (45 kg/h) of steam at 40 lbf/in² gauge (372 KPa) or equivalent compressed air.
†Widely variable.
‡Maximum volume of solids that the bowl can contain, ft³.
NOTE: To convert inches to millimeters, multiply by 25.4; to convert revolutions per minute to radians per second, multiply by 0.105; to convert gallons per minute to liters per second, multiply by 0.063; to convert tons per hour to kilograms per second, multiply by 0.253; and to convert horsepower to kilowatts, multiply by 0.746.

hub end and is accelerated to speed before introducing to the separation pool. The solids settle out to the bowl wall and the clarified liquid overflows the ring weir or discharges through a skimmer pipe. In some designs, internal baffles in the bowl are required to stop wave action primarily along the axial direction. When sufficient thick solid layer has built up inside the bowl, the supernatant liquid is skimmed off by moving the opening of the skimmer pipe radially inward. After the liquid is sucked out, the solids are knifed out as with centrifugal filters. However, unlike centrifugal filters, the cake is always fully saturated with liquid, $S = 100\%$. These centrifuges are used for coarse, fast-settled solids. When greater clarification effectiveness is required, the operation may be totally batchwise with prolonged spinning of each batch. If the solids content in the feed is low, several batches may be successively charged and the resulting supernatant liquor skimmed off before unloading of the accumulated solids.

Commercial centrifuges of this type have bowl diameters ranging from 0.3 to 2.4 m (12 to 96 in). The large sizes are used on heavy-duty applications such as coal dewatering and are limited by stress considerations to operate at 300g. The intermediate sizes for chemical process service develop up to 1000g (see Table 18-10).

Disc Stack Centrifuges One of the most common types of commercially utilized centrifuges is a vertically mounted disc machine, one type of which is shown in Fig. 18-153. Feed is introduced proximate to the axis of the bowl, accelerated to speed typically by a radial vane assembly, and flows through a stack of closely spaced conical discs in the form of truncated cones. Generally 50 to 150 discs are used. They are spaced 0.4 to 3 mm (0.015 to 0.125 in) apart to reduce the distance for solid/liquid separation. The angle made by conical discs with the horizontal is typically between 40 to 55° to facilitate solids conveyance. Under centrifugal force the solids settle against the underside of the disc surface and move down to the large end of the conical disc and subsequently to the bowl wall. Concurrently, the clarified liquid phase moves up the conical channel. Each disc carries several holes spaced uniformly around the circumference. When the disc stack is assembled, the holes provide a continuous upward passage for the lighter clarified liquid released from each conical channel. The liquid collects at the top of the disc stack and discharges through overflow ports. To recover the kinetic energy and avoid foaming due to discharging of a high-velocity jet against a stationary casing, the rotating liquid is diverted to a stationary impeller from which the kinetic energy of the stream is converted to hydrostatic pressure. Unlike most centrifuges operating with a slurry pool in contact with a free surface, disc centrifuges with a rotary seal arrangement can operate under high pressure. The settled solids at the bowl wall are discharged in different forms, depending on the type of disc centrifuges.

Manual Discharge Disc Stack Centrifuges In the simplest design shown in Fig. 18-154*a*, the accumulated solids must be removed manually on a periodic basis, similar to that for the tubular-bowl centrifuge. This requires stopping and disassembling the bowl and removing the disc stack. Although the individual discs rarely require cleaning, manual removal of solids is economical only when the fraction of solids in the feed is very small.

Self-Cleaning Disc Centrifuges More commonly known as clarifiers (two-phase) and separators (three-phase), these centrifuges, which also contain a conical disc stack inside the bowl, automatically discharge accumulated solids on a timed cycle while the bowl is at full speed. Feed is introduced into the bowl via a nonrotating feed pipe and into a distributor which evenly distributes the slurry to the appropriate disc stack channels. Slurry is forced up through the disc stack where solids accumulate on the underside of the discs and slide down the discs, where they are forced to the sludge holding area just inside the maximum diameter of the double cone-shaped bowl, as shown in Fig. 18-153. When the solids chamber is full, the bottom of the bowl, which is held closed to the top portion hydraulically, drops by evacuating the hydraulic operating fluid. The solids are discharged at full speed in a very short time into an outer housing where they are diverted out of the machine. The liquid or liquids (in a three-phase separator) are normally discharged via stationary impellers under pressure. These types of centrifuges are commonly used in the clarification of beverages and the purification of mineral and edible oils.

Disc Nozzle Centrifuges In the nozzle-discharge disc centrifuge, solids are discharged continuously, along with a portion of the liquid phase, through nozzles spaced around the periphery of the bowl, which are tapered radially outward, providing a space for solids

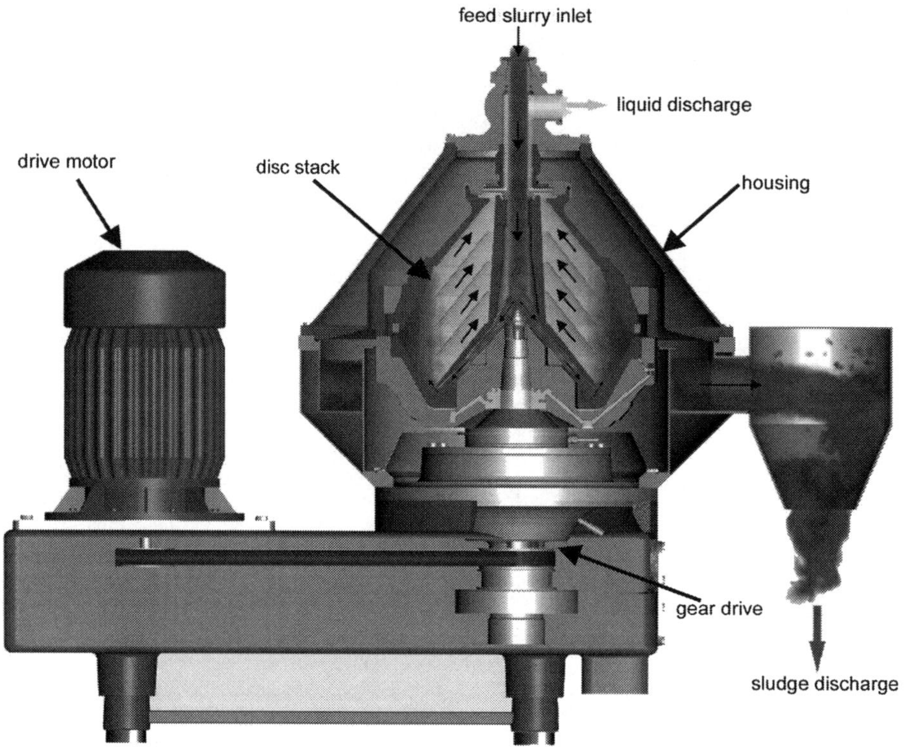

FIG. 18-153 Disc Stack Centrifuge. *(Flottweg Separation Technology.)*

storage (see Fig. 18-154*b*). The angle of repose of the sedimented solids determines the slope of the bowl walls for satisfactory operation. Clarification efficiency is seriously impaired if the buildup of solids between nozzles reaches into the disc stack. The nozzle diameter should be at least twice the diameter of the largest particle to be processed, and prescreening is recommended of extraneous solids. Typically, nozzle diameters range from 0.6 to 3 mm (0.25 to 0.125 in). Large disc centrifuges may have as much as 24 nozzles spaced out at the bowl.

For clarification of a single liquid phase with controlled concentration of the discharged slurry, a centrifuge which provides recirculation is used (see Fig. 18-154*b*). A fraction of the sludge discharged out of the machine is returned to the bowl to the area adjacent to the nozzles through lines external to the machine as well as built-in annular passages at the periphery of the bowl. This has the effect of preloading the nozzles with sludge which has already been separated and reduces the net flow of liquid with the newly sedimented solids from the feed. Increased concentration can be obtained alternatively by recycling a portion of the sludge to the feed, but this increases solids loading at the disc stack, with a corresponding sacrifice in the effluent clarity for a given feed rate.

With proper rotary seals, the pressure in the machine can be contained up to 1.1 MPa (150 psig) or higher. Also, operating temperature can be as high as 315°C (about 600°F). The rotating parts are made of stainless steel with the high-wear nozzles made of tungsten carbide. The bowls may be underdriven or suspended and range from several centimeters to over 1 m (3.3 ft) in outer diameter. The largest size capable of clarifying up to 1920 L/m (500 gpm) requires 112 kW (150 hp). These types of centrifuges are commonly used in applications including corn wet milling (starch separation, gluten thickening), the classification of kaolin clay particles, washing of terephthalic acid crystals, and dewaxing of lube oils.

Decanter Centrifuges The decanter centrifuge (also known as the solid-bowl or scroll centrifuge) consists of a solid exterior bowl with an internal screw or scroll conveyor (see Fig. 18-155). Both the bowl and the conveyor rotate at a high speed, yet there is a difference in speed between the two, which is responsible for conveying the sediment along the machine from the cylinder to the conical discharge end. The rotating assembly is commonly mounted horizontally with bearings on each end. Some centrifuges are vertically mounted with the weight of the rotating assembly supported by a single bearing at the bottom or with the entire machine suspended from the top. With the former configuration, the weight of the rotating assembly provides a good sealing surface at the bearing for high-pressure applications. The bowl may be conical in shape or, in most instances, it has combined conical and cylindrical sections (see Fig. 18-155).

Slurry is fed through a stationary pipe into the feed zone located near the center of the scroll. The product is then accelerated circumferentially and passes through distribution ports into the bowl. The bowl has a cylindrical/conical shape and rotates at a preset speed optimal for the application. The slurry rotates with the bowl at the operating speed and forms a concentric layer at the bowl wall, as shown in Fig. 18-156*a*. In the separation pool or pond, under centrifugal gravity the solids which are heavier compared to the liquid settle toward the bowl wall, while the clarified liquid moves radially toward the pool surface. Subsequently, the liquid flows along the helical channel (or channels, if the screw conveyor has multiple leads) formed by adjacent blades of the conveyor to the liquid bowl head, from which it discharges over the weirs. The annular pool/pond height can be changed by adjusting the radial position of the weir openings, which take the form of circular holes or crescent-shaped slots or by adjusting a stationary impeller, which will discharge the liquid under pressure (see Fig. 18-156*b*).

The solids contained in the slurry are deposited against the bowl wall by centrifugal force. The length of the cylindrical bowl section and the cone angle are selected to meet the specific requirements of an application. The scroll conveyor rotates at a slightly different speed from the bowl, and conveys the deposited solids toward the conical

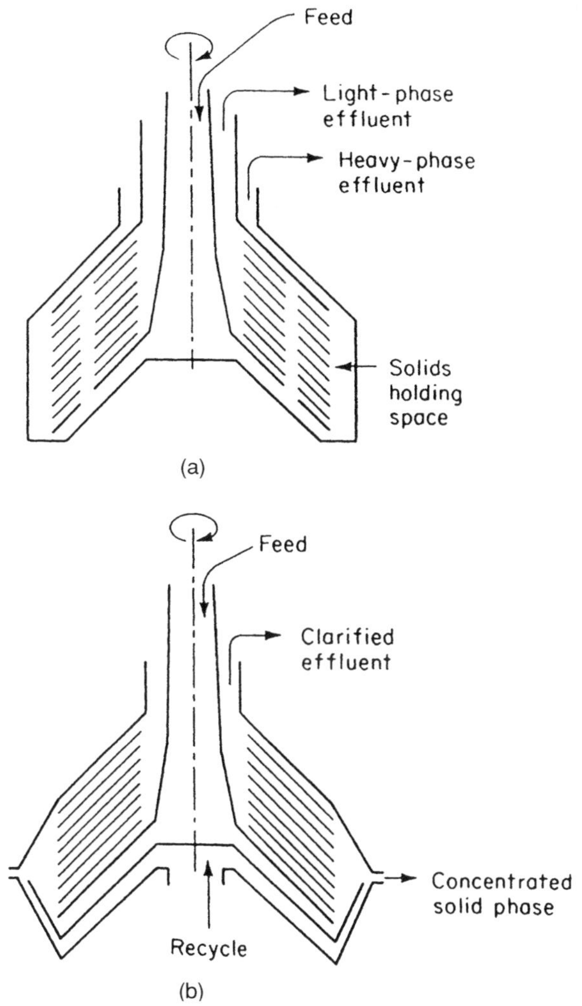

FIG. 18-154 Disc stack-centrifuge bowls: (*a*) separator, solid wall; (*b*) recycle clarifier, nozzle discharge.

end of the bowl, also known as the beach. The half cone angle ranges between 5° and 20°. The cake is submerged in the pool when it is in the cylinder and at the beginning of the beach. In this region, liquid buoyancy helps to reduce the effective weight of the cake under centrifugal gravity, resulting in lower conveyance torque. Farther up the beach, the cake emerges above the pool and moves along the "dry beach," where buoyancy force is absent, resulting in more difficult conveyance and higher torque. But it is also in this section that the cake is dewatered, with expressed liquid returned back to the pool. The centrifugal force helps to dewater, yet at the same time hinders the transport of the cake in the dry beach. Therefore, a balance in cake conveyance and cake dewatering is the key in setting the pool and the *G*-force for a given application. Also, clarification is important in dictating this decision.

The cylindrical section provides clarification under high centrifugal gravity. In some cases, the pool should be shallow to maximize the *G*-force for separation. In other cases, when the cake layer is too thick inside the cylinder, the settled solids—especially the finer particles at the cake surface—entrain into the fast-moving liquid stream above, which eventually ends up in the centrate. A slightly deeper pool becomes beneficial in these cases because there is a thicker buffer liquid layer to ensure settling of resuspended solids. This can be at the expense of cake dryness due to reduction of the dry beach. Consequently, there is again a compromise between centrate clarity and cake dryness. Another reason for the tradeoff of centrate clarity with cake dryness is that, in losing fine solids to the centrate (i.e., classification), the cake with larger particles, having less surface-to-volume ratio, can dewater more effectively, resulting in drier cake. It is best to determine the optimal pool for a given application through tests.

The speed with which the cake transports is controlled by the differential speed. High differential speed facilitates high solids throughput where the cake thickness is kept to a minimum so as not to impair centrate quality due to entrainment of fine solids. Also, cake dewatering is improved due to a reduction in the drainage path with smaller cake height; however, this is offset by the fact that higher differential speed also reduces cake residence time, especially in the dry beach. The opposite holds for low differential speed. Therefore, an optimal differential speed is required to balance centrate clarity and cake dryness. The desirable differential speed is usually maintained using a two-stage planetary gearbox, the housing of which rotates with the bowl speed, with a fixed first-stage pinion shaft. In some applications, the pinion is driven by an electrical backdrive (dc or ac), hydraulic backdrive, or braked by an eddy-current device at a fixed rotation speed. The differential speed is then the difference in speed between the bowl and the pinion divided by the gear ratio. This also applies to the case when the pinion arm is held stationary, in which the pinion speed is zero. The torque at the spline of the conveyor, conveyance torque, is equal to the product of the pinion torque and the gear ratio. Higher gear ratio gives lower differential speed, and vice versa; lower gear ratio gives higher differential for higher solids capacity. The torque at the pinion shaft has been used to control the feed rate or to signal an overload condition by shearing of a safety pin. Under this condition, both the bowl and the conveyor are bound to rotate at the same speed (zero differential) with no conveyance torque and no load at the pinion.

Soft solids, most of which are biological waste such as sewage, are difficult to convey up the beach. Annular baffles or dams have been commonly used to provide a pool-level difference wherein the pool is

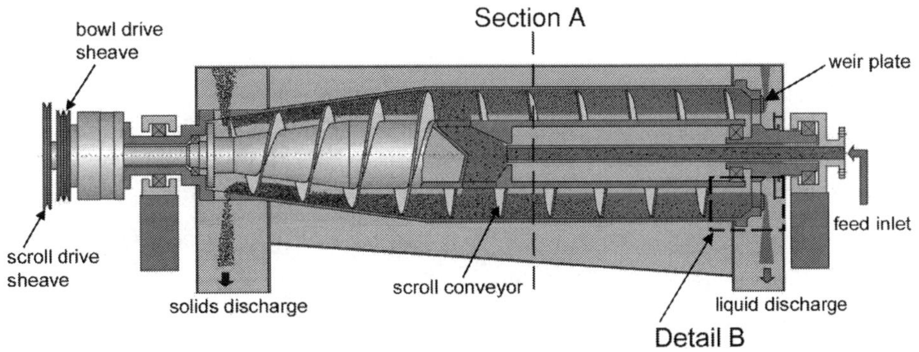

FIG. 18-155 Two-phase decanter centrifuge—gravity liquid discharge. (*Flottweg Separation Technology.*)

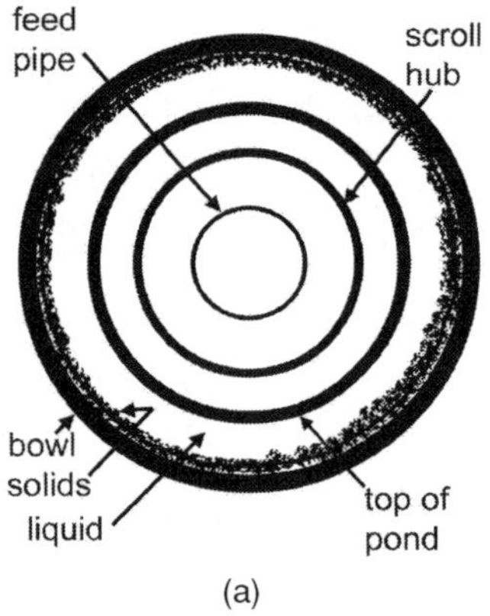

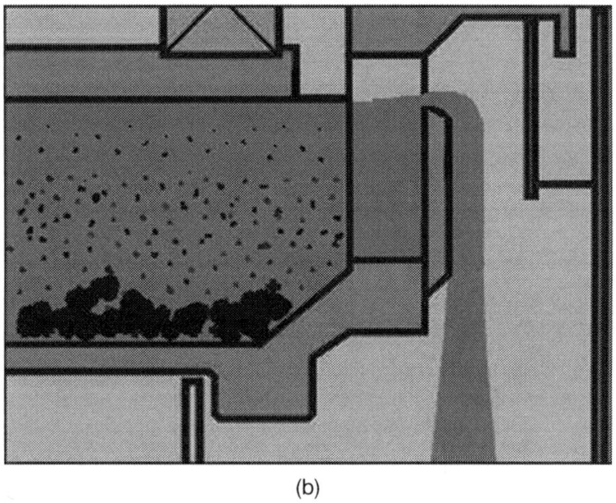

FIG. 18-156 (*a*) Section A from Fig. 18-155. (*b*) Detail B from Fig. 18-155. (*Flottweg Separation Technology.*)

deeper upstream of the baffle toward the clarifier and lower downstream of the baffle toward the beach. The pool-level difference across the baffle, together with the differential speed, provide the driving force to convey the compressible sludge up the beach. This has been used effectively in thickening of waste-activated sludge and in some cases of fine clay with dilatant characteristics.

High solids decanter centrifuges have been used to dewater mixed raw sewage sludge (with volume ratio of primary to waste-activated sludge such as 50 percent to 50 percent or 40 percent to 60 percent, etc.), aerobically digested sludge, and anaerobically digested sludge. Cake solids as dry as 28 percent to 35 percent by weight are obtained for raw mixed sludge and 20 percent to 28 percent for the digested sludges, with the aerobic sludge at the lower end of the range. The typical characteristics of high-solids applications are: low differential speed (0.5 to 3 r/min), high conveyance torque, high polymer dosage (10 to 30 lb dry polymer/ton dry solids, depending on the feed sewage), and slightly lower volumetric throughput rate. An electrical (dc or ac with variable-frequency drive) or hydraulic backdrive on the conveyor with high torque capacity is essential to operate these conditions at steady state.

The horizontal decanter centrifuge is operated below its critical speed. The bowl is mounted between fixed bearings anchored to a rigid frame. The gearbox is cantilevered outboard of one of these bearings, and the feed pipe enters the rotating assembly through the other end. The frame is isolated from the support structure by spring-type or rubber vibration isolators. In the vertical configuration, the bowl and the gearbox are suspended from the drive head, which is connected to the frame and casing through vibration isolators. A clearance bushing at the bottom limits the excursion of the bowl during start-up and shutdown but does not provide the radial constraint of a bearing under normal operating conditions.

Decanter centrifuges with mechanical shaft-to-casing seals are available for pressure containment up to 1.1 MPa (150 psig), similar to the nozzle-disc centrifuge. They can be built to operate at temperatures from −87 to +260°C (−125 to +500°F).

When abrasive solids are processed, the points of wear are protected with replaceable inserts/tiles made from silica carbide, tungsten carbide, ceramic, or other abrasive-resistant materials. These high-wear areas include the feed zone including feed ports; the conveyor blade tip, especially the pressure or pushing face; the conical beach; and the solids discharge ports. Transport of solids is encouraged in some applications by longitudinal strips or grooves at the inner diameter of the bowl, especially at the beach, to enhance the frictional characteristics between the sediment and the bowl surface, and by polished conveyor faces to reduce frictional drag. For fluidlike sediment cake, by using the strips in the beach, a much tighter gap between the conveyor blade tip and the bowl surface is possible with a cake heel layer trapped by the strips. This reduces leakage of the fluid sediment flowing through an otherwise larger gap opening to the pool. Gypsum coating on the bowl wall at the beach section has been used to achieve the same objective.

Various bowl configurations with a wide range of aspect ratios—i.e., length-to-diameter ratio—from less than 1 to 4 are available for specific applications, depending on whether the major objective is maximum clarification, classification, or solids dryness. Generally, the movement of liquid and solids is in countercurrent directions, but in the cocurrent design the movement of liquid is in the same direction as that of the solids. In this design, the feed is introduced at the large end of the machine and the centrate is taken by a skimmer at the beach-cylinder junction. The settled solids transverse the entire machine and discharge at the beach exit. Compound angle beaches are used in

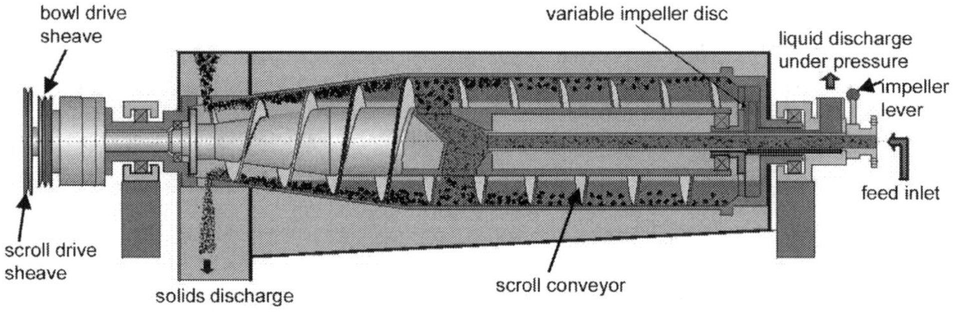

FIG. 18-157 Two-phase decanter centrifuge—pressurized liquid discharge. (*Flottweg Separation Technology.*)

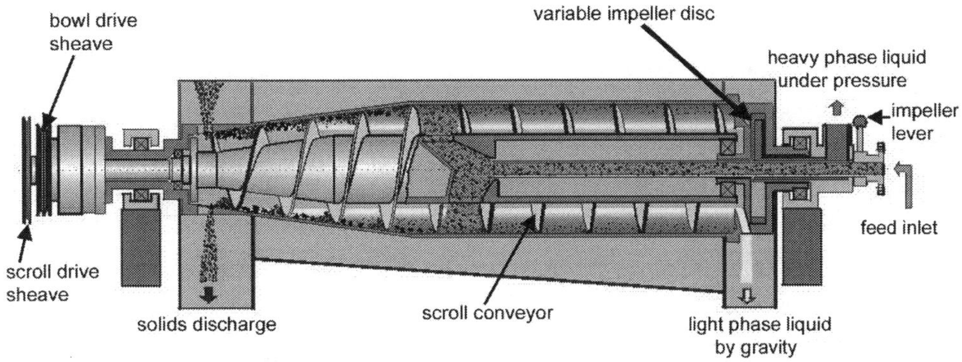

FIG. 18-158 Three-phase decanter centrifuge. (*Flottweg Separation Technology.*)

specific applications such as washing and drying of polystyrene beads. The pool level is located at the intersection of the two angles at the beach, the steeper angle being under the pool and the shallower angle above the pool (i.e., dry beach), allowing a longer dewatering time. The wash is applied at the pool side of the beach-angle intersection and functions as a continuously replenished annulus of wash liquid through which the solids are conveyed. The size of decanter centrifuge ranges from 6 in diameter to 54 in. The larger is the machine, the slower is the speed, and less is the *G*-force (Fig. 18-129). However, it provides a much higher throughput capacity, which cannot be accommodated with smaller machines (see Table 18-10). Decanter centrifuges are utilized in a broad range of industries and applications where a large amount of solids separation from liquids is required on a continuous basis. These industries include but are not limited to food, beverage including dairy, chemical, pharmaceutical, oil, edible oil, industrial, and municipal wastewater.

Three-Phase Decanter (Tricanter) Centrifuges Tricanter centrifuges are similar in principle to decanter centrifuges (see Fig. 18-158), but separate three phases that contain two immiscible liquids and one sedimenting/suspended solids phase. The sedimenting solids that collect on the bowl wall are conveyed out of the centrifuge and discharged similarly to a decanter centrifuge. The two liquid phases are discharged either via gravity over two sets of adjustable weir plates or rings or via a dual discharge system where the heavy liquid phase (typically water) is discharged via a stationary impeller under pressure and the light liquid phase (typically fat or oil) is discharged via gravity over a ring dam. The benefit of the dual discharge system is that the liquid interface zone (and ultimately the pool/pond height) is adjustable while the machine is operating at full speed. These types of centrifuges are commonly used in the fish, animal by-products, oil sludge, and edible oil (i.e., olive and palm) industries.

Specialty Decanter Centrifuges Decanter technology has evolved over the past 10 years to include machines suitable for separation applications normally not effective in standard decanter centrifuges. These specialty decanters/tricanters utilize the same basic premise of solids discharge via an internal scroll, but with specific machine geometries that allow for specialty separations. These specialty decanters include the Sedicanter (shown in Fig. 18-159) and Sorticanter (shown in Fig. 18-160). The Sedicanter, which has a double-cone cocurrent bowl design and specialized scroll geometry, is capable of achieving higher rotational speeds (up to 7750 rpm and 10,000*G*) and can, therefore, increase clarification efficiency and effectively discharge fine, pasty solids where a normal decanter is inefficient and ineffective. The Sedicanter is commonly used in certain biotechnology, vitamin, soy, and yeast separations.

The Sorticanter has a scroll with reversing pitch on one side, which skims a floating solids layer off the top of the carrier liquid, usually an aqueous brine of intermediate density. Sinking solids are scrolled out similarly to normal decanter centrifuges, and liquid is discharged under pressure via a stationary but variable impeller. The Sorticanter is utilized in the plastics recycling industry.

Screenbowl Centrifuges The screenbowl centrifuge consists of a solid-bowl decanter to which, at the smaller conical end, a cylindrical screen has been added (see Fig. 18-161). The scroll spans the entire bowl, conforming to the profile of the bowl. It combines a sedimenting centrifuge together with a filtering centrifuge. Therefore, the solids which are processed are typically larger than 23 to 44 μm.

As in a decanter, an accelerated feed is introduced to the separation pool. The denser solids settle toward the bowl wall and the effluent escapes through the ports at the large end of the machine. The sediment is scrolled toward the beach, typically with a steeper angle compared to the decanter centrifuge. As the solids are conveyed to the

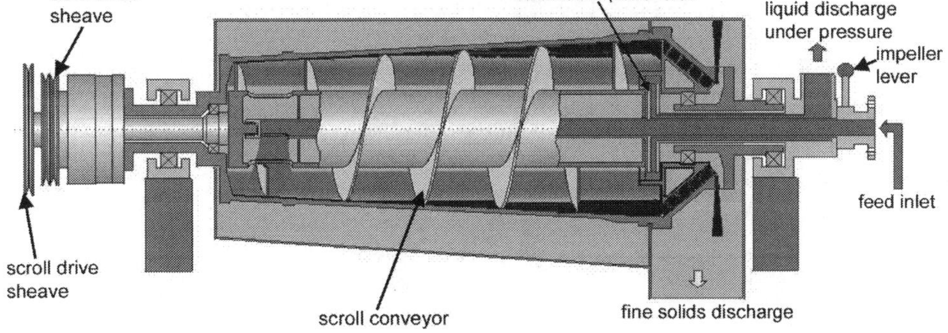

FIG. 18-159 Sedicanter centrifuge. (*Flottweg Separation Technology.*)

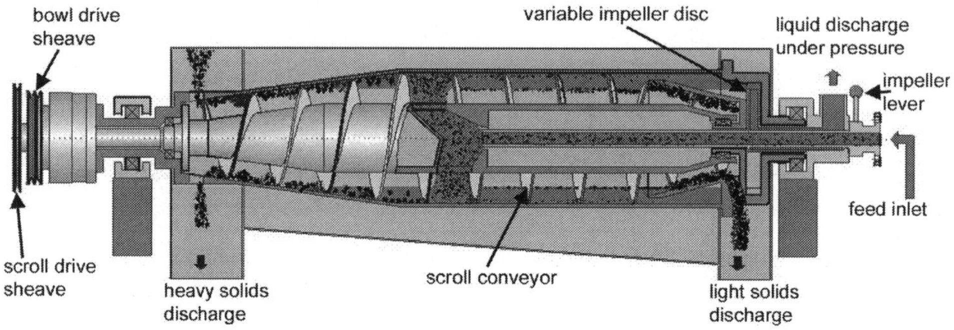

FIG. 18-160 Sorticanter centrifuge. (*Flottweg Separation Technology.*)

screen section, the liquid in the sediment cake further drains through the screen, resulting in drier cake. Washing of the sediment in the first half of the screen section is very effective in removing impurities, with the second half of the screen section reserved for dewatering of mother liquor and wash liquid.

The screen is typically constructed of a wedge-bar with an aperture between adjacent bars, which opens up to a larger radius. This prevents solids from blinding the screen as well as reduces conveyance torque. For abrasive materials such as coal, the screens are made of wear-resistant materials such as tungsten carbide.

Continuous Centrifugal Sedimentation Theory The Stokes settling velocity of a spherical particle under centrifugal field is given by Eq. (18-103). Useful relationships have been established on continuous sedimentation by studying the kinematics of settling of a spherical particle of diameter d in an annular rotating pool. Equating the time rate of change in a radial position to the settling velocity, and the rate of change in an axial position to bulk-flow velocity, thus gives

$$\frac{dr}{dt} = crd^2 \tag{18-111}$$

$$\frac{dx}{dt} = \frac{Q}{\pi\,(r_b^2 - r_p)} \tag{18-112}$$

where $c = (\rho_s - \rho_L)\,\Omega^2/18\mu$, x is distance along the axis of the bowl, Q is the volumetric feed rate, r_b and r_p are, respectively, the bowl and pool surface radii. For a particle located at one end of the bowl at radius r with $r_p < r < r_b$, after transversing the full bowl length, it settles out and is captured by the bowl wall. Solving the above equations with these boundary conditions, the limiting trajectory is:

$$\frac{r}{r_b} = \exp\left[\frac{-\pi\, cL(r_b^2 - r_p^2)d^2}{Q}\right] \tag{18-113}$$

If the same size particle d is located at an initial starting radius less than r given by Eq. (18-113) it is assumed to escape from being captured by the bowl, whereas it would have been captured if it had been at an initial radius greater than r. Assuming that the number of particles with size d is uniformly distributed across the annular pool, the recovery Rec_d (known also as **grade efficiency**) is the differential of the cumulative recovery $Z = 1 - Y$, with Y given in Eq. (18-95) for particles with size d, as the ratio of the two annular areas:

$$\text{Rec}_d = \frac{r_b^2 - r^2}{r_b^2 - r_p^2} \tag{18-114}$$

Combining Eqs. (18-113) and (18-114), the maximum Q to the centrifuge, so as to meet a given recovery Rec_d of particles with diameter d, is

$$\frac{Q_d}{2V_{gd}} = \left(\frac{\pi\,\Omega^2 L}{g}\right)\left(\frac{r_p^2 - r_b^2}{\ln\{1 - \text{Rec}_d[1 - (r_p/r_b)^2]\}}\right) = \Sigma_{\text{Rec}_d} \tag{18-115}$$

Note in Eq. (18-115) that V_{gd} is the settling rate under 1g, and it is a function of the particle size and density and fluid properties. The ratio $Q_d/2V_{gd}$ is then related only to the operating speed and geometry of

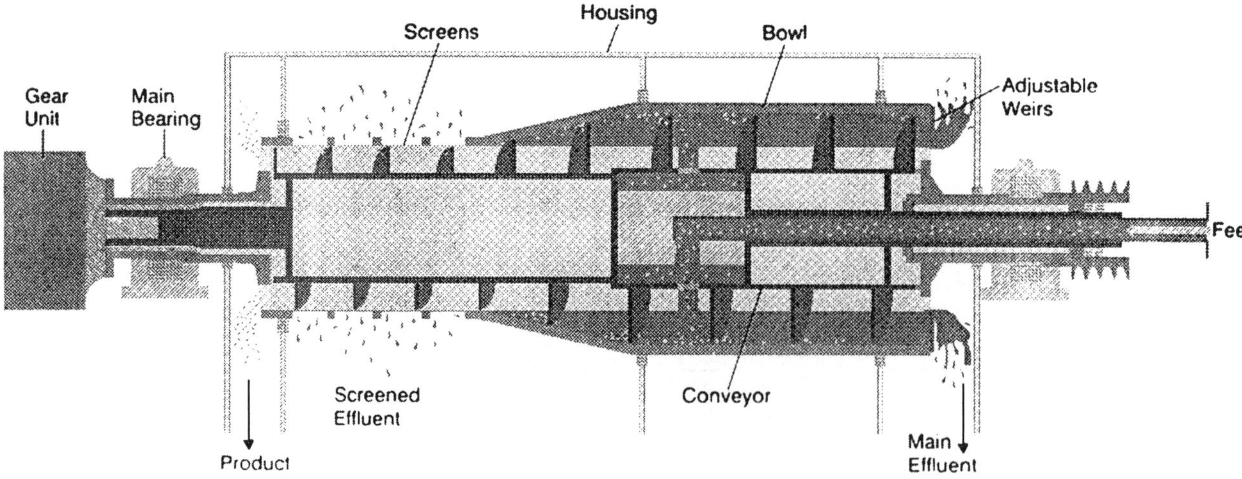

FIG. 18-161 Cylindrical-conical screen-bowl centrifuge. (*Bird Machine Co.*)

the centrifuge, as well as to the size recovery. It measures the *required* surface area for settling under centrifugal gravity to meet a specified Rec_d. When the size recovery Rec_d is set at 50 percent, the general result, Eq. (18-115), reduces to the special case, which is the well-known Ambler's Sigma factor, which for a straight rotating bowl (applicable to bottle centrifuge, decanter centrifuge, etc.) is:

$$\Sigma = \left[\frac{\pi\,\Omega^2 L}{g}\right]\left[\frac{r_b^2 - r_p^2}{\ln\,[2r_b^2/(r_b^2 + r_p^2)]}\right] \tag{18-116}$$

It can be simplified to:

$$\Sigma = \pi\,\Omega^2 L \frac{(3r_b^2 + r_p^2)}{2g} \tag{18-117}$$

For a disc centrifuge, a similar derivation results in

$$\Sigma = \frac{2\pi\,\Omega^2\,(N-1)\,(r_2^3 - r_1^3)}{3g\,\tan\,\theta} \tag{18-118}$$

where N is the number of discs in the stack, r_1 and r_2 are the outer and inner radii of the disc stack, and θ is the conical half-angle.

Typical Σ factors for the three types of sedimenting centrifuges are given in Table 18-13. In scale-up from laboratory tests, sedimentation performance should be the same if the value of Q/Σ is the same for the two machines. This is a widely used criterion for the comparison of centrifuges of similar geometry and liquid-flow patterns developing approximately the same G; however, it should be used with caution when comparing centrifuges of different configurations. In general, the shortcomings of the theory are due to the oversimplified assumptions being made, such as (1) there is an idealized plug-flow pattern; (2) sedimentation abides by Stokes law as extended to many g's; (3) feed solids are uniformly distributed across the surface of the bowl head at one end of the clarifier and capture implies that the particles' trajectory intersect the bowl wall; (4) the feed reaches full tangential speed as it is introduced to the pool; (5) the recovery of given-size particles is at 50 percent; (6) this does not account for possible entrainment of already settled particles in the liquid stream; (7) there is absence of entrance and exit effects.

Experience in using the Σ concept has demonstrated that the calculated Σ factor should be modified by an efficiency factor to account for some of the aforementioned effects which are absent in the theory and, as such, this factor depends on the type of centrifuge. It is nearly 100 percent for simple spin-tube bottle centrifuge, 80 percent for tubular centrifuge, and less than 55 percent for disc centrifuges. The efficiency varies widely for decanter centrifuges, depending on cake conveyability and other factors.

FILTERING CENTRIFUGES

Filtering centrifuges are broadly categorized as continuous operating and batch operating. Both continuous and batch filtering centrifuges use some type of filtration media fitted against the basket (bowl) wall. As the solid-liquid mixture is introduced, the liquid filters through the solids, through the filter media, and typically through perforations in the basket shell (except with rotational siphon designs, discussed later). Filtering centrifuges are primarily chosen over sedimenting types where high cake purity through cake washing is a requirement, or where minimal residual cake moisture is desired. Typical solids retention times range from 5 to 45 s for continuous operating filtering centrifuges and 5 to 180 min for batch operating filtering centrifuges.

Usually, the solids phase is of a higher specific gravity than the liquid; but unlike with sedimenting centrifuges, this is not an absolute requirement. In the nontypical case where the opposite is true, then filtration, be it centrifugal, vacuum, or pressure, is the only option for solid-liquid separation. However, for batch filtering centrifuges with the solids phase lighter than the liquid, care must be taken that liquid is not allowed to build in the basket during the feed step, or else buoyancy forces may float the settled solids, resulting in uneven filtration and high vibration. This is not of concern for continuous filtering centrifuges since the filtration rate must inherently exceed the liquid feed rate for stable operation. Refer to Table 18-14 for typical operating ranges of filtering centrifuges.

BATCH FILTERING CENTRIFUGES

Although continuous centrifuges are often preferred for reasons of lowest capital cost, high unit capacity, and ease of integration into continuous upstream and downstream processes, batch filtering centrifuges with cyclic operation will always have a role in the CPI for reasons of highest possible product purity, lowest possible cake moisture, and highest product recovery. The slurry's physical properties such as particle size distribution or liquid viscosity may require long

TABLE 18-13 Scale-up Factors for Sedimenting Centrifuges

Type of centrifuge	Inside diameter, in	Disc diameter, in/number of discs	Speed, r/min	Σ value, units of 10^4 ft^2	Recommended scale-up factors*
Tubular	1.75	—	23,000	0.32	1†
Tubular	4.125	—	15,000	2.7	21
Tubular	4.90	—	15,000	4.2	33
Disc	—	4.1/33	10,000	1.1	1
Disc	—	9.5/107	6,500	21.5	15
Disc	—	12.4/98	6,250	42.5	30
Disc	—	13.7/132	4,650	39.3	25
Disc	—	19.5/144	4,240	105	73
Scroll conveyor	6	—	6,000	0.27	1
Scroll conveyor	14	—	4,000	1.34	5
Scroll conveyor	14‡	—	4,000	3.0	10
Scroll conveyor	18	—	3,450	3.7	12.0
Scroll conveyor	20	—	3,350	4.0	13.3
Scroll conveyor	25	—	3,000	6.1	22
Scroll conveyor	25	—	2,700	8.6	31

*These scale-up factors are relative capacities of centrifuges of the same type but different sizes when performing at the same level of separation achievement (e.g., same degree of clarification). These factors must not be used to compare the capacities of different types of centrifuges.

†Approaches 2.5 at rates below mL/min.

‡Long bowl configuration.

NOTE: To convert inches to millimeters, multiply by 25.4; to convert revolutions per minute to radians per second, multiply by 0.105; and to convert 10^4 square feet to square meters, multiply by 929.

TABLE 18-14 Operating Range of Filtering Centrifuges

Type of centrifuge	G/g	Minimum feed solids concentration by wt.	Minimum mean particle size, μm	Minimum V_{fo}, m/s
Vibratory	30–120	50	300	5×10^{-4}
Tumbler	50–300	50	200	2×10^{-4}
Screen scroll	500–2000	35	75	1×10^{-5}
Pusher	300–800	40	120	5×10^{-5}
Screen bowl	500–2000	20	45	2×10^{-6}
Peeler	300–2000	5	10	2×10^{-7}
Vertical	200–1000	5	5	1×10^{-7}
Inverting filter	300–1000	5	2	5×10^{-8}

retention time and drive the selection from continuous to batch. Generally, when the value of the product is high, batch operating centrifuges are often preferred.

A development over the last 10 to 20 years is the introduction of true cGMP designs suitable for high-purity fine chemicals and pharmaceutical applications. Requirements in this sector usually favor batch operation due to demands for batch identity; typical properties of the slurry dictate batch operation, high wash requirements, cleanability and inspectability, avoidance of cross-contamination for multipurpose applications, and elimination of operator exposure to the process.

Most modern batch filtering centrifuges are provided with ac variable-frequency drives (VFDs) for speed variation, often with power regenerative braking. Even in the case of peeler centrifuges which are capable of constant-speed operation throughout the cycle, in most cases they are still fitted with VFDs to meet starting requirements (accelerating a high inertial load), operating flexibility, and regenerative braking in hazardous areas. However, there are still units available that use either a two-speed motor (generally considered obsolete) or hydraulic drive.

The main subcategories of batch filtering centrifuges are vertical basket centrifuges (both top unloading and bottom unloading of various configurations), horizontal peeler centrifuges, and horizontal inverting filter centrifuges.

Vertical Basket Centrifuge—Operating Method and Mechanical Design The vertical basket centrifuge is equipped with a cylindrical basket rotating about the vertical axis. The basket shell is normally perforated and lined with filtration media consisting of filter cloth, and backing screens to provide liquid drainage paths to the basket holes. Securing the filter cloth may be by hook and loop attachment or with snap-in retainers.

Feed slurry is introduced into the basket through either single or multiple feed pipes, or by other means such as rotating feed cone to help distribute the solids on the basket wall. In most cases, feed slurry is introduced at an intermediate speed, although in some applications (FGD gypsum, for one), feeding is done at full speed. There are several methods available to control the feed and cake level such as mechanical, paddle-type feelers, capacitance probes, ultrasonic sensors, feed totalizer, or load cells.

The solids distribution profile may tend to be parabolic with thicker cakes near the bottom of the basket, tapering down toward the top, since the G-field is perpendicular to the force of gravity. This is especially true with fast-sedimenting solids that will settle toward the basket bottom before the slurry is fully accelerated by the basket. The coarser solids can settle toward the basket bottom, while the finer solids deposit preferentially toward the top. This can result in uneven filtration resistance in the cake, affecting the wash pattern and efficiency of the wash. In cases where this is a concern, a rotating feed cone may be better for even distribution, or a horizontal peeler centrifuge may be better suited to the application.

During and after feeding, filtrate passes through the cake, filter media, and out through the basket shell and is collected in a housing surrounding the basket and discharged through a tangential nozzle. The solids build up on the basket wall during the feed step until the desired loading is achieved.

After a spin time to filter the mother liquor through the solids, wash liquor is commonly applied in either a single step or various combinations, typically via a wash pipe with nozzles. The cake is spun for a time at high speed, then the machine ramps down to discharge. Solids removal can be accomplished by one of several methods.

Top Unloading Vertical Basket Centrifuges This is one of the oldest types of centrifuges, dating back to about 1900 or even earlier. With this design, the perforated basket is fitted with either a filter cloth or a filter bag. The basket has a solid bottom. After the dry spin portion of the cycle is completed, the machine is stopped. Solids removal is either by manually digging out the cake or by removing (lifting) the filter bag from the top of the unit. Except for small pilot-scale units and some specialty applications, this design is no longer commonly marketed or desired due to labor intensiveness, incompatibility with solvent wet or toxic products (operator exposure), and overall inefficiency of operation. See Fig. 18-162.

One modern top unloading design is used for bulk pharmaceutical processing as part of a complete closed-loop dewatering and drying system including heaters, blowers, and sterile filters. This involves a two-motion discharge "mouth" to pneumatically remove thin layers of cake while blowing sterile, dehumidified hot nitrogen into the housing and into the discharge pipe. This method both pneumatically conveys and flash-dries the solids en route to the waiting conical mixer/dryer for subsequent vacuum drying.

Bottom Unloading Vertical Basket Centrifuges Most common for modern machines is the bottom discharge design, incorporating a swiveling scraper mechanism, typically cutting the cake in a single motion, or with a two-motion, oscillating scraper in the case of finer or stickier cakes or pharmaceuticals. In every case, solids discharge must be at low speed, which necessitates ramping the machine up and down every cycle. After discharge, a thin cake layer or heel remains on the filter cloth. See Fig. 18-163.

Heel removal can be automated by dissolving the heel, flushing the heel out the solids discharge chute with subsequent downstream diverting away from solids handling equipment, or pneumatically removing the heel with blowoff nozzles, discharging it out the solids discharge. Pneumatic heel removal can be accomplished either from within the basket (often incorporated with the knife) or from the outside of the basket.

There are different cover arrangements to access the interior, such as hinged or pivoting manway, half or full opening covers. Filter cloth maintenance or component adjustments usually require entering the unit in any case, except for small sizes. See Fig. 18-164.

Isolation of the load imbalances from the structure has historically been by link, or three-column suspension. This system is relatively inefficient and transmits substantial dynamic forces to the building foundation, limiting operating speeds and performance. In response to the common shortcomings of this design, some manufacturers redesigned the suspension/isolation system to a massive inertial baseplate and housing supported on tuned coil springs and dampers at each corner. This greatly improved the dynamic force attenuation and stability of the machine. This system has allowed designing very large machines (1600-mm-diameter by 1250-mm-high baskets) processing materials with high solids density such as wallboard-grade gypsum from FGD systems with a unit capacity in excess of 10 Mtons/h.

Top Suspended Vertical Centrifuges A special type of top suspended centrifuge is widely used in sugar processing and is shown in Fig. 18-165. Conventionally, the drive is suspended from a horizontal bar supported at both ends from two A-frames. The drive head, which is connected to the motor or a driven pulley through a flexible coupling, carries the thrust and radial bearings that support the basket,

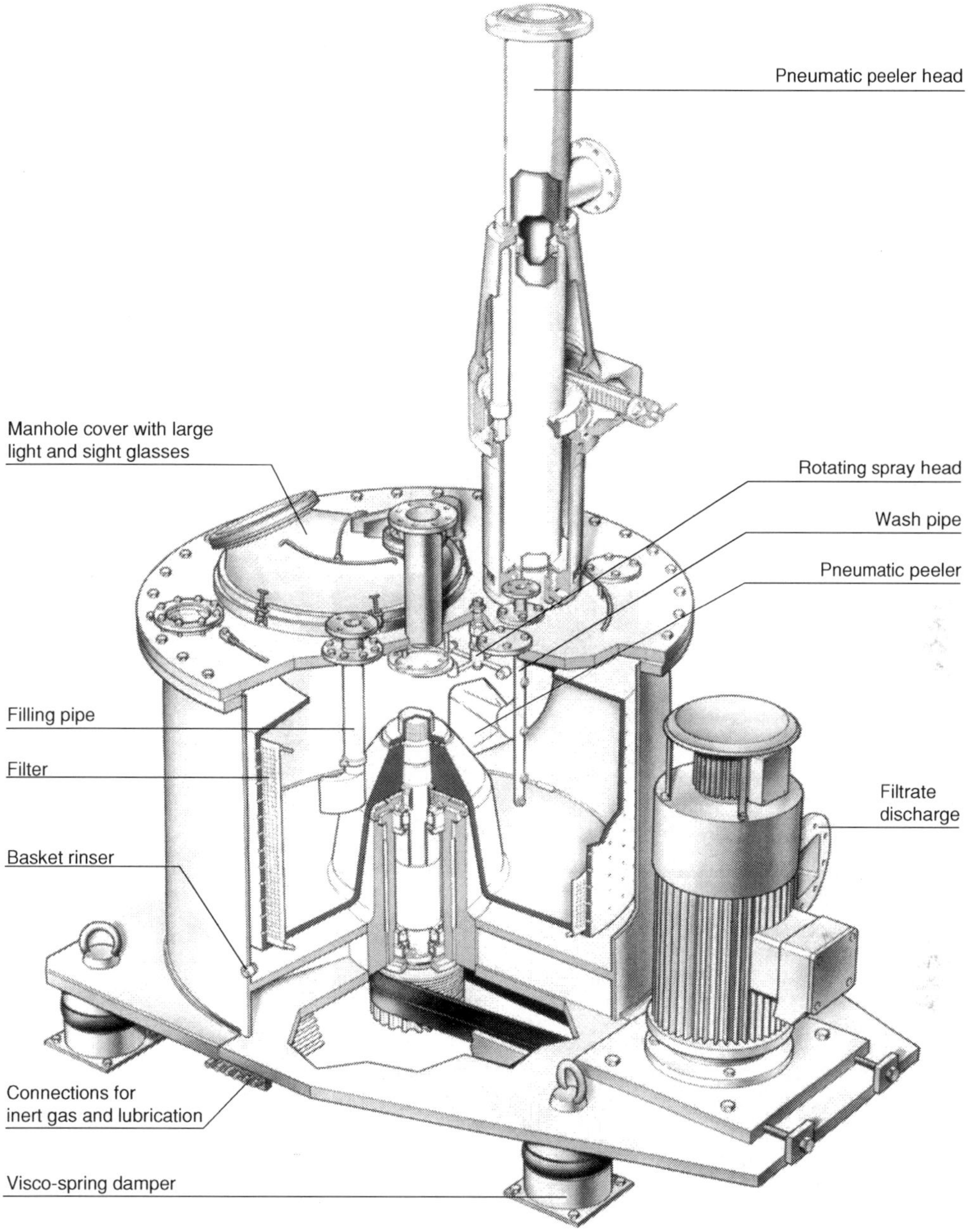

FIG. 18-162 Top unloading vertical basket centrifuge. (*Krauss Maffei Process Technology.*)

shaft, and load. These units can be equipped with large 112-kW (150-hp) drive motors, and on white sugar can process about 350 kg/cycle with 24 cycles/h.

Horizontal Peeler Centrifuge—Operating Method and Mechanical Design The chemical design peeler centrifuge was developed in the 1920s and has wide areas of application (Fig. 18-166). Like the vertical basket, it has a rotating filtration basket, except that it rotates about the horizontal axis. Early machines supported the basket from both ends, but virtually all modern machines are cantilever-supported for purposes of accessibility to the basket and internal components. Some machines are equipped with a full opening door which swings away with internal components, while other designs incorporate a full opening housing that also provides access to the basket exterior. This unit has an extremely rugged construction compared to vertical baskets, required due to the full-speed feeding and discharge capability of the peeler centrifuge. It is often provided with high-power ac VFD drives for accelerating the feed slurry at full speed and for optimum operating flexibility. Gastight construction to 400-mm (16-in) water column is usually standard, and higher pressure ratings can easily be accomplished.

By reorienting the axis to horizontal, many advantages become possible such as superior wash capability with the more uniform solids distribution compared to the vertical basket, with the potential for uneven, parabolic cake profile resulting in uneven wash penetration.

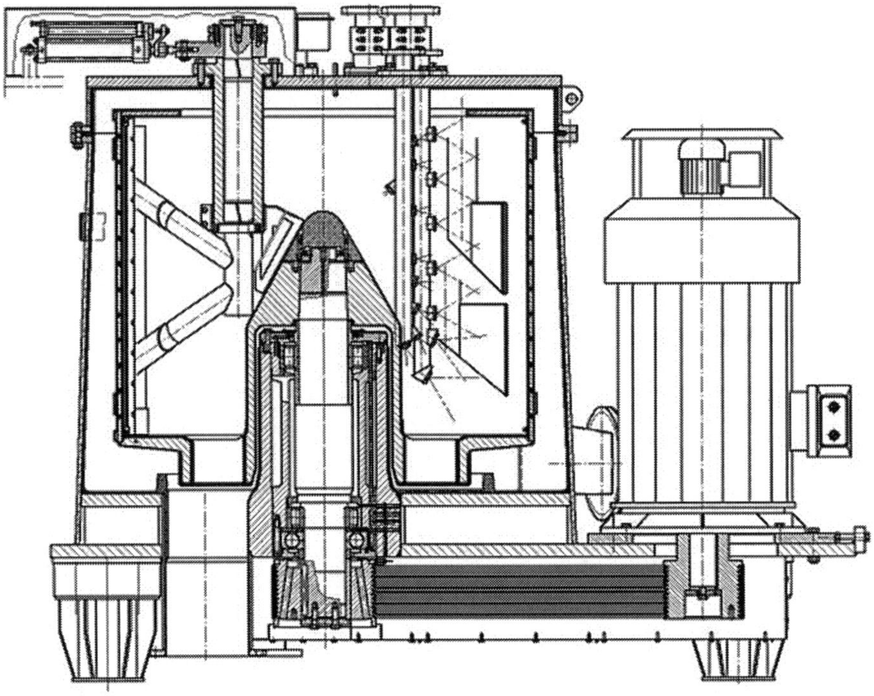

FIG. 18-163 Typical bottom unloading vertical basket centrifuge. (*Krauss Maffei Process Technology.*)

The peeler centrifuge costs more than a comparable basket size vertical machine, although often a smaller peeler centrifuge can outperform a larger vertical basket. In addition, the higher capital cost is offset by numerous process and mechanical advantages, such as these:

- Full opening door contains the feed, wash, feed control, and solids discharge components. Easily swung open, it then provides complete access to the basket interior and all internal components mounted on the door. Filter cloth exchange does not require vessel entry.
- Isolation of dynamic forces is far superior with horizontal machines compared to vertical.
- The peeler centrifuge will distribute the solids more evenly since it is not feeding perpendicular to gravity as is the vertical basket. This

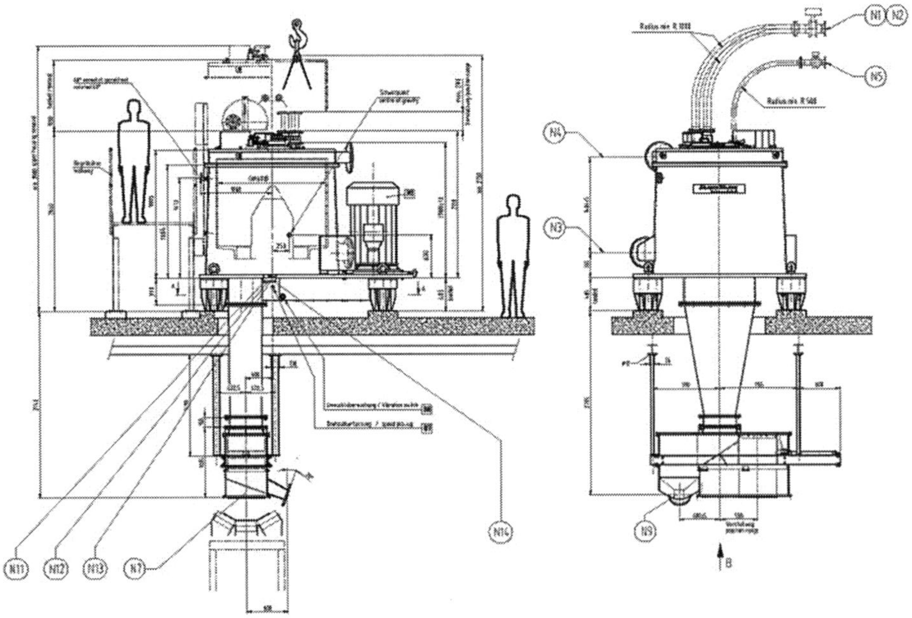

FIG. 18-164 Typical vertical basket centrifuge installation. (*Krauss Maffei Process Technology.*)

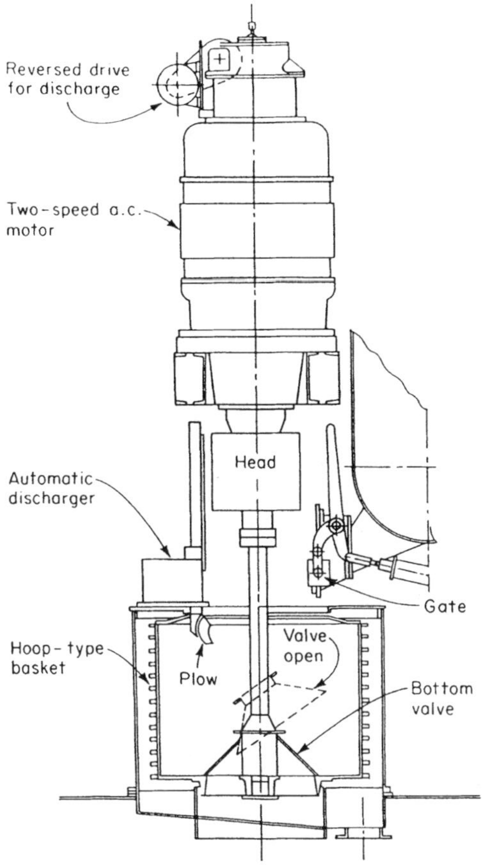

FIG. 18-165 Top-suspended vertical centrifuge. (*Western States Machine Co.*)

provides smoother operation, better wash effect, and an ability to handle faster-draining materials.

- Ability to discharge at high speed eliminates or minimizes dead cycle time required for acceleration and braking for higher capacity, lower power consumption, and lower wear and tear. The cycle time savings is particularly beneficial with short-cycle (fast-filtering) requirements.
- The peeler centrifuge can provide higher centrifugal forces than can vertical baskets, for increased performance and flexibility.
- Peeler centrifuges are available in larger sizes than vertical centrifuges—up to 2100-mm (83-in) diameter.
- The peeler centrifuge is also capable of automatic heel removal by several methods: dissolving the heel, reslurrying and discharging the heel wet and diverting downstream, or dry heel removal pneumatically. Pneumatic heel removal can be accomplished either from within the basket or from outside the basket.

Siphon Peeler Centrifuge The siphon peeler centrifuge (Fig. 18-167) was developed and patented by Krauss-Maffei in the 1970s. Instead of utilizing only centrifugal pressure as the driving force, as do all perforated units both vertical and horizontal, the rotational siphon centrifuge provides an increased pressure gradient by reducing the pressure behind the filter media and thereby increasing the driving force for filtration.

As in the perforated basket design, the liquid filters through the cake and filter media, but instead of discharging through perforations in the basket shell, the basket wall is solid and the liquid flows axially to the basket rear and into a separate chamber. At this point, the filtrate is skimmed out with a radially adjustable skimmer.

In perforated baskets, the driving force for filtration is approximately the hydrostatic pressure established by the liquid column. The driving force diminishes as the liquid column height decreases, often causing a wet layer near the base of the cake due to capillary pressure balancing the centrifugal pressure. In siphon baskets, in addition to the centrifugal pressure, by skimming at a radius greater than the filter cloth, a rotational siphon is established. Due to the gravitational field in which it is working, a height difference Δh of only 20 to 30 mm is sufficient to lower the pressure behind the cloth to the vapor pressure of the liquid. This additional vacuum remains in place until all the interstitial liquid is drawn through the cake, and will overcome the cake capillary pressure, thus preventing this wet layer. Once the supernatant and interstitial

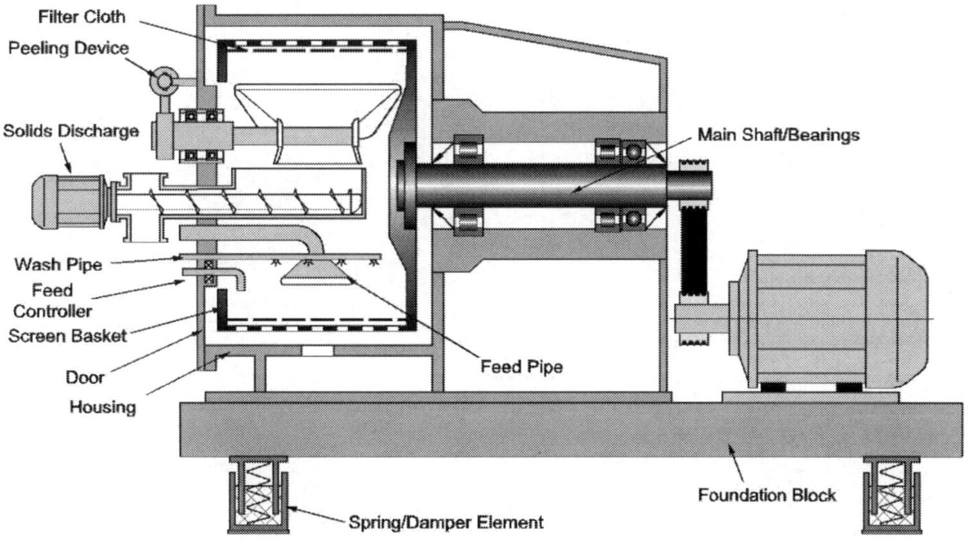

FIG. 18-166 Peeler centrifuge cross section. (*Krauss Maffei Process Technology.*)

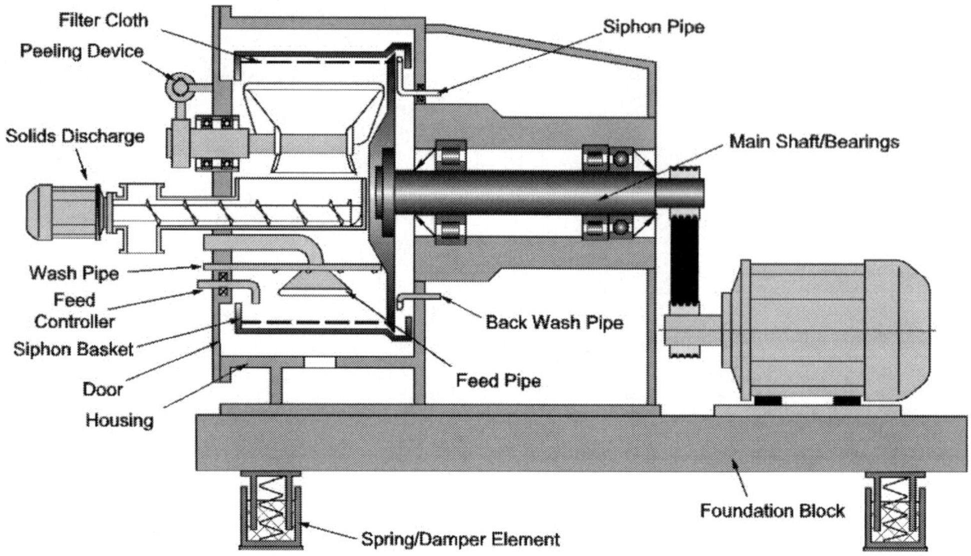

FIG. 18-167 Siphon peeler centrifuge cross section. (*Krauss Maffei Process Technology.*)

liquid drains from the cake, the siphon chamber behind the cloth drains and filtration characteristics are like those of a perforated basket. See Fig. 18-168.

To reestablish the siphon for the next cycle, a priming step precedes feeding where the siphon skimmer is pivoted inward near the rim of the siphon chamber, and liquid is introduced into the siphon chamber that backflows up through the heel cake, displacing gas from the chamber. Feeding then begins with the heel submerged. After a time delay, the siphon swivels downward to the working position (δh of +20 to 30 mm) where it remains for the remainder of the cycle. Besides increased driving force for filtration, other benefits of the rotational siphon include:

- Accurate control of the filtration rate is useful during feed and wash. For fast-filtering products, filtration rates can be throttled, ensuring even solids distribution.
- Backwashing the residual heel between each cycle rejuvenates the heel to maintain good permeability. Heel life is often extended.
- Feeding into a liquid bath helps lay down a more porous heel layer since the larger particles sediment faster than finer particles.
- Separate discharge of filtrate from splash/overflow provides better product yield.
- Deep siphon chambers with cake backwashing capability have been successfully utilized to completely submerge the cake and indefinitely increase wash contact time.

Pressurized Siphon Peeler Centrifuge Theoretically, the same principle can further increase driving force with overpressure in the process housing; for example, 3-bar overpressure would produce up to 4-bar pressure gradient across the cake. In practice, to date this has not been utilized due to the complexity of the installation.

Pharma Peeler Centrifuge For applications requiring hygienic operation, a special type of peeler centrifuge was developed in the 1990s (Fig. 18-169).

Primary applications for this type of machine are in fine chemicals and pharmaceuticals, often in multipurpose use where cross-contamination must be avoided. It provides for ease of cleanability and inspectability with automatic CIP/SIP, access to every wetted surface, pressure-tight construction suitable for steam sterilization, automatic heel removal, and separation of mechanical components from the

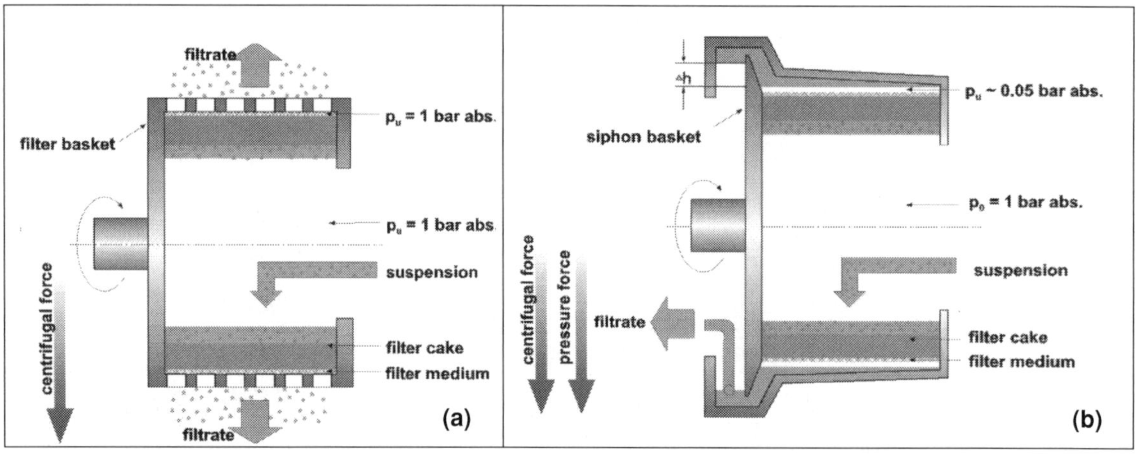

FIG. 18-168 Schematic representation of (*a*) perforate versus (*b*) siphon centrifuge. (*Krauss Maffei Process Technology.*)

FIG. 18-169 Pharma peeler centrifuge. (*Krauss Maffei Process Technology.*)

process end, making it suitable for through-the-wall clean-room installation. Operation is contained, thereby eliminating operator exposure.

Inverting Filter Centrifuge The inverting filter centrifuge was introduced in the late 1970s to provide a means of ensuring that all the filter cake is discharged from the filter medium. By turning the cloth inside out to achieve solids discharge, the problems of operator exposure and variable product quality associated with manual cake removal or residual-heel blinding were largely eliminated. By the late 1980s, this style of centrifuge had found widespread application in pharmaceutical and agricultural chemical production. See Figs. 18-170 to 18-172.

The inverting filter design comprises a horizontal axis shaft with a two-part bowl attached. The perforated cylindrical bowl remains in a fixed axial position throughout the operation, while a bowl insert is able to move along the horizontal axis. The filter cloth is attached at one end to the axially fixed bowl and at the other to the axially movable bowl insert. Therefore, moving the bowl insert causes the filter cloth to turn inside out. In the filtering position, the bowl insert sits inside the bowl, with the filter cloth covering perforations. As with other filtering centrifuges, the cake builds up on the cloth during filling. It is washed using a true positive-displacement, plug-flow wash, and cake dewatering can be achieved simply by spinning (often at the maximum speed) for a time.

As this cake-discharging mechanism involves little or no risk of cloth blinding, inverting filter centrifuges usually operate at optimum conditions with relatively thin cakes and frequent discharges. (Cake thicknesses are typically 1 to 3 in, and cycle times are typically 8 to 14 min.) This style of operation is particularly effective with compressible materials where the filtration rate drops off dramatically with increasing cake thickness. By operating with thin cakes and short cycle times, the average filtration flux throughout the batch operation is maximized for these difficult applications.

Inverting filter centrifuges come in bowl diameters ranging from 300 to 1300 mm and achieve *g*-forces of 3000 – 900 × gravity.

CONTINUOUS FILTERING CENTRIFUGES

Where processing conditions and objectives allow, continuous filtering centrifuges offer the combination of high processing capacities and good wash capabilities. Inherently they are less flexible than batch filtering centrifuges, primarily constrained by much shorter retention time, and in some cases liquid handling capacity requires upstream

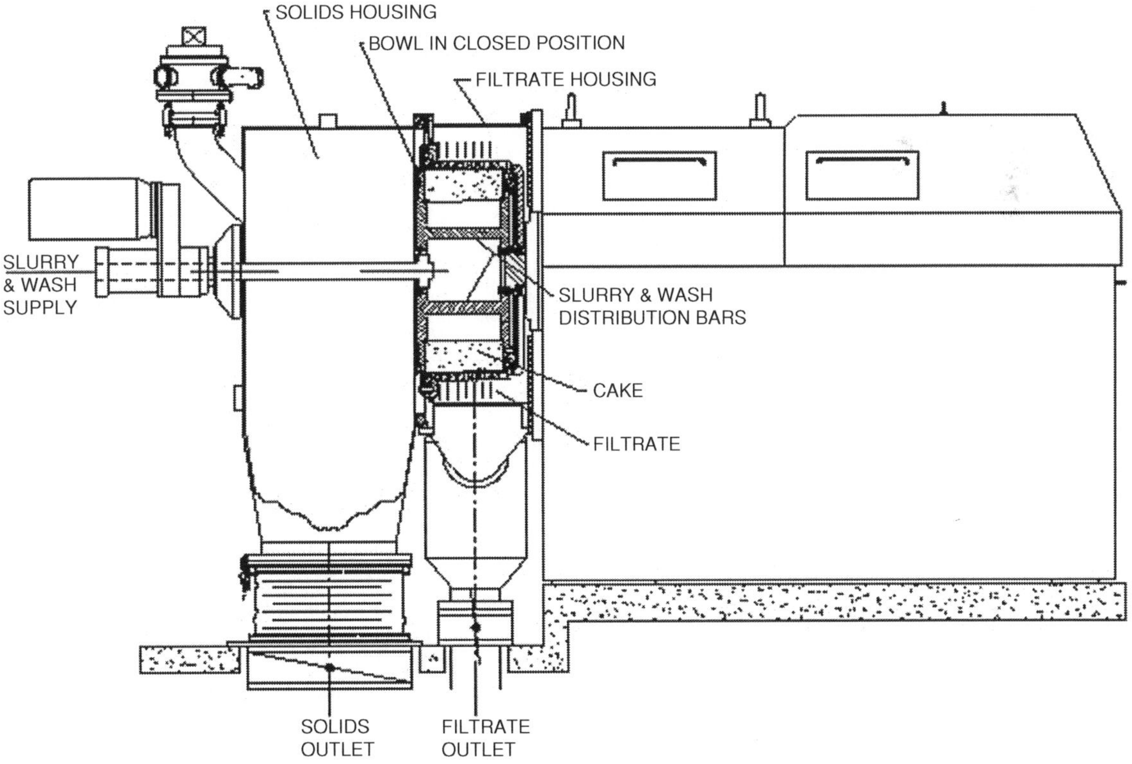

FIG. 18-170 Inverting filter centrifuge. (*Heinkel USA.*)

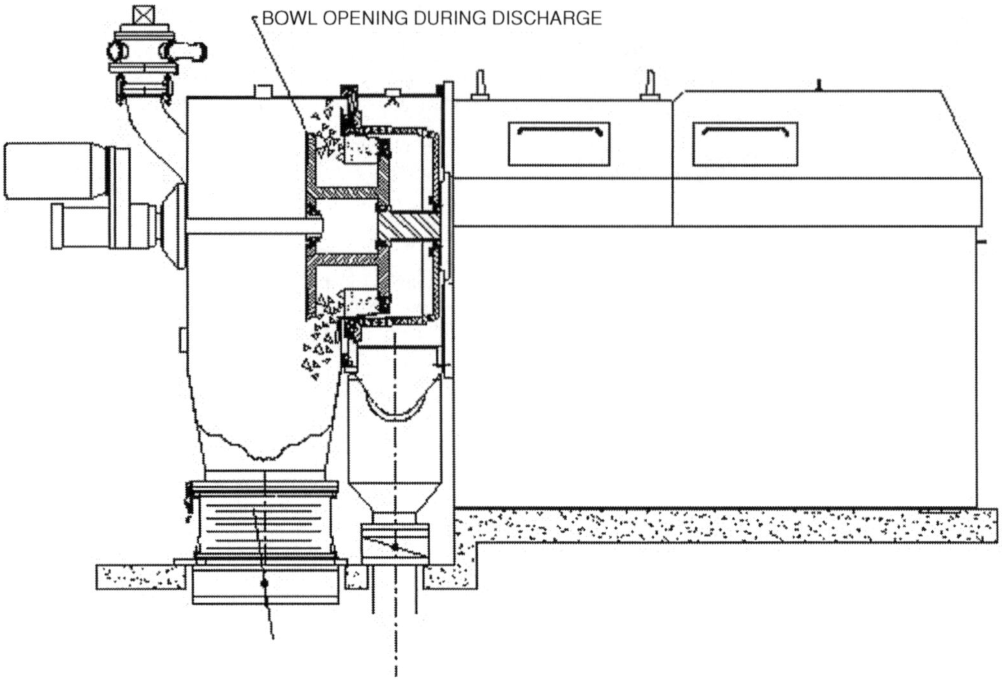

FIG. 18-171 Inverting filter centrifuge. (*Heinkel USA.*)

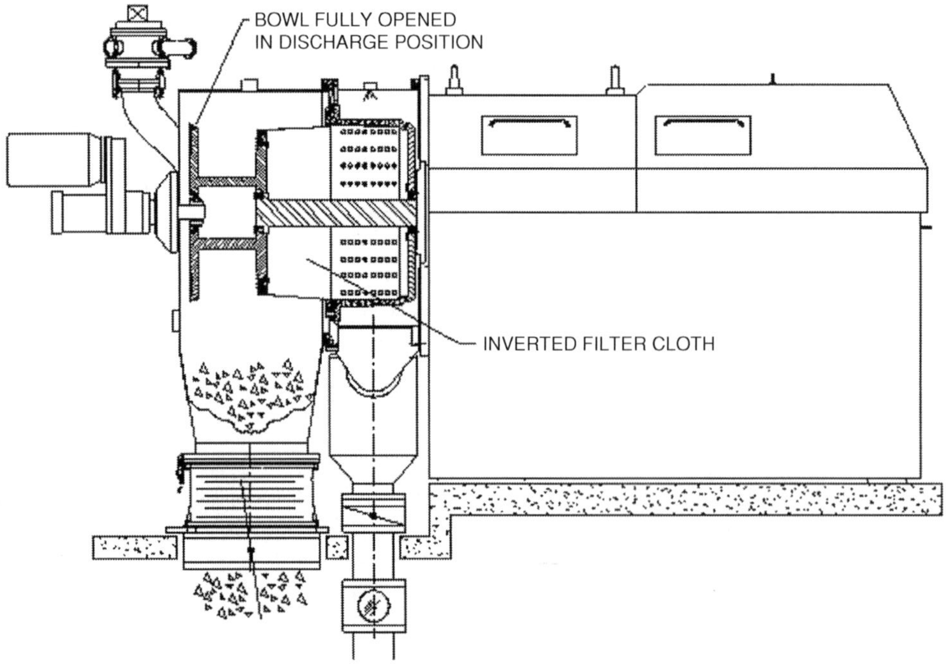

FIG. 18-172 Inverting filter centrifuge. (*Heinkel USA.*)

preconcentration of the slurry. Fines loss to the filtrate is also greater with continuous designs compared to batch.

Conical-Screen Centrifuges When a conical screen in the form of a frustum is rotated about its axis, the component of the centrifugal force normal to the screen surface impels the liquid to filter through the cake and the screen, whereas the component of the centrifugal force parallel to the screen in the longitudinal direction conveys the cake to the screen at a larger diameter. The sliding of the solids on the cone is favored by smooth perforated plates or wedge-wire sections with slots parallel to the axis of rotation, rather than woven wire mesh.

Wide-angle conical screen centrifuges. If the half-angle of the cone screen is greater than the angle of repose of the solids, the solids will slide across it with a velocity which depends on frictional properties of the cake but not on feed rate. The frictional property of the cake depends on the solid property, such as shape and size, as well as on moisture content. If the half-angle of the cone greatly exceeds the angle of repose, the cake slides across the screen at a high velocity, thereby reducing the retention time for dewatering. The angle selected is therefore highly critical with respect to performance on a specific application. Wide-angle and compound-angle centrifuges are used to dewater coarse coal and rubber crumb and to dewater and wash crude sugar and vegetable fibers such as from corn and potatoes.

Shallow-angle conical screen centrifuges. By selecting a half-angle for the conical screen that is less than the angle of repose of the cake and providing supplementary means for the controlled conveyance of the cake across the conical screen from the small to large diameter, longer retention time is available for cake dewatering. Three methods are in common use for cake conveyance:

1. *Vibrational conveyance.* This is referred as the vibratory centrifuge. A relatively high frequency force is superimposed on the rotating assembly. This can be either in-line with the axis of rotation or torsional, around the driveshaft. In either case, the cake under inertial force from the vibration is partly "fluidized" and propelled down the screen under a somewhat steady pace toward the large end, where it is discharged.

2. *Oscillating or "tumbling" conveyance.* This is commonly known as the tumbling centrifuge. The driveshaft is supported at its lower end on a pivot point. A supplementary power source causes the shaft and the rotating bracket it carries to gyrate about the pivot at a controlled amplitude and at a frequency lower than the rate of rotation of the basket. The inertia force generated also provides partial fluidization of the bed of solids in the basket, causing the cake to convey toward the large end, as in the vibrational conveyance.

3. *Scroll conveyance.* Another type of continuous filtering centrifuge is the scroll screen centrifuge, as shown in Fig. 18-173. The scroll screen centrifuges are also sometimes called worm screen centrifuges. The design consists of a fixed-angle rotating basket and a concentric screw conveyor to control the transport and discharge of solids. Common applications include crystal, fiber, and mineral separations.

Scroll screen centrifuges are typically used for continuous feeds of slurries of at least 10 percent solids by volume, of materials with an average size of 100 μm or greater. This design offers some residency to process variation and typically removes the bulk of surface moisture.

The scroll and the screen are rotating in the same direction with a small differential speed of typically less than 100 rpm. The feed is deposited into the acceleration cone of the scroll, then passes through the feed openings of the scroll. The solids are retained on the screen; as the liquid migrates through, the cake passes the screen media and the basket. The discharge housing collects the liquid, and the solids are conveyed to the large diameter of the rotating basket and are continuously discharged.

An internal product wash is also available in the scroll screen centrifuges. Wash liquid is added in a chamber midway along the basket, and the wash liquid migrates through the cake prior to final drying and discharge.

The rotating basket is used to retain the screen media. Wedge-wire as well as sheet metal screens are available, but are typically limited to a minimum opening size of 70 μm or larger. Common basket designs include 10, 15 and 20 degrees.

The scroll acts as a screw conveyor and discharges the solids. The typical solids retention time in the centrifuge ranges from 0.5 to 6 s. A close tolerance, 0.3 to 1 mm, is common between the scroll and screen; therefore little material remains on the screen. This minimizes the potential for imbalances.

Pusher Centrifuges—Operating Method and Mechanical Design Pusher centrifuges (Fig. 18-174) are continuous filtering centrifuges used for dewatering and washing free-draining bulk crystalline, polymer, or fibrous materials. Where suited, they provide the best washing characteristics of any continuous centrifuge due to control of retention time, uniform cake bed, and essentially

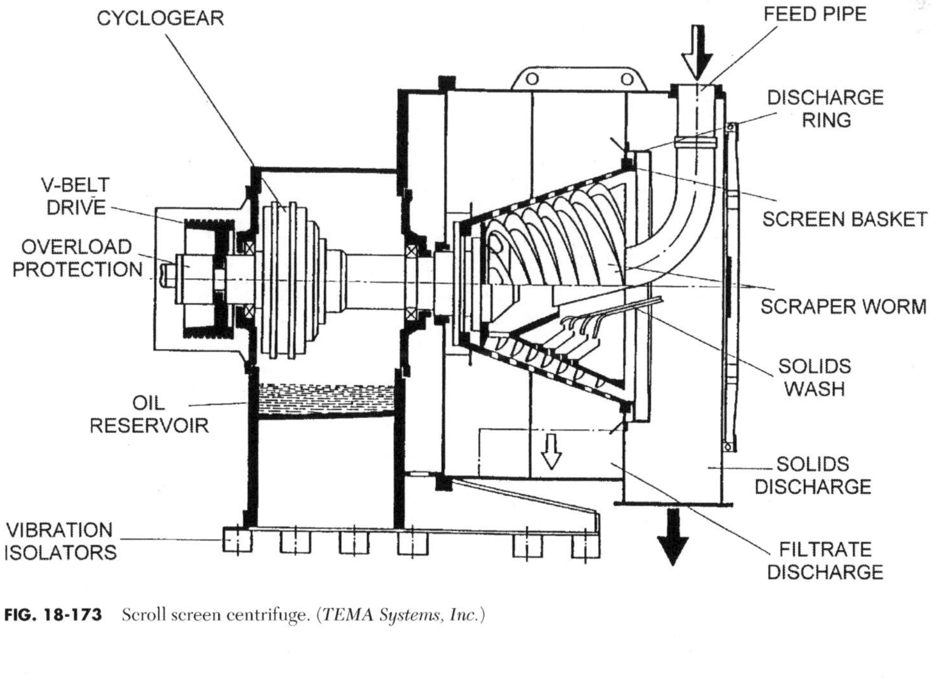

FIG. 18-173 Scroll screen centrifuge. (*TEMA Systems, Inc.*)

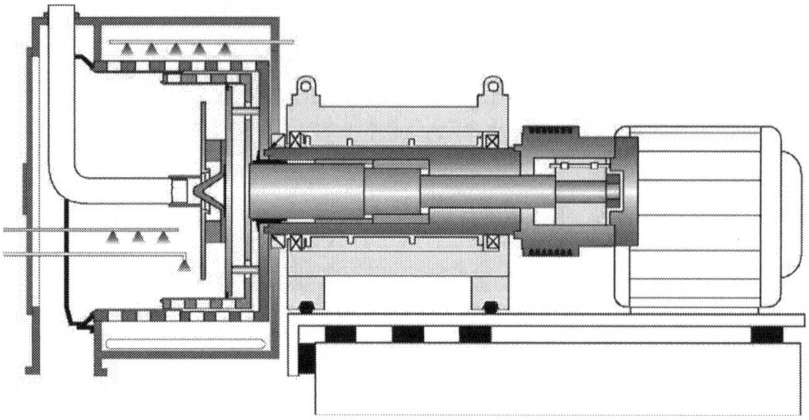

FIG. 18-174 Pusher centrifuge cross section. (*Krauss Maffei Process Technology.*)

plug flow of solids through the unit. For a typical application such as salt, they range in capacity from about 1 ton/h for small (250-mm-diameter) units up to about 120 tons/h or more in the largest units (1250-mm-diameter). They are generally applied where the mean particle size is at least 150 μm. Typical solids retention time is between 10 and 30 s. Normally the machine is fed by a feed pipe, but it can also be used as posttreatment of a prior dewatering step such as a vacuum filter. In this case it is fed by a feed screw. Due to the gentle handling of the product, pusher centrifuges are better suited for fragile crystals than are other types of continuous filtering centrifuges.

Generally there are three limitations to capacity in pusher centrifuges: (1) solids volumetric throughput, (2) liquid filtration capacity, and (3) retention time necessary to achieve desired objectives regarding cake purity and residual cake moisture. In most cases (2) dictates; therefore to optimize capacity and performance, preconcentrating the feed slurry as high as possible is desired. Some designs have a short conical section at the feed end for prethickening within the unit, but generally it is preferable to thicken ahead of the centrifuge with gravity settlers, hydrocyclones, or inclined screens.

As depicted schematically in Fig. 18-174, the rotating assembly consists of a belt-driven outer rotor that rotates at constant speed. The outer shaft (hollow shaft) is fixed to the main or outer basket. Within the hollow shaft is the pusher shaft which is keyed together with the hollow shaft but also oscillates. The reciprocal motion is provided by a mechanical gearbox for smaller units (400-mm-diameter and less) or hydraulically in larger units. The depicted schematic is of a two-stage design in which the pusher shaft is fixed to the inner basket and the pusher plate is attached to the outer basket by posts. The stroke length is between 30 and 85 mm depending on machine size, and stroke frequency is usually between 45 and 90 strokes per minute.

The feed slurry enters through a stationary central pipe into a feed accelerator/distributor, then is introduced onto the (in this case) oscillating inner basket just in front of the pusher plate. In the feed zone, most of the liquid is drained, forming a cake sufficiently stiff to transfer the push force through the bed of solids and transport the cake without shearing. This is why it requires fast-draining materials and is liquid-limited, since it must form a cake within the period of one stroke.

With designs that use a simple feed cone or plate for feed distribution, most of the slurry acceleration takes place on the screen surface, with lower effective slurry speed and driving force for filtration. More advanced designs utilize an impeller-type feed accelerator that largely preaccelerates the slurry prior to introduction on the screen for higher capacity and lower screen wear.

With each stroke of the pusher, the material in the feed zone is pushed up to a certain height primarily depending on the friction coefficient between the solids and the screen and the screen deck length, and secondarily depending on *G*-force and loading. Once the cake in the feed zone is compressed and has formed a ring with this height, it transmits the push force to the stationary bed of cake in the basket which begins to move the cake bed forward until the forward end of the stroke. This cake height is often referred to as the natural cake height. The schematic in Fig. 18-175 shows what is taking place in the feed zone.

The distance the cake ring moves forward divided by the stroke length is defined as the *push efficiency*. The push efficiency varies with solids volumetric loading, resulting in a self-compensating control of varying rates. Depending on the cake properties, primarily compressibility, up to about 90 percent push efficiency is achievable. In some cases volumetric throughput can be further increased beyond the volumetric push capacity at the natural cake height, in which case the push efficiency remains almost constant and the cake height increases with increasing load, commonly referred to as the forced cake height. This realm of operation is usually only possible with multistage designs.

As the cake bed is transported through the basket, it passes through the various process steps shown in Fig. 18-176 with product moisture gradient as shown in Fig. 18-177.

Usually, cake wash ratios of about 0.1 to 0.3 kg wash/kg solids are possible within the normal residence time of the wash zone. This usually can displace at least 95 percent of the mother liquor and impurities. In some cases higher wash ratios or even multistage countercurrent washes are utilized, in which case sufficient residence time via throughput reduction must be considered.

Single-Stage versus Multistage Pusher centrifuges can be single-stage configuration with a single long basket and screen, two-stage (as shown schematically in Fig. 18-177), three-, or four-stage designs. Cake height and push force are primarily influenced by screen deck length and cake friction coefficient.

Single-Stage Where single-stage units are appropriate (ammonium sulfate is one example due to very large crystal size and good cake shear strength), the solids volumetric capacity can be maximized. However, because the push force requirement increases with screen length, cake shear or buckling can be the result with unstable operation. Because the average cake thickness in the feed zone is higher, filtration capacity may be slightly less than with multistage units. Fines losses can be slightly less with single-stage units since a smaller proportion of the cake bed is in contact with the slotted screen and there is no reorientation of crystals between stages. These units are often limited to low-speed operation for stability.

Two-Stage The majority of pusher centrifuges sold today are of this type. It provides greater flexibility compared to single stages in terms of greater filtration capacity, lower tendency for cake shear, and higher speed capability. When it is possible to operate with a forced cake, capacities can approach those of single-stage designs. Wash typically is applied on the latter portion of the first stage and through the

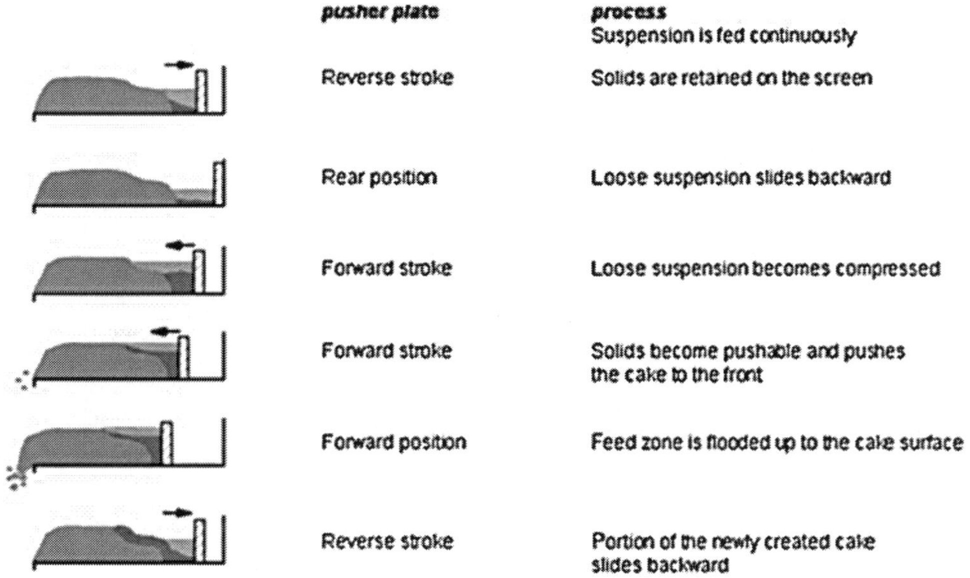

FIG. 18-175 Pusher centrifuge solids transport. (*Krauss Maffei Process Technology.*)

transition onto the second stage. During this transition the crystals are reoriented and the capillaries opened, which can enhance the wash effect. With even stage units, the feed acceleration system is not oscillating relative to the feed pipe. Some advanced designs of feed acceleration systems incorporating impellers benefit from this constant relationship.

Three- and Four-Stage These designs are generally reserved for the largest sizes that have long baskets that need to be subdivided into reasonable-length stages as well as for very special applications with very high friction coefficients, low internal cake shear strength, or fairly high compressibility. For example, in processing high rubber ABS, four-stage units have been utilized; but the deck lengths are so short, with the corresponding thin cakes and short retention time, that capacity and performance are severely reduced. Other types of machine (such as peeler centrifuges or cylindrical conical pushers discussed below) can be better suited.

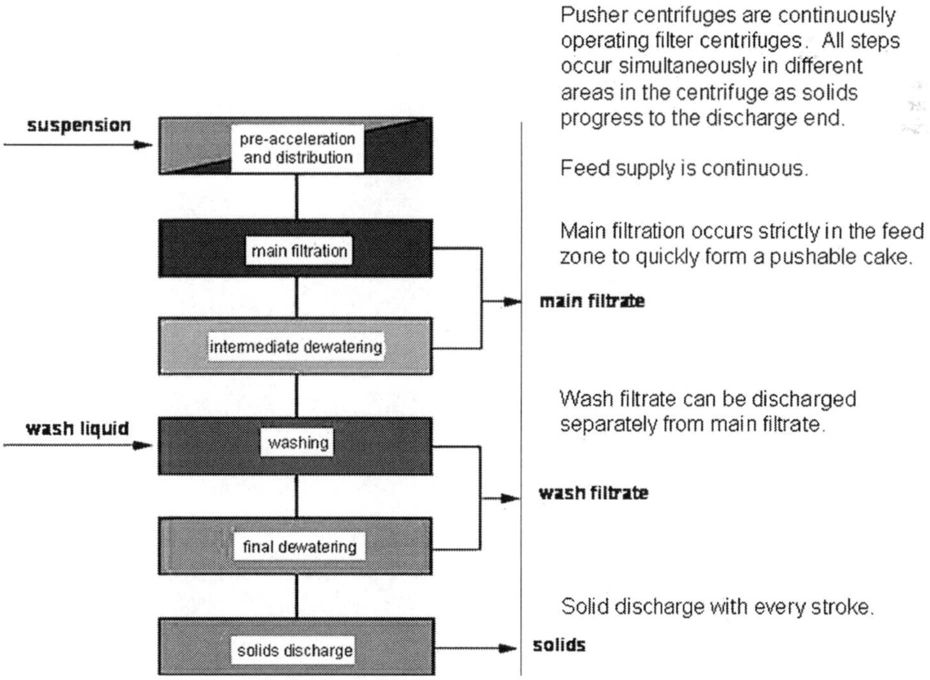

FIG. 18-176 Pusher centrifuge process steps. (*Krauss Maffei Process Technology.*)

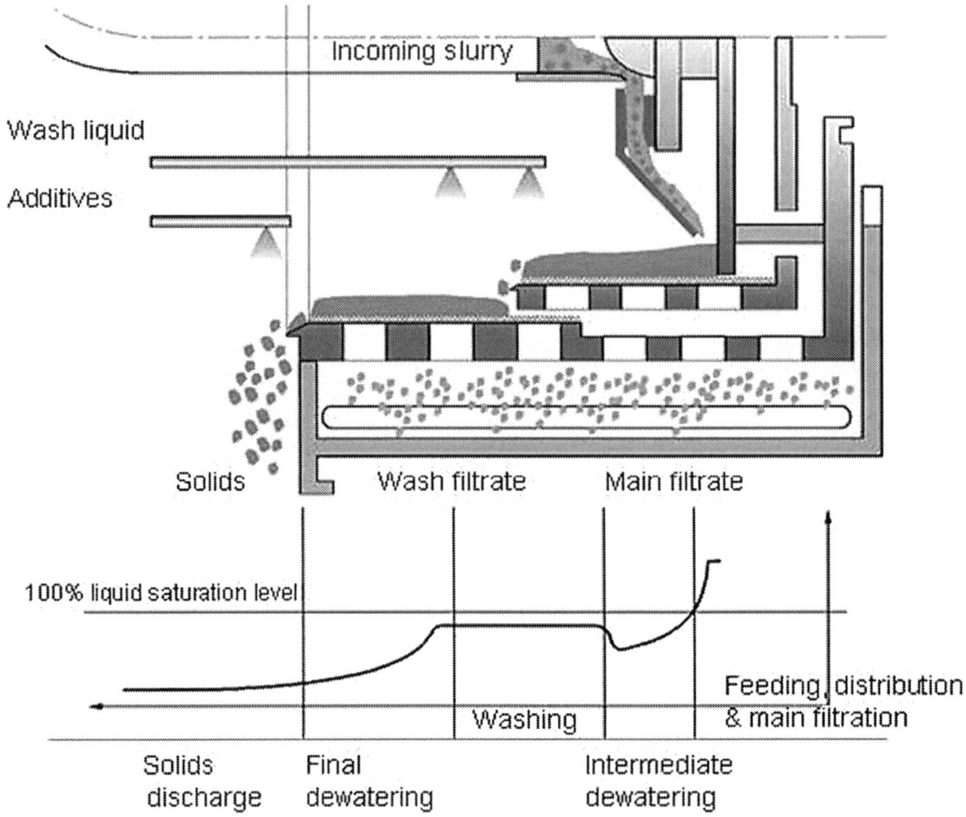

FIG. 18-177 Pusher centrifuge product moisture gradient. (*Krauss Maffei Process Technology.*)

Cylindrical/Conical A variation of single- and two-stage designs utilizes a cylindrical section or stage at the feed end followed by a conical section or stage sloping outward to the discharge end. The benefit of this design is that the axial component of force in the conical end assists with solids transport. Care must be taken that the cone angle not exceed the sliding friction angle of the cake, or else the cake will short-circuit the zone, resulting in poor performance and high vibration. Fabrication costs of the baskets are higher than those of cylindrical designs, and slotted screen construction is complicated with high replacement costs.

Theory of Centrifugal Filtration Theoretical predictions of the behavior of solid-liquid mixtures in a filtering centrifuge are more difficult compared to pressure and gravity filtration. The area of flow and driving force are both proportional to the radius, and the specific resistance and porosity may also change markedly within the cake. Filtering centrifuges are nearly always selected by scale-up from lab tests on materials to be processed, such as using bucket centrifuges where a wide range of test conditions (cake thickness, time, and G-force) can be controlled. Although tests with the bucket centrifuge provide some quantitative data to scale-up, the results include wall effect from buckets, which are not representative of actual cylindrical basket geometry, bucket centrifuges are not useful in quantifying filtration rates. A modified version of the buckets or even a cylindrical perforated basket can be used. In the latter, there is less control of cake depth and circumferential uniformity. The desired quantities to measure are filtration rate, washing rate, spinning time, and residual moisture. Also, with filtering centrifuges such as the screen-bowl centrifuge, screen-scroll centrifuge, and to some extent in multistage pushers, the cake is constantly disturbed by the scroll conveyor or conveyance mechanism; liquid saturation due to capillary rise as measured in bucket tests is absent. While bucket centrifuge tests are very useful for first-look feasibility, it is always recommended to follow with pilot-scale testing of the actual equipment type being considered.

Filtration Rate When the centrifuge cake is submerged in a pool of liquid, as in the case of a fast-sedimenting, solids-forming cake almost instantly, and the rate of filtration becomes limiting, the bulk filtration rate Q for a basket with axial length b is:

$$Q = \frac{\pi b \rho K \Omega^2 (r_b^2 - r_p^2)}{\mu \left(\ln \left[\frac{r_b}{r_c} \right] + \frac{K R_m}{r_b} \right)} \qquad (18\text{-}119a)$$

where μ and ρ are, respectively, the viscosity and density of the liquid; Ω is the angular speed; K is the average permeability of the cake and is related to the specific resistance α by the relationship $\alpha K \rho_s = 1$, with ρ_s being the solids density; r_p, r_c, and r_b are, respectively, the radius of the liquid pool surface, the cake surface, and the filter medium adjacent to the perforated bowl. Here, the pressure drop across the filter medium, which also includes that from the cake heel, is $\Delta p_m = \mu R_m(Q/A)$ with R_m being the combined resistance. The permeability K has a unit m^2, α m/kg, and R_m m^{-1}. The driving force is due to the hydrostatic pressure difference across the bowl wall and the pool surface—i.e., the numerator of Eq. (18-119a), and the resistance is due to the cake layer and the filter medium—i.e., denominator of Eq. (18-119a). Fig. 18-178 shows the pressure distribution in the cake and the liquid layer above. The pressure (gauge) rises from zero to a maximum at the cake surface; thereafter, it drops monotonically within the cake in overcoming resistance to flow. There is a further pressure drop across the filter medium, the magnitude dependent on the combined resistance of the medium and the heel at a given flow rate. This

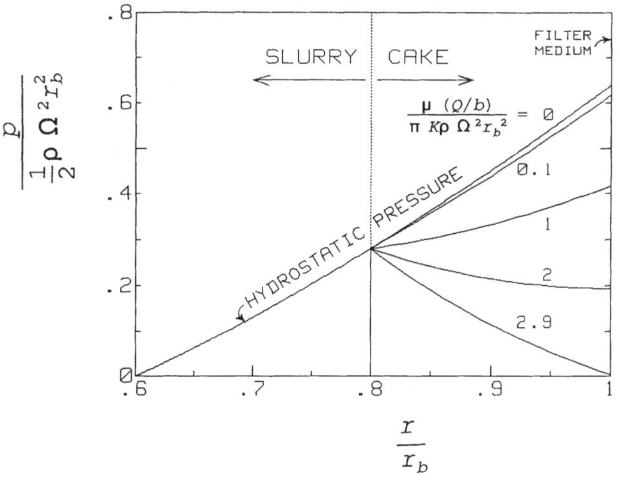

FIG. 18-178 Pressure distribution in a basket centrifuge under bulk filtration.

scenario holds, in general, for incompressible as well as for compressible cake. For the latter, the pressure distribution also depends on the compressibility of the cake.

For incompressible cake, the pressure distribution and the rate depend on the resistance of the filter medium and the permeability of the cake. Figure 18-178 shows several possible pressure profiles in the cake with increasing filtration rates through the cake. It is assumed that $r_c/r_b = 0.8$ and $r_p/r_b = 0.6$. The pressure at $r = r_b$ corresponds to pressure drop across the filter medium Δp_m with the ambient pressure taken to be zero. The filtration rate as well as the pressure distribution depend on the medium resistance and that of the cake. High medium resistance or blinding of the medium results in greater penalty on filtration rate.

In most filtering centrifuges, especially the continuous-feed ones, the liquid pool above the cake surface should be minimum to avoid liquid running over the cake. Setting $r_p = r_c$ in Eq. (18-119*a*), the dimensionless filtration flux is plotted in Fig. 18-179 against r_c/r_b for different ratios of filter-medium resistance to cake resistance, KR_m/r_b. For negligible medium resistance, the flux is a monotonic decreasing function with increasing cake thickness, i.e., smaller r_c. With finite medium resistance, the flux curve for a range of different cake thicknesses has a

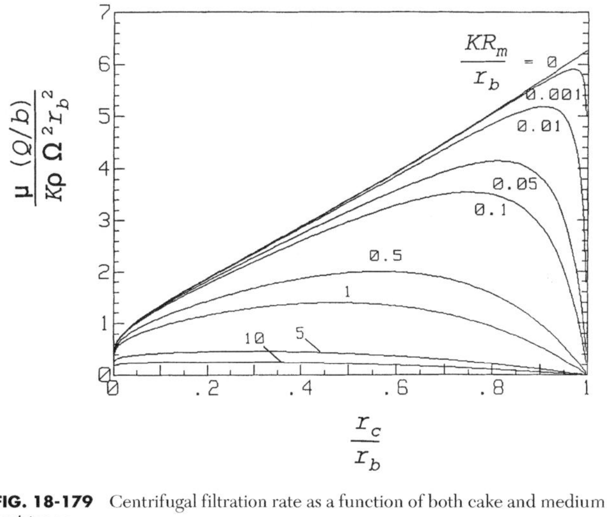

FIG. 18-179 Centrifugal filtration rate as a function of both cake and medium resistance.

maximum. This is because for thin cake the driving liquid head is small and the medium resistance plays a dominating role, resulting in lower flux. For very thick cake, despite the increased driving liquid head, the resistance of the cake becomes dominant; therefore, the flux decreases again. The medium resistance to cake resistance should be small, with $KR_m/r_b < 5$ percent. However, the cake thickness, which is directly proportional to the throughput, should not be too small, despite the fact that the machine may have to operate at somewhat less than the maximum flux condition.

It is known that the specific resistance for centrifuge cake, especially for compressible cake, is greater than that of the pressure or vacuum filter. Therefore, the specific resistance has to be measured from centrifuge tests for different cake thicknesses so as to scale up accurately for centrifuge performance. It cannot be extrapolated from pressure and vacuum filtration data. For cake thickness that is much smaller compared to the basket radius, Eq. (18-119*a*) can be approximated by

$$V_f = V_{fo}\left(\frac{h}{h_c}\right) \qquad (18\text{-}119b)$$

where $h = r_b - r_p$ is the liquid depth, $h_c = r_b - r_c$ is the cake thickness, and $v_{fo} = (\rho GK/\mu)$ is a characteristic filtration velocity. Table 18-14 shows some common filtering centrifuges and the application with respect to the G-level, minimum feed-solids concentration, minimum mean particle size, and typical filtration velocity. The vibratory and tumbler centrifuges have the largest filtration rate of 5×10^{-4} m/s (0.02 in/s) for processing 200-μm or larger particles, whereas the pendulum has the lowest filtration rate of 1×10^{-7} m/s (4×10^{-6} in/s) for processing 5-μm particles with increased cycle time. The screen-scroll, pusher, screen-bowl, and peeler centrifuges are in between.

Film Drainage and Residual Moisture Content Desaturation of the liquid cake ($S < 1$) begins as the bulk filtration ends, at which point the liquid level starts to recede below the cake surface. Liquids are trapped in: (1) cake pores between particles that can be drained with time (free liquid), (2) particle contact points (pendular liquid), (3) fine pores forming continuous capillaries (capillary rise), (4) particle pores or they are bound by particles (bound liquid). Numbers (1) to (3) can be removed by centrifugation, and, as such, each of these components depends on G to a different extent. Only desaturation of the free liquid, and to a much lesser entent the liquid at contact points, is a function of time. The wet cake starts from a state of being fully saturated, $S = 1$, to a point where $S < 1$, depending on the dewatering time. At a very large amount of time, it approaches an equilibrium point S_∞, which is a function of G, capillary force, and the amount of bound liquid trapped inside or externally attached to the particles.

The following equations, which have been tested in centrifugal dewatering of granular solids, prove useful:

Total saturation:

$$S_{\text{total}} = S_\infty + S_T(t) \qquad (18\text{-}120)$$

Equilibrium component:

$$S_\infty = S_c + (1 - S_c)(S_p + S_z) \qquad (18\text{-}121)$$

Transient component:

$$S_T(t) = (1 - S_c)(1 - S_p - S_z)S_t(t) \qquad (18\text{-}122)$$

The details of the mathematical model of these four components are given below.

Drainage of free liquid in thin film:

$$S_t(t) = \left(\frac{4}{3}\right)\left(\frac{1}{t_d^n}\right), \quad t_d > 0 \qquad (18\text{-}123)$$

where for smooth-surface particles, $n = 0.5$, and for particles with rough surfaces, n can be as low as 0.25.

Bound liquid saturation:

$$S_p = \text{function(particle characteristics)} \qquad (18\text{-}124)$$

Pendular saturation:

$$S_z = 0.075, \quad N_c \le 5 \tag{18-125a}$$

$$S_z = \frac{5}{(40 + 6N_c)}, \quad 5 \le N_c \le 10 \tag{18-125b}$$

$$S = \frac{0.5}{N_c}, \quad N_c \ge 10 \tag{18-125c}$$

Frequently when $N_c < 10$, S_p and S_z are combined for convenience, the sum of which is typically 0.075 for smooth particles and can be as high as 0.35 for rough-surface particles. This has to be determined from tests.

Saturation due to capillary rise:

$$S_c = \frac{4}{B_o} \tag{18-126}$$

where the dimensionless time t_d, capillary number N_c, and Bond number B_o are, respectively:

$$t_d = \frac{\rho\, G d_h^2 t}{\mu\, H} \tag{18-127}$$

$$N_c = \frac{\rho\, G d_h^2}{\sigma \cos\theta} \tag{18-128}$$

$$B_o = \frac{\rho\, G H d_h}{\sigma \cos\theta} \tag{18-129}$$

where ρ and μ are, respectively, the density and viscosity of the liquid; θ is the wetting angle of the liquid on the solid particles; σ is the interfacial tension; H is the cake height; d is the mean particle size; and t is the dewatering time. The hydraulic diameter of the particles can be approximated by either $d_h = 0.667\, \varepsilon d/(1 - \varepsilon)$ or $d_h = 7.2(1 - \varepsilon)K^{1/2}/\varepsilon^{3/2}$, where ε is the cake porosity.

Example
Given: ρ = 1000 kg/m^3, ρ_s = 1200 kg/m^3, μ = 0.004 N·s/m^2, $\sigma \cos\theta$ = 0.068 N/m, H = 0.0254 m, d = 0.0001 m, ε = 0.4, G/g = 2000, t = 2 s, S_p = 0.03.

Calculate: $d_h = 4.4 \times 10^{-5}$ m, t_d = 748, N_c = 0.56, B_o = 322, S_t = 0.048, S_z = 0.075, S_c = 0.012, S_∞ = 0.116, S_T = 0.043, S_{total} = 0.158, W = 0.919.

Note that W is the solids fraction by weight and is determined indirectly from Eq. (18-89). The moisture weight fraction is 0.081.

The transient component depends not only on G, cake height, and cake properties, but also on dewatering time, which ties to solids throughput for a continuous centrifuge and cycle time for batch centrifuge. If the throughput is too high or the dewatering cycle is too short, the liquid saturation can be high and becomes limiting. Given that time is not the limiting factor, dewatering of the liquid lens at particle contact points requires a much higher G-force. The residual saturation depends on the G-force to the capillary force, as measured by N_c, the maximum of which is about 7.5%, which is quite significant. If the cake is not disturbed (scrolled and tumbled) during conveyance and dewatering, liquid can be further trapped in fine capillaries due to liquid rise, the amount of which is a function of B_o, which weighs the G-force to the capillary force. This amount of liquid saturation is usually smaller as compared to capillary force associated with liquid-lens (also known as pendular) saturation. Lastly, liquid can be trapped by chemical force at the particle surface or physical capillary or interfacial force in the pores within the particles. Because the required desaturating force is extremely high, this portion of moisture cannot be removed by mechanical centrifugation. Fortunately, for most applications it is a small percentage, if it exists.

SELECTION OF CENTRIFUGES

Table 18-15 summarizes the several types of commercial centrifuges, their manner of liquid and solids discharge, their unloading speed, and their relative volumetric capacity. When either the liquid or the solids discharge is not continuous, the operation is said to be cyclic. Cyclic or batch centrifuges are often used in continuous processes by providing appropriate upstream and downstream surge capacity.

Sedimentation Centrifuges These centrifuges frequently are selected on the basis of tests on tubular, disc, or helical-conveyor centrifuges of small size. The centrifuge should be of a configuration similar to that of the commercial centrifuge it is proposed to be used for. The results in terms of capacity for a given performance (effluent clarity and solids concentration) may be scaled up by using the sigma concept of Eqs. (18-117) to (18-119). Spin-tube tests may be used for information on systems containing well-dispersed solids. Such tests are totally unreliable on systems containing a dispersed phase that agglomerates or flocculates during the time of centrifugation.

Filtering Centrifuges These filters often can be selected on the basis of batch tests on a laboratory unit, preferably one at least 12 in (305 mm) in diameter. A bucket centrifuge test would be helpful to study the effect of G, cake height, and dewatering time, but not filtration rates. It is always recommended to follow bucket tests with pilot-scale testing of the actual equipment type being considered. Caution has to be taken in correcting for capillary saturation, which may be absent in large continuous centrifuges with scrolling conveyances.

Unless operating data on similar material are available from other sources, continuous centrifuges should be selected and sized only after tests on a centrifuge of identical configuration.

It seems needless to state but is frequently overlooked that test results are valid only to the extent that the slurry and the test conditions duplicate what will exist in the operating plant. This may involve testing on a small scale (or even on a large one) with a slipstream from an existing unit, but the dependability of the data is often worth the extra effort involved. Most centrifuge manufacturers provide testing services and demonstration facilities in their own plants and maintain a supply of equipment for field-testing in the customer's plant, such as with a pilot centrifuge module with associated peripheral equipment. Larger-scale pilot equipment provides better scale-up accuracy, e.g., in evaluating the effect of cake thickness in batch filtering centrifuges.

COSTS

Neither the investment cost nor the operating cost of a centrifuge can be directly correlated with any single characteristic of a given type of centrifuge. The costs depend on the features of the centrifuge tailored toward the physical and chemical nature of the materials being separated, the degree and difficulty of separation, the flexibility and capability of the centrifuge and its auxiliary equipment, the environment in which the centrifuge is located, and many other nontechnical factors, including market competition. The cost figures presented herewith represent centrifuges only for use in the process industries as of 2004. In any particular installations, the costs may be somewhat less or much greater than those presented here. The prices presented herewith are for rough guidance only. Substantial variations will be found due to volatility of currency exchange and material costs.

The useful parameter for value analysis is the installed cost of the number of centrifuges required to produce the demanded separative effect (end product) at the specified capacity of the plant. The possible benefits of adjustments in the upstream and downstream components of the plant and the process should be carefully examined in order to minimize the total overall plant costs; the systems approach should be used.

Purchase Price Typical purchase prices, including drive motors, of tubular and disc sedimenting centrifuges are given in Table 18-16. The price will vary upward with the use of more exotic materials of construction, the need for explosion-proof electrical gear, the type of enclosure required for vapor containment, and the degree of portability, and this holds for all types of centrifuges.

The average purchase prices of continuous-feed, solid-bowl centrifuges made, respectively, of 316 stainless steel and steel are shown in

TABLE 18-15 Characteristics of Commercial Centrifuges

Method of separation	Rotor type	Centrifuge type	Manner of liquid discharge	Manner of solids discharge or removal	Centrifuge speed for solids discharge	Capacity*
Sedimentation	Batch	Ultracentrifuge				1 mL
		Laboratory, clinical	Batch	Batch manual	Zero	To 6 L
	Tubular	Supercentrifuge	Continuous†	Batch manual	Zero	To 1,200 gal/h
		Multipass clarifier	Continuous†	Batch manual	Zero	To 3,000 gal/h
	Disc	Solid wall	Continuous†	Batch manual	Zero	To 30,000 gal/h
		Light-phase skimmer	Continuous	Continuous for light-phase solids	Full	To 1,200 gal/h
		Peripheral nozzles	Continuous	Continuous	Full	To 24,000 gal/h
		Peripheral valves	Continuous	Intermittent	Full	To 3,000 gal/h
		Peripheral annulus	Continuous	Intermittent	Full	To 12,000 gal/h
	Solid bowl	Constant-speed horizontal	Continuous†	Cyclic	Full (usually)	To 60 ft^3
	(scroll	Variable-speed vertical	Continuous†	Cyclic	Zero or reduced	To 16 ft^3
	conveyor)	Continuous decanter	Continuous	Continuous screw conveyor	Full	To 54,000 gal/h To 100 tons/h solids
Sedimentation and filtration		Screen-bowl decanter	Continuous	Continuous	Full	To 60,000 gal/h To 125 tons/h solids
Filtration	Conical screen	Wide-angle screen	Continuous	Continuous	Full	To 40 tons/h solids
		Differential conveyor	Continuous	Continuous	Full	To 80 tons/h solids
		Vibrating, oscillating, and tumbling screens	Continuous	Essentially continuous	Full	To 250 tons/h solids
		Reciprocating pusher	Continuous	Essentially continuous	Full	Limited data
	Cylindrical screen	Reciprocating pusher, single and multistage	Continuous	Essentially continuous	Full	To 100 tons/h solids
		Horizontal	Cyclic	Intermittent, automatic	Full (usually)	To 25 tons/h solids
		Vertical, underdriven	Cyclic	Intermittent, automatic, or manual	Zero or reduced	To 10 tons/h solids
		Vertical, suspended	Cyclic	Intermittent, automatic, or manual	Zero or reduced	To 10 tons/h solids

*To convert gallons per hour to liters per second, multiply by 0.00105; to convert tons per hour to kilograms per second, multiply by 0.253; and to convert cubic feet to cubic meters, multiply by 0.0283.

†Feed and liquid discharge interrupted while solids are unloaded.

Fig. 18-180. The average purchase prices of continuous-feed filtering centrifuges are shown in Fig. 18-181. This chart includes a comparison of prices on screen-bowl, pusher, screen-scroll, and oscillating conical baskets. On average, the screen bowl is approximately 10 percent higher in price than the solid bowl of the same diameter and length. This incremental cost results from the added complexity of the screen section, bowl configuration, and casing differences. Prices for both the solid-bowl and the filtering centrifuges do not include the drive motor, which typically adds another 5 to 25 percent to the cost. The higher end of this range represents a variable-speed-type drive. If a variable-speed backdrive is used instead of the gear unit, the incremental cost is about another 10 to 15 percent, depending on the capability.

The average prices of the batch centrifuge are shown in Fig. 18-182. All the models include the drive motor and control. In Fig. 18-182, the inverting filter, horizontal peeler, and the advanced vertical peeler are the premium baskets especially used for specialty chemicals and pharmaceuticals. Control versatility with the use of programmable logic control (PLC), automation, and cake-heel removal are the key features which are responsible for the higher price. The underdriven, top-driven, and pendulum baskets are less expensive with fewer features.

Installation Costs Installation costs of centrifuges vary over an extremely wide range, depending on the type of centrifuge, on the area and kind of structure in which it is installed, and on the details of installation. Some centrifuges, such as portable tubular and disc oil purifiers, are shipped as package units and require no foundation and a minimum of connecting piping and electrical wiring. Others, such as large batch automatic and continuous scroll-type centrifuges, may require substantial foundations and even building reinforcement, extensive interconnecting piping with required flexibility, auxiliary feed and discharge tanks and pumps and other facilities, and elaborate electrical and process-control equipment. Minimum installation costs, covering a simple foundation and minimum piping and wiring, are about 5 to 10 percent of purchase price for tubular and disc centrifuges;

TABLE 18-16 Typical Purchase Prices, Including Drive Motors, of Tubular and Disc Sedimenting Centrifuges, 2004

Type	Bowl diameter, in (mm)	Approximate Σ value, units of 10^4 ft^2 (10^3 m^2)	Designation	Purchase price, 2004 $
Tubular	4 (102)	2.7 (2.5)	Oil purifier	60,000–80,000
	4 (102)	2.7 (2.5)	Chemical separation	60,000–80,000
	5 (127)	4.2 (3.9)	Blood fractionation	100,000–140,000
Manual discharge disc	13.5 (343)	21 (20)	Hermetic	100,000–130,000
	24 (610)	95 (88)	Centripetal pump	150,000–300,000
Continuous nozzle-discharge disc	12 (305)	12 (11)	Clarifier	100,000–130,000
	18 (457)	25 (23)	Separator	150,000–200,000
	30 (762)	100 (93)	Recycle clarifier	270,000–300,000
Self-cleaning disc	14 (356)	13 (12)	Centripetal pump	130,000–150,000
	18 (457)	22 (20)	Centripetal pump	170,000–200,000
	24 (610)	38 (35)	Centripetal pump	250,000–300,000

*NOTE: All prices quoted are for stainless steel construction with the exception of the oil purifier noted.

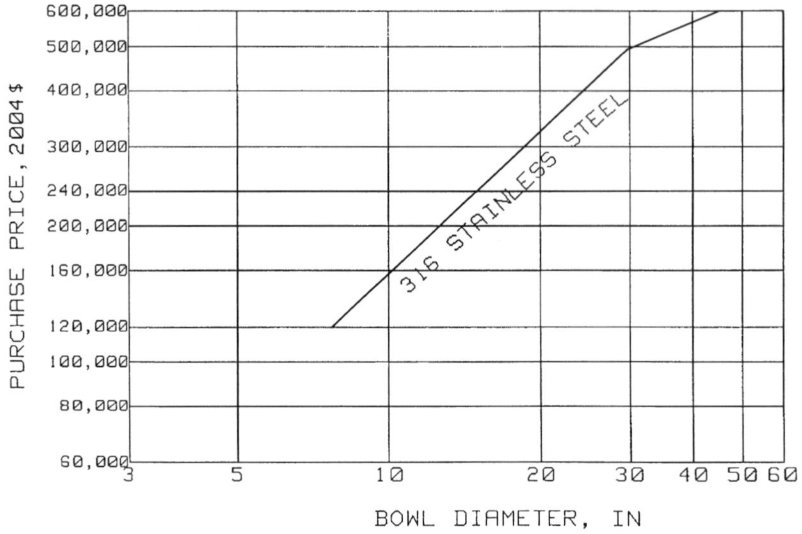

FIG. 18-180 Costs of continuous-feed solid-bowl.

10 to 25 percent for bottom drive, batch automatic, and continuous-scroll centrifuges; and up to 30 percent for top-suspended basket centrifuges. If the cost of all auxiliaries—special foundations, tanks, pumps, conveyors, electrical and control equipment, etc.—is included, the installation cost may well range from 1 to 2 times the purchase price of the centrifuge itself.

Maintenance Costs Because of the care with which centrifuges are designed and built, their maintenance costs are in line with those of other slower-speed separation equipment, averaging in the range of 1 to 4 percent for batch machines, 3 to 8 percent for pusher centrifuges, and 5 to 10 percent for decanters and disc centrifuges per year of the purchase price for centrifuges in light to moderate duty. For centrifuges in severe service and on highly corrosive fluids, the maintenance cost may be several times this value. Maintenance costs are likely to vary from year to year, with lower costs for general maintenance and periodic large expenses for major overhaul. Centrifuges are subject to erosion from abrasive solids such as sand, minerals, and grits. When these solids are present in the feed, the centrifuge components are subject to wear. Feed and solids discharge ports, unloader knives, helical scroll blade tips, etc., should be protected with replaceable wear-resistant materials. Excessive out-of-balance forces strongly contribute to maintenance requirements and should be avoided.

Operating Labor Centrifuges run the gamut from completely manual control to fully automated operation. For the former, one operator can run several centrifuges, depending on their type and the application. Fully automatic centrifuges usually require little direct operation attention. In most production environments, PLC- or DCS-based automatic controls are the norm.

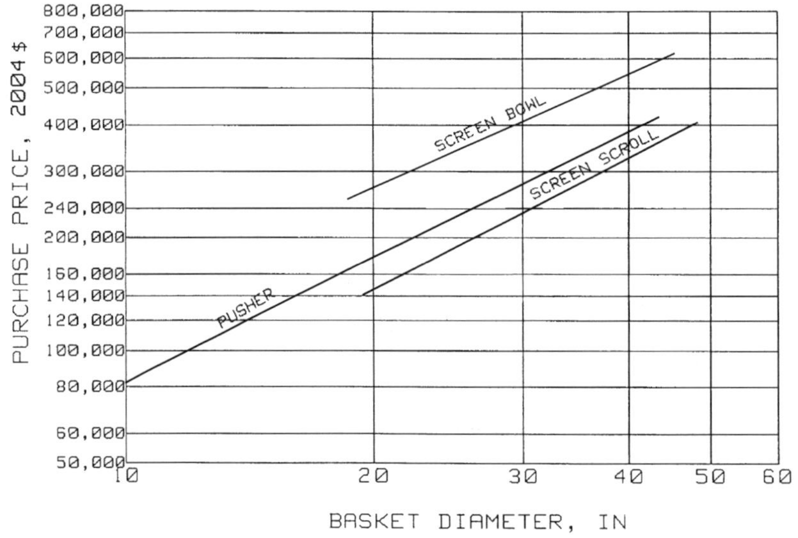

FIG. 18-181 Costs of continuous baskets (316 stainless steel).

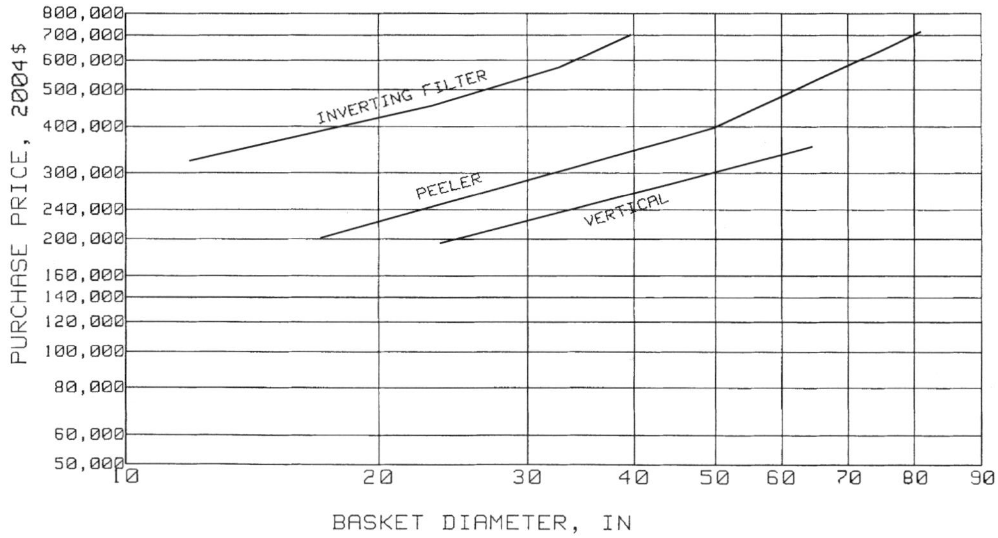

FIG. 18-182 Costs of batch baskets (316 stainless steel).

EXPRESSION

General References: F. M. Tiller and L. L. Horng, "Hydraulic Deliquoring of Compressible Filter Cakes," *AIChE J.*, **29** (2) (1983). F. M. Tiller and C. S. Yeh, The Role of Porosity in Filtration VI: Filtration Followed by Expression," *AIChE J.*, **33** (1987). F. M. Tiller and W. Li, "Dangers of Lab-Plant Scaleup for Solid/Liquid Separation Systems," *Chem. Eng. Commun.*, **190** (1) (2003). F. M. Tiller and T. C. Green, "Role of Porosity in Filtration IX: Skin Effect with Highly Compressible Materials," *AIChE J.*, **19** (1973). F. M. Tiller and W. Li, "Determination of the Critical Pressure Drop for Filtration of Supercompactible Cakes," *Water Sci. and Technol.*, **44** (10) (2001). M. Shirato, T. Murase, and T. Aragaki, "Slurry Deliquoring by Expression," *Progress in Filtration*, vol. 4, Elsevier, 1986. M. Shirato et al., "Internal Flow Mechanism in Filter Cakes," *AIChE J.*, **15** (1969). M. Shirato et al., "Analysis of Consolidation Process in Filter Cake Dewatering by Use of Difficult to Filter Slurries," *J. Chem. Eng. Japan*, **19** (6) (1986). F. M. Tiller and W. F. Leu, "Basic Data Fitting in Filtration," *J. Chinese Inst. Chem. Engr.*, **11** (1980). W. Chen, F. J. Parma and W. Schabel, "Testing Methods for Belt Press Biosludge Dewatering," *Filtration J.*, **5** (1) (2005).

FUNDAMENTALS OF EXPRESSION

Definition Deliquoring of filter cakes is one of the last stages of solid-liquid separations. It has been widely applied in a variety of fields, e.g., in food industries to increase product yield, in wastewater treatment plants to reduce transportation and disposal cost by decreasing sewage sludge moisture content, and in chemical processes to eliminate liquid content in the solid product prior to drying. The energy required to express liquid from solid-liquid mixtures is negligible compared to that of any thermal method. Deliquoring operations include hydraulic expression, mechanical expression, air or gas blowing, and gravity or centrifugal drainage. Hydraulic expression is provided by direct pump pressure or reversed or right-angled flow of liquid at the end of filtration (Tiller and Horng, 1983). The term *expression* used in this presentation refers to mechanical compression of a solid-liquid mixture by applying diaphragms, rolls, pistons, or screw presses on the surface of cakes.

Filtration and Expression of Compactible Filter Cakes

Filtration A filter cake can be incompactible, moderately compactible, highly compactible, or supercompactible (Tiller and Li, 2003). Tiller and Green (1973) showed that porosity or solidosity (volume fraction of cake solids ε_s, solidosity + porosity = 1) is not uniformly distributed in a compactible cake, and a skin cake of high resistance to liquid flow is developed next to the filter medium, as shown in Fig. 18-183 (Tiller and Li, 2001). The skin deters frictional forces necessary to consolidate the cake and increase solidosity in a large portion of the cake. As a result, as illustrated by Fig. 18-184, increasing filtration pressure on highly compactible filter cakes cannot attain substantial deliquoring (flocculated latex) while increasing filtration pressure does help to make a dryer cake on a less compactible material (Kaolin Flat D).

Expression Mechanical expression applies pressure directly on filter cakes rather than relying on flow frictions generated by hydraulic pressure drop to deliquor the cake. The effects of stress distribution in a compactible filter cake by these two different mechanisms are shown in Fig. 18-185. The stress distribution of an expression is more uniform than that of a pressure filtration, leading to a more uniform filter cake. Expression is therefore a better choice for deliquoring of compactible filter cakes.

Fundamental Theory A theoretical model was developed by Shirato (1969, 1986) based on Terzaghi's and Voigt's consolidation model in

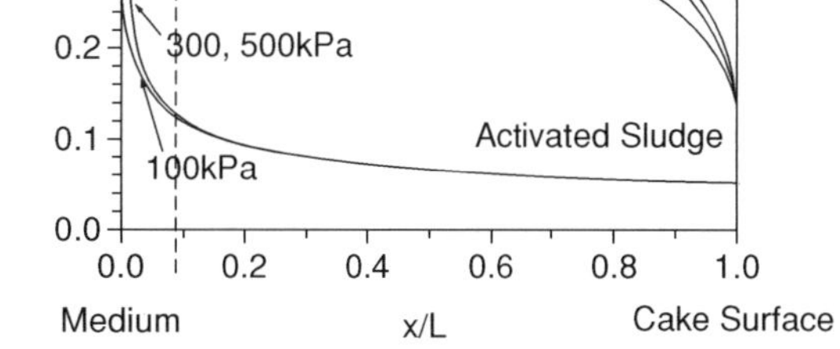

FIG. 18-183 Solidosity ε_s variations as a function of fractional distance throughout filter cake thicknesses.

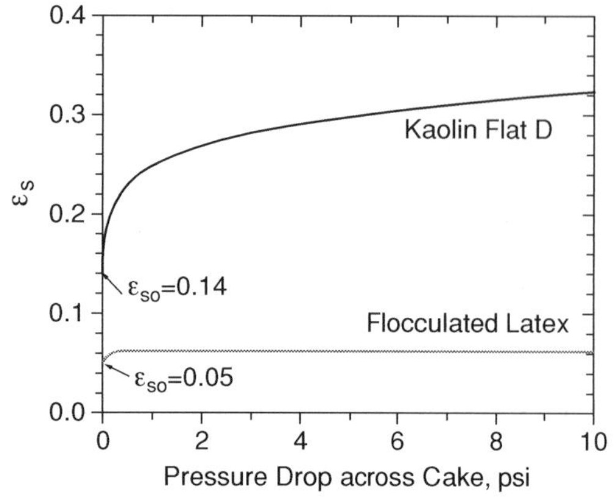

FIG. 18-184 Effect of filtration pressure on average solidosity ε_s.

soil mechanics. Shirato's expression theory includes a filtration stage followed by a consolidation. Average consolidation ratio U_c is given as a function of consolidation time θ_c and other characteristic parameters of an expression process including true solids density, liquid density, liquid viscosity, specific resistance (or permeability) versus pressure, porosity versus pressure, and frictional stress on solids throughout cake thickness versus applied pressure (Shirato et al., 1986). The relationships of specific resistance, and porosity versus pressure, and local frictional stress on solids throughout cake thickness during the primary consolidation stage are given by empirical constitutive equations (Tiller and Leu, 1980), and can be determined by a compression-permeability cell test (Tiller, 1977, 1980), as shown in Fig. 18-186.

Factors Affecting Expression Operations Based on fundamental theory, variables affecting expression include characteristics of suspending particles, properties of liquid, properties of filter cake, and expression operation conditions as summarized in Fig. 18-187. Expression efficiency is determined by the properties of the filter cake, which very much depend on characteristics of the suspending particle, properties of liquid, and operation conditions. Interrelationships of the above parameters are described by empirical equations covering restrictive ranges.

EXPRESSION EQUIPMENT

This type of equipment uses mechanical expression rather than pump pressure for cake compression. Dryer cakes and faster cycle rate can

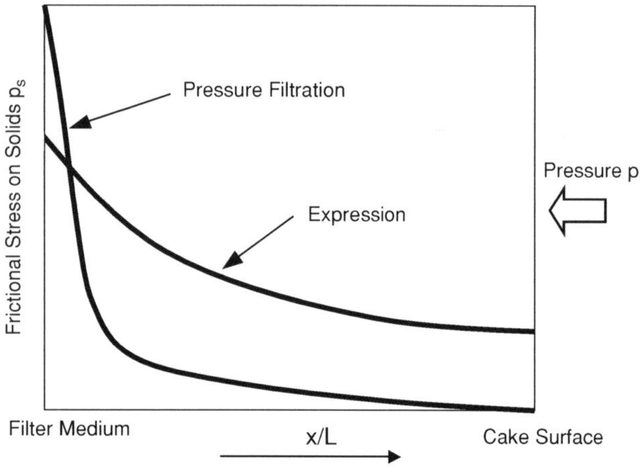

FIG. 18-185 Comparisons of frictional stress distributions in expression and pressure filtration.

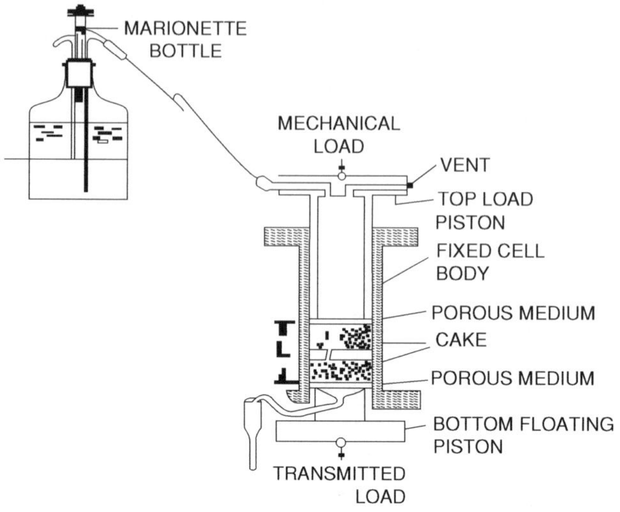

FIG. 18-186 Compression-permeability (C-P) cell.

be achieved compared to pressure filters. Low- to high-pressure (up to 2000 psi) units are available for expression equipment. They can be divided into two categories, batch expression equipment, which allows higher compression pressure and has lower slurry handling capacity, and continuous expression equipment, which uses lower compression pressure but offers higher slurry handling capacities.

Batch Expression Equipment In batch expression equipment, the cake is initially formed by pressure filtration just as in other pressure filters. After the filtration stage, a squeezing device such as a diaphragm is inflated with gas or liquid to compress the cake. Batch expression equipment allows longer compression time and higher compression pressure. The cake can be very dry.

Diaphragm Presses Diaphragm presses, also called membrane presses, are derived from filter presses, which were described in the pressure filtration section. In a diaphragm press, a diaphragm (Fig. 18-188*a*) is attached to the recessed chamber plate. The operation of a diaphragm press is the same as that of a chamber press during the filtration step. At the end of filtration, the diaphragm is inflated (Fig. 18-188*b*) to squeeze the filter cake to achieve the mechanical expression. After the squeezing, the diaphragm is deflated and the filter chamber opened to discharge the cake.

The diaphragm can be made of polypropylene or rubber, but polypropylene is most often used today. Both air and water can be used as the inflating medium for the diaphragm. As the inflating medium needs to be brought into the filter plates by hoses, a dangerous condition can exist if a hose is broken with air flowing in it. Therefore, hydraulic fluid (mostly water) is used to inflate the diaphragm to squeeze the cake. Air is only used occasionally in small pilot units.

As in filter presses, one disadvantage of the diaphragm press is the manual operation for filter cake discharge. With recent development, automatic cake discharge devices are available from most filter manufacturers. However, the reliability of an automatic cake discharge device needs to be verified by actual field operation. Normally, automatic cake discharge has a better chance of success in diaphragm presses than filter presses as the cakes are normally dryer in diaphragm presses.

The cake deliquoring is primarily done during the expression step so the cake formation period is normally carried out under low pressure and a high-pressure slurry pump is not necessary; it helps to reduce floc damage during pumping. The normal expression pressure used in a diaphragm press is 110 or 220 psi; in some designs pressure up to 800 psi can be used.

Diaphragm presses are superior to filter presses in deliquoring compactible cakes (such as biological sludge, pulps, or highly flocculated materials). As a diaphragm press is more expensive than a regular filter press, the use of a diaphragm press may not be advantageous

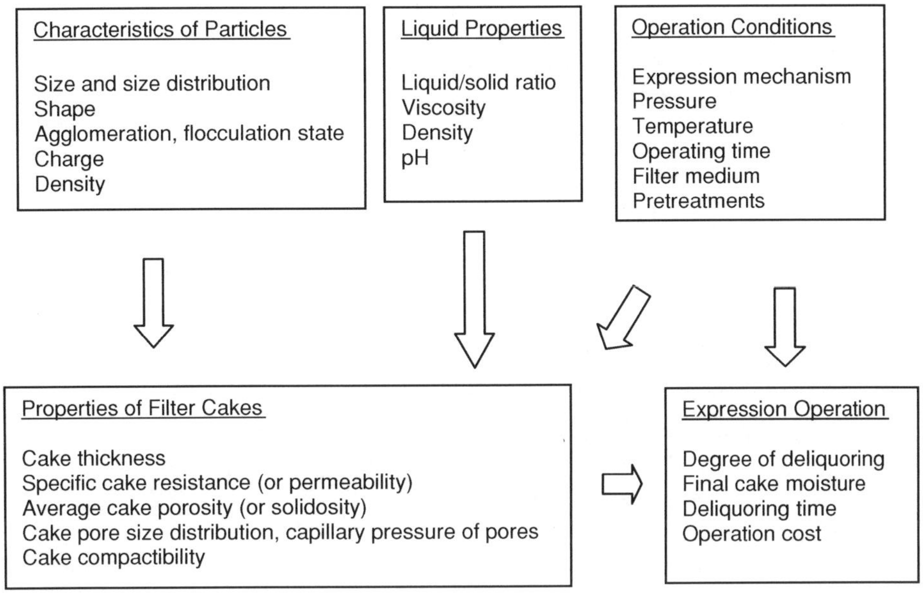

FIG. 18-187 Variables affecting expression.

if solids are not very compactible. There are laboratory and pilot tests available to determine the need for a diaphragm press.

The best way to evaluate diaphragm presses for an application is to run tests with a small pilot unit. Although smaller test units are available, pilot units with 1-ft^2 filter plate area are more common and are recommended.

A laboratory pressure filter (Fig. 18-189) equipped with a piston can provide a simple feasibility test. In this kind of device, the suspension is poured into the filter cylinder, and the first stage of the test is just like a pressure filtration test. After the filtration, compressed air or water is used to push the piston down to squeeze the filter cake. The filtration rate, final cake thickness and dryness are recorded for evaluation and comparison with the same test without the compression by the piston.

Horizontal Diaphragm Presses This is similar to the diaphragm press except the filter plates lay horizontally (while in diaphragm press, the filter plates are operated vertically). The press can be a single-chamber unit, or multiple chambers can be stacked to achieve greater filtration area.

In each filter plate, the filter medium is attached to a moving belt (Fig. 18-190). An elastomer seal is used at the edge of the filter chamber. The slurry is fed into the filter chamber, and the operation starts as a pressure filtration. After filtration, the diaphragm is inflated to squeeze on the cake. At the end of expression, the filter chamber opens and the belt moves the cake out of the filter chamber for discharging. The filter chamber is then closed and ready for the next filtration cycle. Permanent filter belt or disposable medium can be used as filter media. The disposable media are especially useful when handling particles which have high tendency to foul the filter media. With the moving belt, the press operation is fully automatic and is another advantage of this equipment.

The testing for evaluating the horizontal diaphragm press is the same as that described above for the (vertical) diaphragm presses. To ensure automatic operation, the cake solids should not stick to the seal of the filter chamber and need to be carefully evaluated during testing.

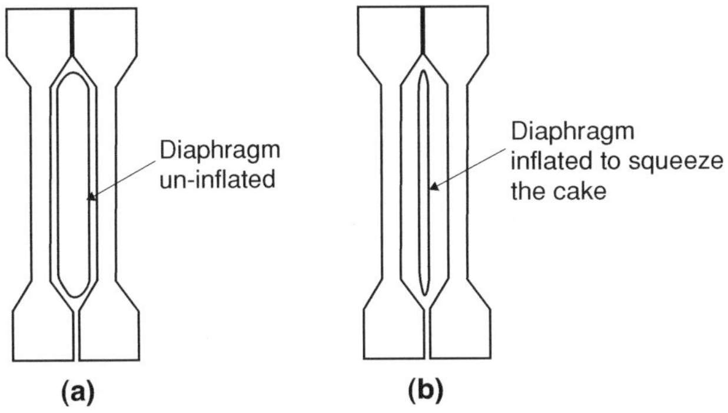

FIG. 18-188 Diaphragm press plate.

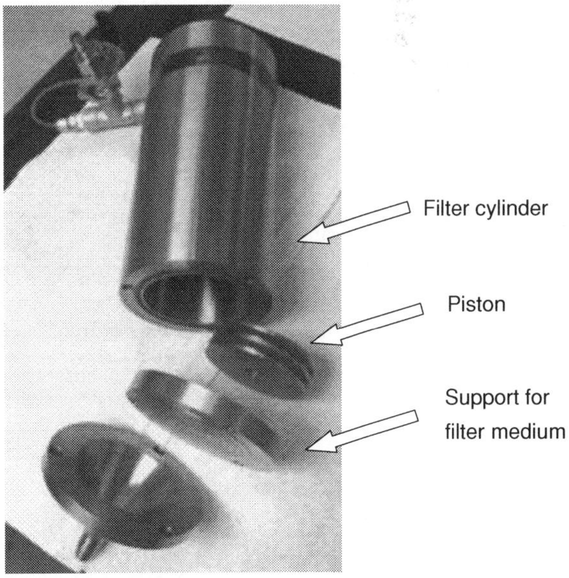

FIG. 18-189 Laboratory pressure filter with a piston to compress the cake.

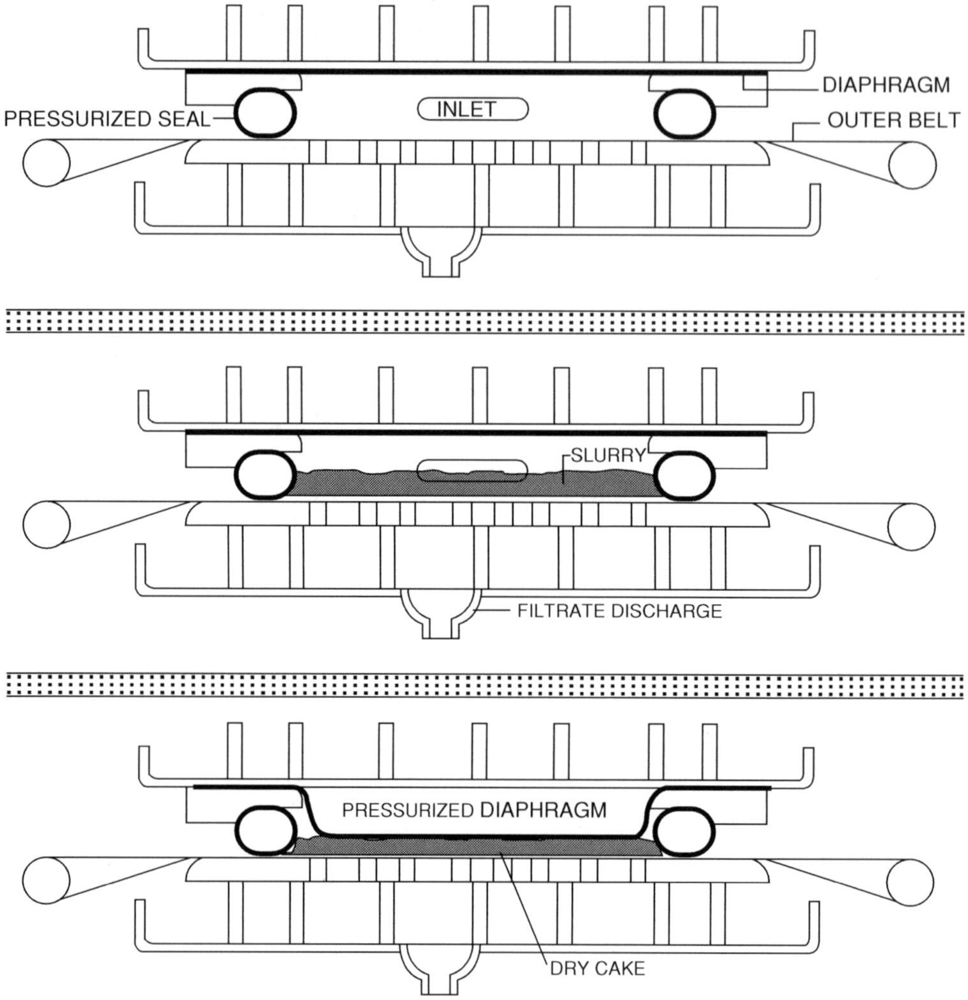

FIG. 18-190 A horizontal diaphragm press. (*Courtesy of Filtra Systems.*)

Tower Presses This press is similar to the stacked horizontal diaphragm presses, but only one filter belt is used (Fig. 18-191). The operation is also fully automatic. The primary applications are in chemical, mineral and pharmaceutical industries. The testing method is the same as the diaphragm presses. One important factor in designing a tower press is the solids need to be able to be cleared from the chamber seal, otherwise leakage will occur in the following filtration cycle.

Tubular Presses As the name implies, this press is composed of a candle filter inside a cylindrical hydraulic casing (Fig. 18-192). The filter cloth is wrapped around the filter candle, and a diaphragm is attached to the inner side of the outer casing. During the filtration step, the space in between two cylinders is filled with slurry, and pressure filtration is conducted. At the end of the filtration step, the diaphragm is inflated to squeeze the cake around the filter candle. At the end of expression, the bottom of the hydraulic casing tube is opened and the filter assembly is lowered. Air is then introduced to pulse the cake off the candle. After the cake is discharged, the inner filter candle moves back, and the bottom is closed for the next filtration cycle.

Tubular presses use the highest pressures among all expression equipment. The pressure can be as high as 1500 psi. With the high pressure, the cake can be very dry (> 95 percent dryness). This type of equipment normally has low capacity so multiple units are used.

Typical applications of tubular presses are for fine particle dewatering including minerals, talc, and $CaCO_3$.

The same laboratory testing equipment as in the diaphragm press can be used but with a higher pressure. A commercially available piston press can also be used.

Continuous Expression Equipment Continuous expression equipment has the advantage of large capacity and automatic operation. Compared to batch expression equipment, lower pressure is used to squeeze the cake in the continuous expression equipment. As a result, the cakes are not as dry as those from the batch expression devices.

Belt Filter Presses Belt presses (Fig. 18-193) have two filter belts that move around rollers of different sizes to dewater the slurry. A typical belt press may have one or more of the following stages: a preconditioning zone, a gravity drainage zone, a linear compression zone (low pressure), and a roller compression zone (high pressure).

The conditioned slurry is fed into the belt press at the preconditioning zone (a tank or pipe), where coagulant and flocculant are added to condition the slurry. The slurry then goes to a horizontal section where the slurry is thickened by gravity drainage. At the end of the gravity drainage section, the thickened slurry (or dilute cake) drops into a wedge section where the wet cake starts to be squeezed by both belts under pressure. At the end of the wedge section, both

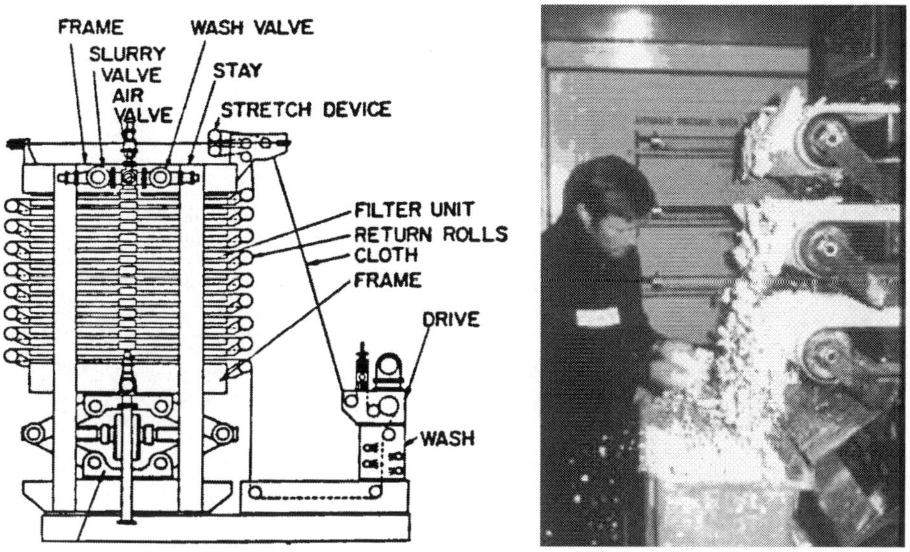

FIG. 18-191 Tower press. (*Courtesy of Larox.*)

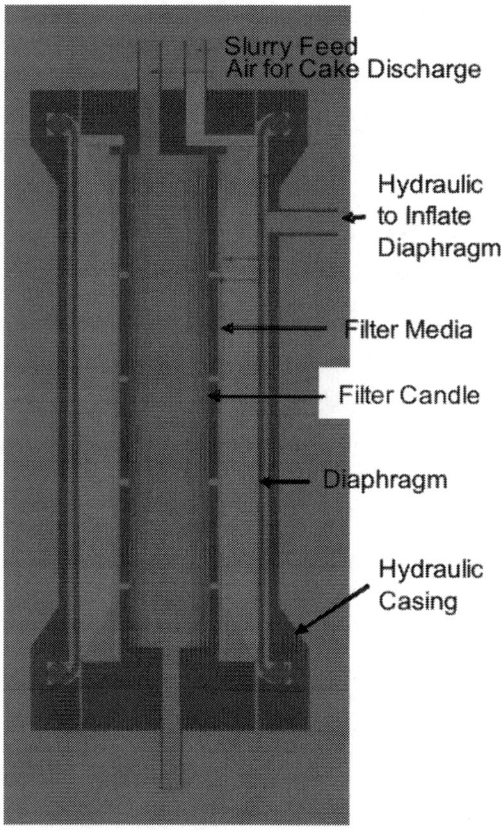

FIG. 18-192 Tubular press. (*Courtesy of Metso Minerals.*)

belts come together with the cake sandwiched in between and move through a series of rollers. The final dewatering is accomplished by moving the cake through these rollers in the order of decreasing roller diameters. While the roller diameter gets smaller, the pressure exerted on the cake gets higher. After the final roller, the two belts are separated to release the cake. Each belt goes through some washing nozzles to clean off any remaining solids on the belt.

It is important to condition the slurry by coagulation and/or flocculation before it is fed into the belt press. An insufficiently flocculated slurry will not dewater properly, and the cake might be squeezed out through the belts or from the side (both sides of a belt press are open). Good conditioned flocs look like cottage cheese, and it is a good field indication for troubleshooting. Most of the challenges in operating a belt press are in the slurry conditioning and the optimization of flocculant dosage. Flocculant consumption can contribute to a significant operation cost if proper control strategy is not used.

The pressure applied on the cake in a belt press operation is low compared to that in other compression filters. The applied pressures are commonly expressed in pli (pound per linear inch) which is not straightforward in translating to a commonly recognized pressure unit. As a rough comparison, the pressures used in belt presses are around 10 to 20 psi. This pressure can be controlled by the belt and roller tension but seldom is adjusted by operators in the field.

Belt presses have the advantage of large capacity and automatic operation. The initial capital cost is also low. They were originally developed in the pulp and paper industry. Any slurry with fibers will do well in a belt press, and high-fiber material can be added to the slurry as a filter aid for belt press operation. Today, in addition to pulp dewatering, the belt press is widely used in wastewater sludge dewatering.

Due to the relatively low pressure used, the final cakes are not very dry. The dryness of biological sludge cakes from a belt press ranges from 10 to 20 wt %. As fiber content goes up, the cake can be as dry as 40 wt % in dryness.

Testing for applications in belt presses is most commonly done by flocculation in beakers and visual observation of the size and strength of the formed flocs. The conditioned slurry can be poured into a filter for a gravity drainage test. These tests can be useful for an experienced person to evaluate if a slurry can be used in belt presses and to optimize an existing belt press. However, the simulation of the final cake dryness is not

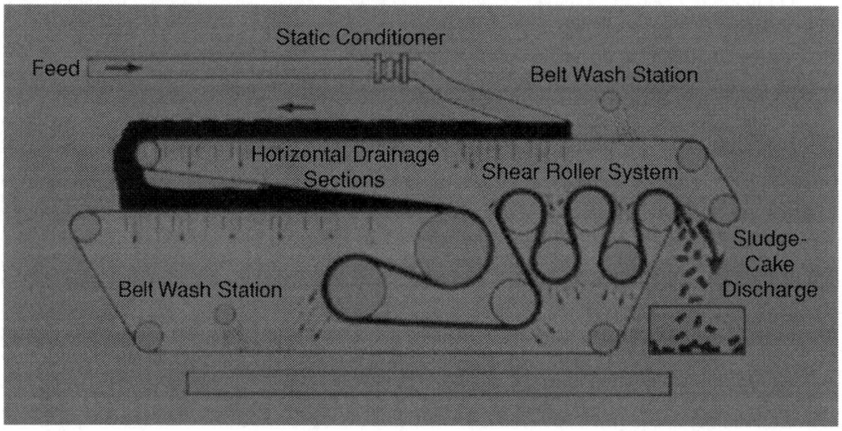

FIG. 18-193 A belt filter press. (*Courtesy of Ashbrook.*)

FIG. 18-194 The crown press.

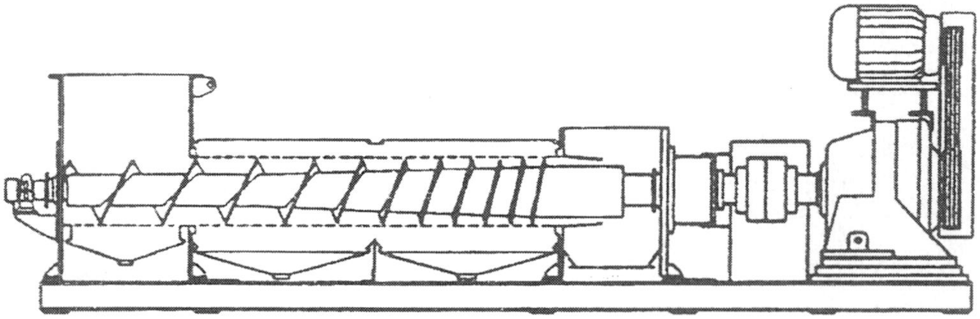

FIG. 18-195 A screw press.

possible with the above methods. The most effective testing is done with a commercially available apparatus called the crown press (Fig. 18-194). This device can simulate the roller actions on the actual belt press and can provide very accurate cake dryness predictions.

Screw Presses A typical screw press is shown in Fig. 18-195, where the slurry is fed into the feed tank at the left-hand side. The core of a screw press is a screw conveyor turning inside a perforated or slotted cylinder. The screw has a smaller diameter at the feed end, and the diameter gradually increases and the screw pitch is shortened toward the discharge end. This design allows gradually decreasing space for slurry/cake and also increasing squeezing pressure on the cake. As the cake moves toward the outlet, the water is squeezed out through the perforated cylinder.

Screw presses also have the advantage of continuous and automatic operation. Screw presses are primarily used in the pulp and paper, citrus, and dairy industries. Applications also exist in many other industries such as dewatering of synthetic rubbers and wastewater sludge. Three pressure (high, medium, and low) ranges are used. High-pressure screw presses are used for vegetable and animal oil; the capacities are relatively smaller. Medium-pressure units are used to dewater deformable particles (such as plastic pellet and synthetic rubber) and paper pulp. Wastewater sludge applications normally use low-pressure options.

SELECTION OF A SOLIDS-LIQUID SEPARATOR

A good solids-liquid separator performs well in service, both initially and over time. It operates reliably day after day, with enough flexibility to accommodate to normal fluctuations in process conditions, and does not require frequent maintenance and repair. Selection of such a separator begins with a preliminary listing of a number of possible devices, which may solve the problem at hand, and usually ends with the purchase and installation of one or more commercially available machines of a specific type, size, and material of construction. Rarely is it worthwhile to develop a new kind of separator to fill a particular need.

In selecting a solids-liquid separator, it is important to keep in mind the capabilities and limitations of commercially available devices. Among the multiplicity of types on the market, many are designed for fairly specific applications, and unthinking attempts to apply them to other situations are likely to meet with failure. The danger is the more insidious because failure often is not of the clean no-go type; rather it is likely to be in the character of underproduction, subspecification product, or excessively costly operation—the kinds of limping failure that may be slowly detected and difficult to analyze for cause. In addition, it should be recognized that the performance of mechanical separators—more, perhaps, than most chemical-processing equipment—strongly depends on preceding steps in the process. A relatively minor upstream process change, one that might be inadvertent, can alter the optimal separator choice.

PRELIMINARY DEFINITION AND SELECTION

The steps in solving a solids-liquid separation problem, in general, are:

1. Define the overall problem, with expert assistance if necessary.
2. Establish process conditions.
3. Identify appropriate separator types; make preliminary selections.
4. Develop a test program.
5. Take representative samples.
6. Make simple tests.
7. Modify process conditions if necessary.
8. Consult equipment manufacturers.
9. Make final selection; obtain quotations.

Problem Definition Intelligent selection of a separator requires a careful and complete statement of the nature of the separation problem. Focusing narrowly on the specific problem, however, is not sufficient, especially if the separation is to be one of the steps in a new process. Instead, the problem must be defined as broadly as possible, beginning with the chemical reactor or other source of material to be separated and ending with the separated materials in their desired final form. In this way the influence of preceding and subsequent process steps on the separation step will be illuminated. Sometimes, of course, the new separator is proposed to replace an existing unit; the new separator must then fit into the current process and accept feed materials of more or less fixed characteristics. At other times the separator is only one item in a train of new equipment, all parts of which must work in harmony if the separator is to be effective.

Assistance in problem definition and in developing a test program should be sought from persons experienced in the field. If your organization has a consultant in separations of this kind, by all means make use of the expertise available. If not, it may be wise to employ an outside consultant, whose special knowledge and guidance can save time, money, and headaches. It is important to do this early; after the separation equipment has been installed, there is little a consultant can do to remedy the sometimes disastrous effects of a poor selection. Often it is best to work with established equipment manufacturers throughout the selection process, unless the problem is unusually sensitive or confidential. Their experience with problems similar to yours may be most helpful and avoid many false starts.

Preliminary Selections Assembling background information permits tentative selection of promising equipment and rules out clearly unsuitable types. If the material to be processed is a slurry or pumpable suspension of solids in a liquid, several methods of mechanical separation may be suitable, and these are classified into settling and filtration methods as shown in Fig. 18-196. If the material is a wet solid, removal of liquid by various methods of expression should be considered.

Settling does not give a complete separation: one product is a concentrated suspension and the other is a liquid which may contain fine particles of suspended solids. However, settling is often the best way to process very large volumes of a dilute suspension and remove most of the liquid. The concentrated suspension can then be filtered with smaller equipment than would be needed to filter the original dilute suspension, and the cloudy liquid can be clarified if necessary. Settlers can also be used for classifying particles by size or density, which is usually not possible with filtration.

Screens may sometimes be used to separate suspensions of coarse particles, but are not widely applicable. For separating fine solids from liquids, cake filtration or the newer systems of crossflow filtration should be considered. Crossflow filtration includes ultrafiltration, where the solids are macromolecules or very fine solids ($D_p \leq 0.1$ µm), and microfiltration, where the particle size generally ranges from 0.1 to 5 µm. In microfiltration a suspension is passed at high velocity of 1 to 3 m/s (3 to 10 ft/s) and moderate pressure (10 to 30 lb_f/in^2 gauge) parallel to a semipermeable membrane in sheet or tubular form. Organic membranes are made of various polymers including cellulose acetate, polysulfone, and polyamide; and they are usually asymmetric, with a thin selective skin supported on a thicker layer that has larger pores. Inorganic membranes of sintered metal or porous alumina are also available in various shapes, with a range of average pore sizes and permeabilities. Most membranes have a wide distribution of pore sizes and do not give complete rejection unless the average pore size is much smaller than the average particle size in the suspension.

In microfiltration, particles too large to enter the pores of the membrane accumulate at the membrane surface as the liquid passes through. They form a layer of increasing thickness that may have appreciable hydraulic resistance and cause a gradual decrease in permeate flow. A decline in liquid flow may also result from small particles becoming embedded in the membrane or plugging some of the pore mouths. The particle layer may reach a steady-state thickness because of shear-induced migration of particles back into the mainstream, or the liquid flux may continue to decline, requiring frequent backwashing or other cleaning procedures. Because of the high velocities the change in solids concentration per pass is small, and the suspension is either recycled to the feed tank or sent through several

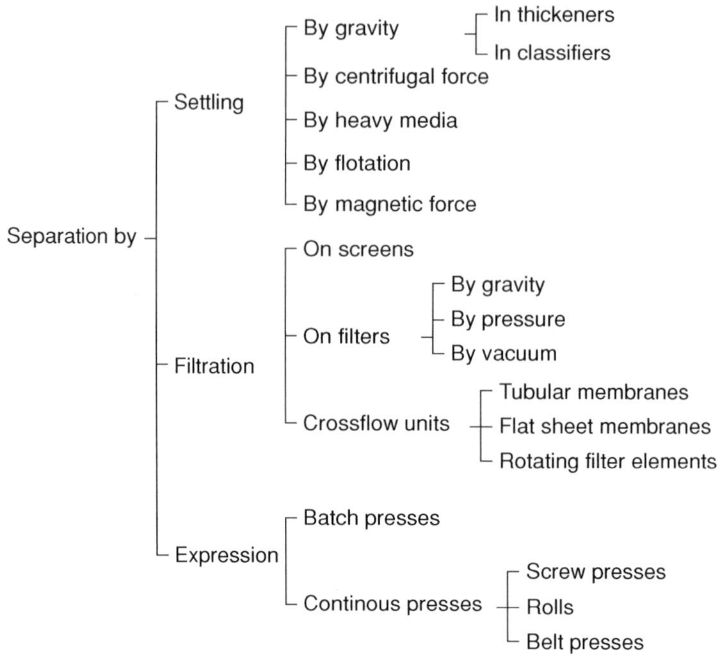

FIG. 18-196 Main paths to solids-liquid separation.

units in series to achieve the desired concentration. The products are a clear liquid and a concentrated suspension similar to those produced in a settling device, but the microfiltration equipment is much smaller for the same production rate.

SAMPLES AND TESTS

Once the initial choice of promising separator types is made, representative liquid-solid samples should be obtained for preliminary tests. At this point, a detailed test program should be developed, preferably with the advice of a specialist.

Establishing Process Conditions Step 2 is taken by defining the problem in detail. Properties of the materials to be separated, the quantities of feed and products required, the range of operating variables, and any restrictions on materials of construction must be accurately fixed, or reasonable assumptions must be made. Accurate data on the concentration of solids, the average particle size or size distribution, the solids and liquid densities, and the suspension viscosity should be obtained *before* selection is made, not after an installed separator fails to perform. The required quantity of the liquid and solid may also influence separator selection. If the solid is the valuable product and crystal size and appearance are important, separators that minimize particle breakage and permit nearly complete removal of fluid may be required. If the liquid is the more valuable product, can minor amounts of solid be tolerated, or must the liquid be sparkling clear? In some cases, partial or incomplete separation is acceptable and can be accomplished simply by settling or by crossflow filtration. Where clarity of the liquid is a key requirement, the liquid may have to be passed through a cartridge-type clarifying filter after most of the solid has been removed by the primary separator.

Table 18-17 lists the pertinent background information that should be assembled. It is typical of data requested by manufacturers when they are asked to recommend and quote on a solid-liquid separator. The more accurately and thoroughly these questions can be answered, the better the final choice is likely to be.

Representative Samples For meaningful results, tests must be run on representative samples. In liquid-solids systems good samples are hard to get. Frequently a liquid-solid mixture from a chemical process varies significantly from hour to hour, from batch to batch, or from week to week. A well-thought-out sampling program over a prolonged period, with samples spaced randomly and sufficiently far apart, under the most widely varying process conditions possible, should be formulated. Samples should be taken from all shifts in a continuous process and from many successive batches in a batch process. The influence of variations in raw materials on the separating characteristics should be investigated, as should the effect of reactor or crystallizer temperature, intensity of agitation, or other process variables.

Once samples are taken, they must be preserved unchanged until tested. Unfortunately, cooling or heating the samples or the addition of preservatives may markedly change the ease with which solids may be separated from the liquid. Sometimes they make the separation easier, sometimes harder; in either case, tests made on deteriorated samples give a false picture of the capabilities of separation equipment. Even shipping of the samples can have a significant effect. Often it is so difficult to preserve liquid-solids samples without deterioration that accurate results can be obtained only by incorporating a test separation unit directly in the process stream.

Simple Tests It is usually profitable, however, to make simple preliminary tests, recognizing that the results may require confirmation through subsequent large-scale studies.

Preliminary gravity settling tests are made in a large graduated cylinder in which a well-stirred sample of slurry is allowed to settle, the height of the interface between clear supernatant liquid and concentrated slurry being recorded as a function of settling time. Centrifugal settling tests are normally made in a bottle centrifuge in which the slurry sample is spun at various speeds for various periods of time, and the volume and consistency of the settled solids are noted. In gravity settling tests in particular, it is important to evaluate the effects of flocculating agents on settling rates.

Preliminary filtration tests may be made with a Büchner funnel or a small filter leaf, covered with canvas or other appropriate medium and connected to a vacuum system. Usually the suspension is poured carefully into the vacuum-connected funnel, whereas the leaf is immersed

TABLE 18-17 Data for Selecting a Solids-Liquid Separator*

1. Process
   a. Describe the process briefly. Make up a flowsheet showing places where liquid-solid separators are needed.
   b. What are the objections to the present process?
   c. Briefly, what results are expected of the separator?
   d. Is the process batch or continuous?
   e. Number the following objectives in order of importance in your problem: (*a*) separation of two different solids ______; (*b*) removal of solids to recover valuable liquor as overflow ______; (*c*) removal of solids to recover the solids as thickened underflow ______ or as "dry" cake ______; (*d*) washing of solids ______; (*e*) classification of solids ______; (*f*) clarification or "polishing" of liquid ______; (*g*) concentration of solids ______.
   f. List the available power and current characteristics.

2. Feed
   a. Quantity of feed:
   Continuous process: ______ gal/min; ______ h/day; ______ lb/h of dry solids.
   Batch process: volume of batch: ______; total batch cycle: ______ h.
   b. Feed properties: temp. ______; pH ______; viscosity ______.
   c. What maximum feed temperature is allowable?
   d. Chemical analysis and specific gravity of carrying liquid.
   e. Chemical analysis and specific gravity of solids.
   f. Percentage of solids in feed slurry.
   g. Screen analysis of solids: wet ______ dry ______
   h. Chemical analysis and concentration of solubles in feed.
   i. Impurities: form and probable effect on separation.
   j. Is there a volatile component in the feed? ______ Should the separator be vapor-tight? ______ Must it be under pressure? ______ If so, how much? ______

3. Filtration and settling rates
   a. Filtration rate on Büchner funnel: ______ gal/(min)(ft^2) of filter area under a vacuum of ______ in Hg. Time required to form a cake ______ in thick: ______ s.
   b. At what rate do the solids settle by gravity?
   c. What percentage of the total feed volume do the settled solids occupy after settling is complete? After how long?

4. Feed preparation
   a. If the feed tends to foam, can antifoaming agents be used? If so, what type?
   b. Can flocculating agents be used? If so, what agents?
   c. Can a filter aid be used?
   d. What are the process steps immediately preceding the separation? Can they be modified to make the separation easier?
   e. Could another carrying liquid be used?

5. Washing
   a. Is washing necessary?
   b. What are the chemical analysis and specific gravity of wash liquid?
   c. Purpose of wash liquid: to displace residual mother liquor or to dissolve soluble material from the solids?
   d. Temperature of wash liquid.
   e. Quantity of wash allowable, in lb/lb of solids.

6. Separated solids
   a. What percentage of solids is desired in the cake or thickened underflow?
   b. Is particle breakage important?
   c. Amount of residual solubles allowable in solids.
   d. What further processing will have to be carried out on the solids?

7. Separated liquids
   a Clarity of liquor: what percentage of solids is permissible?
   b. Must the filtrate and spent wash liquid be kept separate?
   c. What further processing will be carried out on the filtrate and/or spent wash?

8. Materials of construction
   a. What metals look most promising?
   b. What metals must not be used?
   c. What gasket and packing materials are suitable?

*U.S. customary engineering units have been retained in this data form. The following SI or modified-SI units might be used instead: centimeters = inches × 2.54; kilograms per kilogram = pounds per pound × 1.0; kilograms per hour = pounds per hour × 0.454; liters per minute = gallons per minute × 3.785; liters per second·square meter = gallons per minute·square foot × 0.679; and pascals = inches mercury × 3377.

in a sample of the slurry and vacuum is applied to pull filtrate into a collecting flask. The time required to form each of several cakes in the range of 3 to 25 mm (⅛ to 1 in) thick under a given vacuum is noted, as is the volume of the collected filtrate. Properly conducted tests with a Büchner or a vacuum leaf closely simulate the action of rotary vacuum filters of the top- and bottom-feed variety, respectively, and may give the experienced observer enough information for complete specification of a plant-size filter. Alternatively, they may point to pressure-filter tests or, indeed, to a search for an alternative to filtration. Centrifugal filter tests are made in a perforated basket centrifugal filter 254 or 305 mm (10 or 12 in) in diameter lined with a suitable filter medium. Slurry is poured into the rotating basket until an appropriately thick cake—say, 25 mm (1 in)—is formed. Filtrate is recycled to the basket at such a rate that a thin layer of liquid is just visible on the surface of the cake. The discharge rate of the liquor under these conditions is the draining rate. The test is repeated with cakes of other thicknesses to establish the productive capacity of the centrifugal filter.

Batch tests of microfiltration may be carried out in small pressurized cells with a porous membrane at the bottom and a magnetic stirrer to provide high shear at the membrane surface. These tests may quickly show what type of membrane, if any, gives satisfactory separation, but scaling up to large production units is difficult. Small modules with hollow-fiber, tubular, or spiral-wound membranes are available from equipment vendors, so that tests can be made with continuous flow at pressures and velocities likely to be used for large-scale operation. The permeate flux should be measured as a function of time for different slurry concentrations, pressure drops, and solution velocities or Reynolds numbers. Often a limiting flux will be reached as the pressure drop is increased, but operation at a lower pressure drop is often desirable since the flux decline may not be as great and the average permeation rate over a batch cycle may be greater.

More detailed descriptions of small-scale sedimentation and filtration tests are presented in other parts of this section. Interpretation of the results and their conversion into preliminary estimates of such quantities as thickener size, centrifuge capacity, filter area, sludge density, cake dryness, and wash requirements also are discussed. Both the tests and the data treatment must be in experienced hands if error is to be avoided.

Modification of Process Conditions Relatively small changes in process conditions often markedly affect the performance of specific solids-liquid separators, making possible their application when initial test results indicated otherwise or vice versa. Flocculating agents are an example; many gravity settling operations are economically feasible only when flocculants are added to the process stream. Changes in precipitation or crystallization steps may greatly enhance or diminish filtration rates and hence filter capacity. Changes in the temperature of the process stream, the solute content, or the chemical nature of the suspending liquid also influence solids-settling rates. Occasionally it is desirable to add a heavy, finely divided solid to form a pseudo-liquid suspending medium in which the particles of the desired solid will rise to the surface. Attachment of air bubbles to solid particles in a flotation cell, using a suitable flotation agent, is another way of changing the relative densities of liquid and solid.

Consulting the Manufacturer Early in the selection campaign—certainly no later than the time at which the preliminary tests are completed—manufacturers of the more promising separators should be asked for assistance. Additional tests may be made at a manufacturer's test center; again a major problem is to obtain and preserve representative samples. As much process information as tolerable should be shared with the manufacturers to make full use of their experience with their particular equipment. Full-scale plant tests, although expensive, may well be justified before final selection is made. Such tests demonstrate operation on truly representative feed, show up long-term operating problems, and give valuable operating experience.

In summary, separator selection calls for clear problem definition, in broad terms; thorough cataloging of process information; and preliminary and tentative equipment selection, followed by refinement of the initial selections through tests on an increasingly larger scale. Reliability, flexibility of operation, and ease of maintenance should be weighed heavily in the final economic evaluation; rarely is purchase price, by itself, a governing factor in determining the suitability of a liquid-solids separator.

Made in United States
Orlando, FL
04 November 2026

90206121R00085